Molecular Biology of the SARS-Coronavirus

Sunil K. Lal
Editor

Molecular Biology of the SARS-Coronavirus

Springer

Editor
Dr. Sunil K. Lal
Virology Group, ICGEB
P. O. Box 10504
Aruna Asaf Ali Road
New Delhi 110067
India
E-mail: sunillal@icgeb.res.in

ISBN: 978-3-642-03682-8 e-ISBN: 978-3-642-03683-5
DOI 10.1007/978-3-642-03683-5
Springer Heidelberg Dordrecht London New York

Library of Congress Control Number: 2009934478

Cover Illustration: 3D model of SARS-CoV, with a wedge cut out of it to reveal the nucleocapsid (see Chap. 3 by Daniel R. Beniac and Timothy F. Booth)

Background: HL-CZ cells transfected with SARS-CoV Spike construct and incubated with anti-Spike human monoclonal antibody followed by secondary FITC-labeled anti-human antibody (see Chap. 18 by T. Narasaraju, P.L. Soong, J. ter Meulen, J. Goudsmit and Vincent T.K. Chow)

Cover design: WMXDesign GmbH, Heidelberg, Germany

Printed on acid-free paper

Springer is part of Springer Science+Business Media (www.springer.com)

Foreword

SARS was the first new plague of the twenty-first century. Within months, it spread worldwide from its "birthplace" in Guangdong Province, China, affecting over 8,000 people in 25 countries and territories across five continents. SARS exposed the vulnerability of our modern globalised world to the spread of a new emerging infection. SARS (or a similar new emerging disease) could neither have spread so rapidly nor had such a great global impact even 50 years ago, and arguably, it was itself a product of our global inter-connectedness. Increasing affluence and a demand for wild-game as exotic food led to the development of large trade of live animal and game animal markets where many species of wild and domestic animals were co-housed, providing the ideal opportunities for inter-species transmission of viruses and other microbes. Once such a virus jumped species and attacked humans, the increased human mobility allowed the virus the opportunity for rapid spread. An infected patient from Guangdong who stayed for one day at a hotel in Hong Kong led to the transmission of the disease to 16 other guests who travelled on to seed outbreaks of the disease in Toronto, Singapore, and Vietnam, as well as within Hong Kong itself. The virus exploited the practices used in modern intensive care of patients with severe respiratory disease and the weakness in infection control practices within our health care systems to cause outbreaks within hospitals, further amplifying the spread of the disease. Health-care itself has become a two-edged sword.

While SARS exposed the vulnerabilities of the modern human condition, it also highlighted the global capacity for a rapid public health and scientific response to an emerging infectious disease threat. Public health and scientific responses succeeded in identifying the causative agent, developing diagnostic tests, and interrupting the spread of the outbreak. The complete virus genome was fully deciphered within weeks and in the ensuing months and years saw an outpouring of scientific research about the disease and its causative agent, the SARS coronavirus. The natural animal reservoir (bats) and amplifier hosts were defined, the virus receptor on human cells identified and novel antiviral drugs and candidate vaccines developed. This resurgence of attention on coronaviruses led to a much better scientific understanding about the biology of the coronaviruses in general, the discovery of two new coronaviruses that cause human disease (NL-63 and HKU-1) and a range of novel coronaviruses that infect animals.

The precursor of the SARS coronavirus still persists in its natural reservoir host and whether this precursor virus will readapt to humans at some time in the future remains unknown. However, the human adapted SARS coronavirus remains in laboratories and may yet escape, either inadvertently or through malicious action. We thus need to remain vigilant to the re-emergence of a SARS-like disease. What is certain, however, is that we will be confronted with other emerging infectious diseases in the decade ahead and that most of these diseases will arise from an animal reservoir. Thus, the mechanisms of the emergence of SARS serve as an excellent case-study to better understand how viruses jump species-barriers to cause disease outbreaks in humans. The synthetic reconstruction of an infectious bat-SARS-like precursor virus, the largest life form to be created by synthetic biology to date, has provided an excellent model for understanding such mechanisms. This book, which includes the current understanding of the molecular biology of SARS coronavirus and its applications to understanding pathogenesis, host responses, inter-species transmission, therapeutics and vaccine design, is therefore timely.

J.S. Malik Peiris
The University of Hong Kong and HKU-Pasteur Research Centre
Hong Kong Special Administrative Region, China

Preface

The SARS outbreak took the whole world by surprise in November 2002. It was the most unprecedented epidemic outbreak in recorded history and the first major new infectious disease of this century, unusual in its high morbidity and mortality rates and in strategically taking advantage of modern international travel to propagate itself around the world. What followed was a global havoc created by this disease, bringing the healthcare system of affected areas to a grinding halt, affecting healthcare providers, disrupting scheduled emergency surgeries and vital treatment to patients with serious conditions, overloading hospitals with infected cases, forcing public events to be cancelled, and schools, and borders to be closed. The economic impact on individuals and businesses was profound, downregulating tourism, education, and employment.

The epidemic was completely different from all known traditional atypical types of pneumonia because patients experienced lack of oxygen at the onset of the disease and hence required the aid of modern respiratory equipment to breathe. This syndrome was contagious enough to infect a substantial number of people widely and easily. In our days of medical advancement and high technology, which has subsequently led to increased life spans and longevity, a growing confidence had emerged in mankind that it had now achieved the ability to overcome the most complicated life-threatening situations. SARS shattered this confidence and made us realize once again that there are hundreds of dangerous and virulent microorganisms living on the other side of the border that can kill humans. What separates us from them is only the species barrier.

This is not the first time the species barrier has been crossed. The SARS outbreak was just another outbreak in South-East Asia, the breeding ground for notorious viruses. The current novel H1N1 swine-flu outbreak that emerged from Mexico, bird-flu H5N1 influenza in Hong Kong in 1996, human enterovirus 71 in Malaysia, Taiwan, and Singapore in 1977, 1998, and 2000 respectively, and the Nipah virus in Malaysia and Singapore in 1998, are all similar examples. The SARS outbreak was a short-lived near-pandemic situation that originated in the Guangdong province of south China in late 2002 and was efficiently contained by July 2003, with 8,096 known infected cases and 774 deaths (a case-fatality rate of 9.6%) infecting individuals from 37 countries worldwide (mortality by age group: below 1% for people aged 24 or younger, 6% for those aged 25–44, 15% for those aged 45–64 and more than 50% for those over 65). If SARS had not been fully contained, the world would have faced a full-blown pandemic. We must not forget that SARS has not been eradicated (e.g., smallpox). It is still present in its natural host reservoirs and carries the threat and potential to return into the human population any time.

We were able to subvert a potentially explosive spread of the new coronavirus (SARS-CoV) outbreak thanks to WHO's global alert, getting together an emergency network of 11 leading laboratories from 9 countries to investigate this new virus. Within a short span of 1 month, these laboratories did a commendable job by tracing the viral etiology and developing a diagnostic test. Over the years, much has been learnt about this new SARS-CoV; however, our knowledge on the molecular biology of SARS-CoV, its life-cycle, infection, and pathogenesis still remain unclear. This virus is mysterious in its ways and this book looks at various molecular aspects of this virus which help us in understanding these complexities.

Prior to the SARS outbreak, human coronaviruses were only associated with mild diseases. SARS-related CoV became the first coronavirus to cause severe disease in humans. In April 2003, the complete genome sequence of the SARS-CoV was revealed. The genome contains unique 5' and 3' UTRs (untranslated regions) containing higher-order structures which play essential roles in viral transcription and replication, assisted by cellular proteins to perform RNA synthesis, a model elegantly reviewed by Liu and Leibowitz in this book. The SARS-CoV genome contains five major open reading frames (ORFs) that encode the replicase polyprotein, the spike (S), envelope (E), and membrane (M) glycoproteins, and the nucleocapsid protein (N). S binds species-specific host cell receptors and triggers a fusion between the viral envelope and the cell membrane. Lambert's chapter clearly describes the basic cell biology of ACE2 and Pöhlmann's chapter elaborates on the S-ACE2 interface. Receptor binding and the subsequent structural changes that result have been described in detail by Beniac and Booth. The S protein is the virulence factor in many different coronaviruses and the principal viral antigen that elicits neutralizing antibody on behalf of the host. To study this, Chow's lab has undertaken whole transcriptome analysis of S transfected host cells and identified novel pathways that become altered. Replicase proteins have been extensively discussed in the chapters by Ziebuhr and Canard. Immediate early proteins, like the RNA dependent RNA polymerase (RDRP) and proteases, are responsible for preparing the infected cell for virus takeover. Dinman describes programmed -1 ribosomal frameshifting as an essential and unique feature of the virus for the translation of these proteins. The overlapping polyproteins 1a and 1ab are extensively cleaved by the internally encoded SARS-CoV proteases, Mpro, and PLpro and are extensively discussed by Chang in his chapter. The N protein forms the capsid and also plays several regulatory roles during viral pathogenesis which have been described by Surjit and myself. Cell type specific apoptosis induction of host cells by viral proteins has been elegantly described by Hermann Schätzl et al. Three chapters are dedicated to describe the current knowledge on accessory proteins by

Pekosz, Sun, and Tan. Sheahan and Baric's chapter and Li and Xu's chapter describe exhaustively the pathogenesis and protective immunity against SARS-CoV in humans. Cell signaling and associated lung fibrosis due to TGF-/Smad pathways are discussed in the chapters by Mizutani and Chen, respectively. The importance and application of retroviral pseudotypes for highly pathogenic diseases like SARS, using surrogates of the live virus for neutralization assays, has been described by Nigel Temperton.

I wish to congratulate and thank all the contributing authors for the exhaustive coverage of their respective subjects and publication of this book. We hope the readers find this book a consolidated compilation of our current understanding of the molecular biology of SARS-CoV.

International Centre for Genetic
Engineering & Biotechnology,
New Delhi

Sunil K. Lal

Contents

Part I
Viral Entry

Chapter 1
Cellular Entry of the SARS Coronavirus: Implications for Transmission, Pathogenicity and Antiviral Strategies

Ilona Glowacka, Stephanie Bertram, and Stefan Pöhlmann

Abstract A novel coronavirus was identified as the causative agent of the lung disease severe acute respiratory syndrome (SARS). The outbreak of SARS in 2002/2003 was associated with high morbidity and mortality and sparked international research efforts to develop antiviral strategies. Many of these efforts focussed on the viral surface protein spike (S), which facilitates the first indispensable step in the viral replication cycle, infectious entry into target cells. For infectious cellular entry to occur, the S protein must engage a cellular receptor, the carboxypeptidase angiotensin-converting enzyme 2 (ACE2). The interface between ACE2 and S protein, which has been characterized at the structural level, constitutes a key target for vaccines and inhibitors, and is believed to be an important determinant of viral pathogenesis and interspecies transmission. In this chapter, we will discuss how SARS-S mediates cellular entry and we will review the implications of this process for SARS coronavirus (SARS-CoV) transmission, disease development and antiviral intervention.

1.1 Introduction

The emergence of the severe acute respiratory syndrome coronavirus (SARS-CoV) in Guangdong Province, China, in 2002, and its subsequent spread in Asia and Canada clearly exemplified the vulnerability of societies and economies to a novel, highly pathogenic respiratory agent (Stadler et al. 2003; Peiris et al. 2003b). The outbreak, which was halted solely by the quarantine of exposed individuals and the use of conventional prevention measures such as surgical masks, was paralleled by an international, collaborative scientific effort to develop means for therapeutic and

S. Pöhlmann (✉)
Institute of Virology, OE 5230, Hannover Medical School, Carl-Neuberg-Straße 1, 30625 Hannover, Germany
e-mail: poehlmann.stefan@mh-hannover.de

preventive intervention (Peiris et al. 2004; Stadler and Rappuoli 2005). The basis for the development of successful antiviral strategies is a thorough understanding of the molecular biology underlying viral amplification and pathogenesis, and many significant discoveries have been made in the SARS field since the identification of the virus early in 2003 (Drosten et al. 2003; Ksiazek et al. 2003; Peiris et al. 2003a). Several of these findings provided important insights into the structure and function of the viral spike (S) protein, which is used by the virus as the key to bind and enter host cells (Hofmann and Pöhlmann 2004). The most well-known examples are the identification of angiotensin-converting enzyme 2 (ACE2) as the host factor which is engaged by the viral S protein for infectious entry into cells, and the elucidation of the structure of the S protein receptor binding domain (RBD) in complex with ACE2 (Li et al. 2003, 2005a). These findings have major implications not only for vaccine and inhibitor development but also for our understanding of the SARS zoonosis, since adaptation of SARS-S to robust usage of human ACE2 was probably of key importance for efficient SARS-CoV spread in humans (Li et al. 2005a, 2005c). In this chapter, we will discuss how SARS-CoV gains access to target cells and how this process can be inhibited. In addition, we will review how the molecular interactions underlying SARS-CoV entry impact viral pathogenesis and interspecies transmission.

1.2 The Spike Protein: Key to the Host Cell

The SARS-Sprotein is a type I transmembrane protein, which comprises 1,255 amino acids and contains 23 consensus signals for N-linked glycosylation (Hofmann and Pöhlmann 2004). S protein is synthesized in the secretory pathway of infected cells. It contains an N-terminal signal sequence, which mediates import of the nascent protein into the endoplasmatic reticulum, where the protein is folded and modified with mannose-rich carbohydrates. Upon transport of the protein into the Golgi apparatus, most, if not all, of the high-mannose carbohydrates are processed into complex glycans (Nal et al. 2005). Evidence of O-glycosylation of SARS-S has not been reported. A novel dibasic ER retrieval motif in the cytoplasmic tail of SARS-S promotes accumulation of the S protein at the ER–Golgi intermediate compartment and the Golgi region (McBride et al. 2007), the sites where progeny particles are assembled (Stertz et al. 2007; Siu et al. 2008). Formation and budding of new particles are driven by the membrane protein (M), the envelope protein (E) and the nucleocapsid protein (N) (Huang et al. 2004; Hsieh et al. 2005; Siu et al. 2008); interactions with the M protein might facilitate S protein incorporation into particles. Trimers of the S protein protrude from the viral envelope and provide virions with a crown (Lat. *corona*) -like appearance, from which the name "coronaviruses" is derived.

The domain organization of SARS-S resembles that of several well-characterized viral membrane proteins, such as influenza virus hemagglutinin (HA) and human immunodeficiency virus (HIV) envelope protein (Env) (Hofmann and Pöhlmann 2004).

These proteins employ comparable strategies to facilitate fusion of viral and host cell membranes and are termed class I fusion proteins (Kielian and Rey 2006). They are distinguished from class II fusion proteins (Kielian 2006), found, for example, on flavi- and alphaviruses, by their distinct spatial organization and the particular configuration of the functional elements required for fusion with target cells: class I fusion proteins are inserted perpendicular to the viral membrane and contain an N-terminal surface unit (SU) and a C-terminal transmembrane unit (TM). The globular SU interacts with cellular receptors, while the TM promotes

Fig. 1.1 Domain organization of coronavirus S proteins (adapted from Hofmann and Pöhlmann 2004). The position of the S protein open reading frame in the SARS-CoV genome is indicated in the *upper panel*. Coronavirus S proteins exhibit a domain organization characteristic for class I fusion proteins. The domain organization of prototype class I fusion proteins, the HIV envelope protein, and the influenza virus HA is shown below. A signal peptide is located at the N terminus and mediates import of the nascent protein into the secretory pathway of infected cells. The surface unit S1 contains a receptor binding domain (RBD), which allows engagement of cellular receptors for infectious entry. The transmembrane unit (S2) harbors functional elements pivotal to membrane fusion: a fusion peptide, two helical regions, and a transmembrane domain. Proteolytic cleavage into the S1 and S2 subunits by host-cell proteases is indicated by a *triangular arrow*. AIBV: avian infectious bronchitis virus; hCoV: human CoV; HR: helical region; MHV: murine hepatitis virus; SARS: severe acute respiratory syndrome

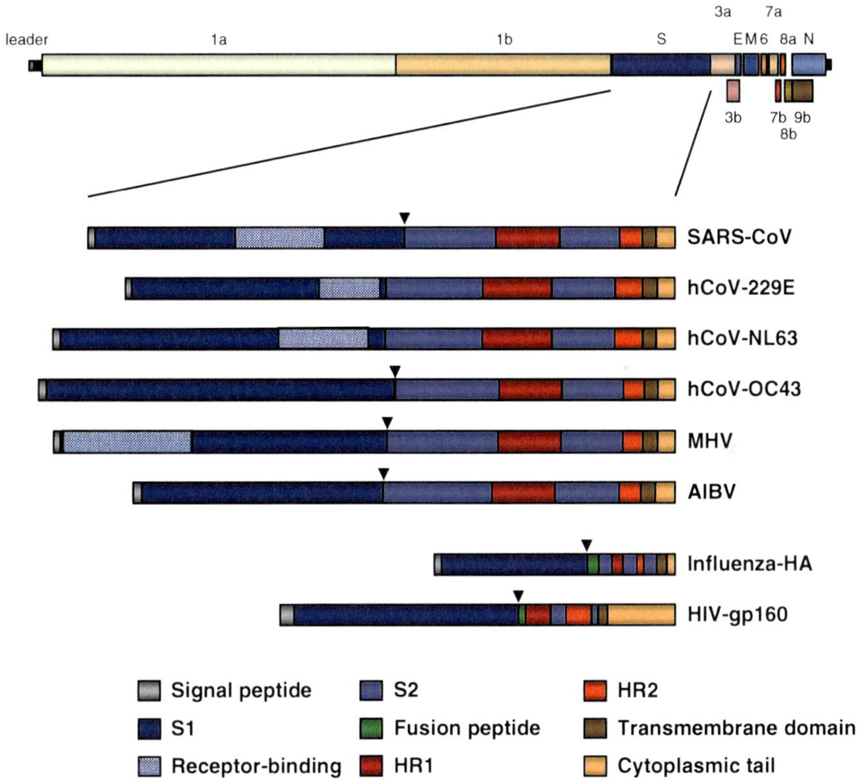

fusion of the viral and host cell membrane (Kielian and Rey 2006). The latter process depends on the presence of a fusion peptide and two helical regions (HR), conserved elements which are intimately involved in the membrane fusion process (Fig. 1.1), as discussed below. The S protein and the aforementioned fusion proteins are adapted to usage by different cellular receptors. Therefore, the SU (termed S1) of SARS-S does not exhibit appreciable sequence homology to the respective sequences of other class I fusion proteins. In contrast, the functional elements in TM, particularly the HRs, are conserved between different class I fusion proteins. Consequently, the TM (termed S2) of SARS-S shares homology with the corresponding sequences of other viral fusion proteins (Hofmann and Pöhlmann 2004), which has important implications for development of antiviral strategies, as discussed below.

1.3 The Attachment Factors DC-SIGN and DC-SIGNR: Enhancers or Inhibitors of SARS-CoV Infection?

The interaction of SARS-S with ACE2 is the first indispensable step in the entry cascade, as discussed below. It needs to be noted, however, that SARS-S also binds to other cell surface factors and these interactions, although being ultimately dispensable for infectious entry, might profoundly alter infection efficiency. Thus, it is well established that the calcium-dependent (C-type) lectin dendritic cell-specific intercellular adhesion molecule-3 grabbing non-integrin (DC-SIGN, CD209; Geijtenbeek et al. 2000) and the related protein DC-SIGNR (L-SIGN, CD209L; Pöhlmann et al. 2001; Bashirova et al. 2001) bind to SARS-S (Marzi et al. 2004; Yang et al. 2004; Jeffers et al. 2004; Khoo et al. 2008) in a glycan-dependent fashion (Shih et al. 2006; Han et al. 2007). DC-SIGNR was detected in the lung on ACE2-positive, SARS-CoV-infected cells and on uninfected bystander cells (Chan et al. 2006), while DC-SIGN expression was found to be induced upon SARS-CoV infection (Yen et al. 2006), suggesting that these lectins might have ample opportunity to capture SARS-CoV in infected individuals.

Despite the potentially important role of DC-SIGN/R in SARS-CoV infection, the consequences of DC-SIGN/DC-SIGNR (collectively referred to as DC-SIGN/R) engagement by SARS-S for viral infectivity are largely unclear. One group suggested that DC-SIGN/R function as bona fide receptors which facilitate viral entry into otherwise nonpermissive HeLa cells (Han et al. 2007). DC-SIGNR was also identified in a functional screen for receptors used by SARS-CoV for cellular entry (Jeffers et al. 2004), further suggesting that DC-SIGNR might support viral entry, at least under certain conditions and probably with low efficiency. In contrast, three other studies failed to detect an appreciable receptor function for DC-SIGN/R but suggested that these proteins might enhance infectious entry into ACE2-expressing cell lines (Marzi et al. 2004; Yang et al. 2004; Shih et al. 2006). Finally, and in contrast to all aforementioned studies, Chan and co-workers provided evidence that DC-SIGNR plays a protective role in SARS-CoV infection (Chan et al. 2006). Thus, it was demonstrated that DC-SIGNR-dependent uptake of SARS-CoV into cell lines might lead to viral degradation and might thus reduce viral infectivity for target cells (Chan et al. 2006). In agreement with this finding, evidence was obtained that the combination of certain DC-SIGNR allelic variants, which resulted in reduced SARS-CoV uptake in cell culture, was associated with increased risk of SARS-CoV infection in humans (Chan et al. 2006), albeit these findings are not undisputed (Tang et al. 2007; Zhi et al. 2007). In any case, most functional studies described above have in common that they were carried out with cell lines, which do not adequately model type II pneumocytes, the major targets of SARS-CoV infection (Hamming et al. 2004; Ding et al. 2004; To and Lo 2004; Mossel et al. 2008), and further work with primary lung epithelium is required to help to elucidate the role of DC-SIGN/R in SARS-CoV infection. Notably, a single study examined the impact of DC-SIGN/R-specific antibodies on viral spread in primary human airway epithelium cultured at the air–liquid interface and observed no inhibition (Sims et al. 2008), although it was not investigated if these lectins were indeed expressed by the cells examined. Finally, it is worth mentioning that SARS-S binds to lectins other than DC-SIGN/R, such as the C-type lectin LSECtin (Gramberg et al. 2005) which is largely co-expressed with DC-SIGNR, and the consequences of these interactions for viral amplification have not been determined. Collectively, it is clear that binding to DC-SIGN/R and related lectins has the potential to modulate viral spread in vivo. It remains to be determined, however, if lectin binding augments or suppresses viral replication. Recently described knock-in mice for human DC-SIGN (Schaefer et al. 2008) or SIGNR1 (a murine homologue of human DC-SIGN) knock-out mice (Lanoue et al. 2004) might be useful to clarify these questions.

1.4 The Two Faces of ACE2: SARS-CoV Receptor and Protector Against Lung Damage

In contrast to attachment factors, cellular receptors are indispensable for infectious viral entry. In order to discover such factors, several laboratories used the soluble SARS-S1 subunit for co-immunoprecipitation of cellular binding partners. A milestone study by Li and colleagues identified the carboxypeptidase ACE2, an integral part of the renin–angiotensin system (see below), as a high-affinity SARS-S interactor (Li et al. 2003). Ectopic expression of ACE2 on barely permissive 293T cells facilitated efficient SARS-S-dependent cell–cell and virus–cell fusion (Li et al. 2003), suggesting that ACE2 might play an important role in SARS-CoV entry. Similar results were obtained by an independent study (Wang et al. 2004), which used a comparable approach to identify cellular binding partners of SARS-S. Subsequently, it was shown that endogenous expression of ACE2 correlates with susceptibility to SARS-CoV infection of cell lines (Nie et al. 2004; Hofmann et al.

2004a) and that ectopic expression of ACE2 facilitates SARS-S-driven infection of otherwise nonsusceptible cells (Mossel et al. 2005). Moreover, it was demonstrated that SARS-CoV infects ACE2-positive type II pneumocytes and ACE2-positive cells in the intestinal epithelium (Hamming et al. 2004; Ding et al. 2004; To and Lo 2004; Chan et al. 2006; Mossel et al. 2008), albeit ACE2-independent infection of target cells has also been suggested (Gu et al. 2005; Gu and Korteweg 2007). Finally, knock-out of ACE2 in mice was found to largely abrogate susceptibility to SARS-CoV infection (Kuba et al. 2005), indicating that ACE2 functions as a bona fide SARS-CoV receptor, which is necessary and sufficient for infectious entry into target cells.

1.4.1 The Structure of the Interface Between SARS-S and ACE2

A thorough understanding of the interface between SARS-S and ACE2 is key to the development of antiviral strategies targeting viral entry. The domains and amino acid residues in SARS-S and ACE2, which contribute to the efficient interaction of these proteins, were initially mapped by mutagenic analyses. These studies identified amino acids 318–510 in SARS-S as an independently folded RBD, which binds to ACE2 with higher affinity than the full length S protein (Xiao et al. 2003; Wong et al. 2004; Babcock et al. 2004). The RBD was also shown to be the major target of neutralizing antibodies (He et al. 2004a, 2004b, 2005), and several residues within amino acids 450–490 were suggested to be critical for optimal ACE2 engagement (Wong et al. 2004; Li et al. 2005c). In addition, exploitation of species-specific differences in murine, rat and human ACE2 allowed the mapping of certain amino acid residues, particularly L353, as important for receptor function (Li et al. 2004, 2005c). These results were supported and extended by the subsequent solution of the structure of the RBD in complex with ACE2 (Li et al. 2005a): the RBD consists of a core (a five-stranded antiparallel β-sheet), and an extended loop, which contains all amino acids making contacts with ACE2. The extended loop, also termed receptor-binding motif (RBM), comprises amino acids 424–494 (Li et al. 2005a), and thus includes the residues defined by mutagenic analysis to be important for SARS-S interactions with ACE2 (Wong et al. 2004; Li et al. 2005c). The RBM contacts the N-terminal helix of ACE2 and the loop between helices α2 and α3. Moreover, a portion of the RBM inserts between a short helix in ACE2 (amino acids 329–333) and a β-hairpin at ACE2 residue L353, supporting the previously postulated contribution of L353 to appropriate spike–receptor interactions (Li et al. 2005c). Conformational changes inherent to the peptidase activity of ACE2 do not impact the availability of the S protein binding site (Li et al. 2005a), in agreement with the observation that an ACE2 inhibitor which blocks peptidase activity and arrests ACE2 in a closed conformation does not inhibit SARS-S-dependent entry (Towler et al. 2004; Li et al. 2005c). Collectively, the functional and structural studies defined amino acids in SARS-S and ACE2, which facilitate the tight association of these proteins. In addition, the results highlighted that natural variation of these sequences might have important implications for SARS-CoV transmission and pathogenicity, as discussed below.

1.4.2 Sequence Variations at the SARS-S/ACE2 Interface Might Impact Viral Transmission and Pathogenicity

Horseshoe bats harbor SARS-CoV-related viruses and might constitute the natural reservoir of SARS-CoV (Lau et al. 2005; Li et al. 2005b). However, the sequence homology between bat and human viruses is limited (Lau et al. 2005; Li et al. 2005b). Thus, the S protein of animal viruses does not contain an RBM-like sequence and does not use ACE2 for cellular entry (Ren et al. 2008). It is therefore probable that SARS-CoV was introduced into the human population via an intermediate host, and palm civets, which harbor viruses with high sequence homology to human SARS-CoV, are possible candidates (Guan et al. 2003; Song et al. 2005). Notably, the S proteins of human viruses from the 2002/2003 epidemic bind human ACE2 with much higher efficiency than their palm civet counterparts (Li et al. 2005c), indicating that efficient spread in humans required adaptation of the SARS-S sequence. Indeed, sequence comparison revealed that the civet RBD contains four amino acid changes relative to the human sequence. Two of these changes are located outside the RBM and do not impact receptor interactions (Li et al. 2005c). In contrast, the remaining two changes, N (human) to K (palm civet) at position 479 and T (human) to S (palm civet) at position 487, afflicted residues making direct contact with ACE2 and significantly decreased binding to human ACE2 (Li et al. 2005a, 2005c). Thus, N479 and T487 might be required for efficient spread in and between humans (Li et al. 2005a, 2005c; Li 2008). Interestingly, viruses isolated from sporadic SARS cases in the winter of 2003/2004, which were not associated with severe disease or human-to-human transmission, contained a serine at position 487 (Li et al. 2005a), further indicating that this amino acid might play a key role in human-to-human transmission and viral pathogenicity.

The potential for zoonotic transmission of SARS-CoV might also be determined by species-specific variations in the ACE2 sequence. Thus, murine and rat ACE2, which do not (rat), or only inefficiently (murine), support SARS-S-driven entry (Li et al. 2004, 2005c), contain a leucine (human) to histidine (mouse, rat) exchange at position 353. This exchange impedes formation of robust contacts with T487 in SARS-S and thereby prevents murine and rat ACE2 from efficiently supporting SARS-S-driven cellular entry (Li et al. 2005a). In addition, the rat but not the murine receptor contains a M82N exchange, which introduces a glycosylation signal. The glycan added to N82 blocks the interaction with L472 in SARS-S and further decreases receptor function, explaining why rat ACE2 is less capable of supporting SARS-S-driven entry than murine ACE2 (Li et al. 2005a). These results, in conjunction with the aforementioned variations in the RBD sequence, highlight that the efficiency of the SARS-S interaction with ACE2 might be a critical determinant of interspecies transmission of SARS-CoV.

1.4.3 The Human Coronavirus NL63 Uses ACE2 for Cellular Entry

A novel human coronavirus, NL63, was discovered by two Dutch groups in the aftermath of the SARS-CoV outbreak (van der Hoek et al. 2004; Fouchier et al. 2004). NL63 is a group I CoV and shows high sequence similarity to the long-known human CoV 229E. The 229E virus, like all other group I viruses described at the time of the NL63 discovery, uses CD13 (aminopeptidase N) as a receptor for cellular entry (Hofmann and Pöhlmann 2004). Considering the specificity of group I viruses for CD13 and taking into account that the spike proteins of 229E and NL63 share 56% sequence identity (van der Hoek et al. 2004; Pyrc et al. 2004), it was surprising that NL63 was shown to use ACE2 and not CD13 for cellular entry (Hofmann et al. 2005). This finding raised the question of whether both viruses use similar strategies to engage ACE2. Mapping studies revealed that an N-terminal unique region in NL63-S, which was suspected to function as RBD, is in fact dispensable for receptor engagement (Hofmann et al. 2006). In contrast, several motifs within amino acids 232 and 684 were found to be required for ACE2 binding within an initial study, and it was suggested that NL63-S might not harbor a single continuous RBD (Hofmann et al. 2006). However, subsequent analyses narrowed the region responsible for ACE2 binding to amino acids 301–643 and 476–616, respectively, and a SARS-S RBM-like motif was identified in NL63-S (Li et al. 2007; Lin et al. 2008). Several amino acid substitutions in ACE2 were found to alter ACE2 usage by SARS-S but not by NL63-S (Hofmann et al. 2006), indicating that both S proteins might interact with different ACE2 surfaces. This interpretation is not undisputed (Li et al. 2007) and solution of the structure of NL63-S in complex with ACE2 might be required to clarify whether SARS-S and NL63-S recognize ACE2 differentially. In any case, it is clear that both viruses employ different mechanisms to activate membrane fusion once the S proteins have bound to ACE2. Thus, it is believed that upon ACE2 engagement SARS-CoV is internalized into endosomal vesicles, where the pH-dependent cellular protease cathepsin L activates SARS-S by cleavage (Simmons et al. 2005). In contrast, low pH and cathepsin activity seem to be largely dispensable for NL63-S-driven entry and it is at present unclear how NL63-S-driven membrane fusion is triggered (Huang et al. 2006; Hofmann et al. 2006).

1.4.4 SARS Versus NL63: A Correlation Between ACE2 Downregulation and Viral Pathogenicity?

NL63 is a globally distributed pathogen which is acquired early in childhood and does not usually cause severe disease (Pyrc et al. 2007). This observation contrasts with the high pathogenicity of SARS-CoV and raises the question of which viral factors determine disease severity. Again, S protein interactions with ACE2 might play a central role. Thus, a milestone discovery by Imai and colleagues indicated that ACE2 expression protects against development of acute respiratory distress syndrome (ARDS) (Imai et al. 2005). ACE2 is an integral component of the renin–angiotensin system (RAS), a key regulator of blood pressure and, as demonstrated by Imai and colleagues (Imai et al. 2005), lung function (Imai et al. 2008; Penninger et al. 2008). Knock-down of ACE2 in a mouse model caused accumulation of angiotensin II, which promoted development of ARDS by signaling via the AT1R receptor (Imai et al. 2005, 2008; Penninger et al. 2008). Conversely, inhibition of AT1R and application of soluble ACE2 protected against ARDS (Imai et al. 2005). Interestingly, a soluble form of the S1 subunit of SARS-S was shown to downregulate ACE2 expression in vitro and in vivo (Kuba et al. 2005), indicating that SARS-S engagement of ACE2 might promote SARS development even in the absence of productive infection. The S protein of NL63 exhibits a markedly reduced affinity for ACE2 compared to SARS-S (Mathewson et al. 2008) and seems to engage the receptor in a different fashion (Hofmann et al. 2006), suggesting that differential ACE2 downregulation by SARS-CoV and NL63 could contribute to the differential pathogenicity of these viruses. However, it is largely unclear how SARS-CoV decreases ACE2 expression and the effect of NL63 on ACE2 levels has not been systematically investigated. Notably, a recent study indicates that SARS-S might promote shedding of the ACE2 ectodomain by inducing ACE2 cleavage by TACE/ADAM17 (Fig. 1.2), a process that seems to be essential for

Fig. 1.2 ACE2 downregulation by SARS-S might promote development of SARS (adapted from Kuba et al. 2006). ACE and ACE2 are key components of the renin–angiotensin system. ACE processes angiotensin I (ANG1) into angiotensin II (ANG2) and accumulation of ANG2 can promote acute lung failure via angiotensin II type 1 receptor (AT1R). This process is prevented by ACE2, which converts ANG2 into angiotensin 1-7 (ANG-(1-7)). The angiotensin II type 2 receptor (AT2R) also exerts a protective function. The interactions of SARS-S with ACE2 drive infectious entry but also induce downregulation of ACE2, possibly by promoting ACE2 cleavage by TACE/ADAM17. Diminished ACE2 expression then facilitates SARS development. ACE: angiotensin-converting enzyme; TACE: TNF-α converting enzyme; ADAM17: ADAM metallopeptidase domain 17

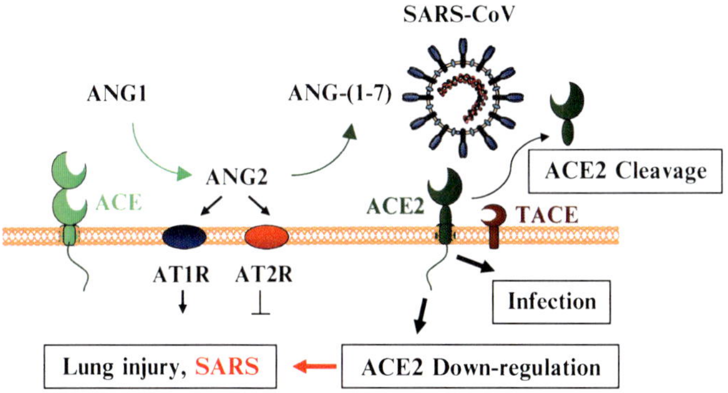

infectious entry (Haga et al. 2008). In contrast, NL63-S did not induce appreciable ACE2 shedding (Haga et al. 2008). Thus, the previously observed ACE2 down-regulation by SARS-S might have been due to proteolytic cleavage and dissociation of the ectodomain rather than ACE2 internalization and degradation. However, it is unclear if shedding of the ACE2 ectodomain actually exacerbates SARS development, considering that soluble ACE2 protects against ARDS in a mouse model (Imai et al. 2005).

1.5 Cleavage by Endosomal Cathepsin Proteases Activates SARS-S

Class I fusion proteins usually require proteolytic cleavage to transit into an activated state (Hofmann and Pöhlmann 2004; Kielian and Rey 2006). However, the strategies to accomplish proteolytic activation can vary. Many fusion proteins are cleaved by subtilisin-like proteases in the secretory pathway of infected cells, and proteolytically processed proteins are incorporated into virions. This applies to the S proteins of most strains of murine hepatitis virus (MHV), a group II coronavirus. The membrane fusion reaction is subsequently triggered by binding of the cleaved S proteins to their cellular receptor, CEACAM-1 (Williams et al. 1991; Nash and Buchmeier 1997; de Haan et al. 2004; Qiu et al. 2006). Consequently, entry is pH-independent and encompasses fusion of the viral membrane with the plasma membrane of target cells (Nash and Buchmeier 1997; de Haan et al. 2004; Qiu et al. 2006). The influenza virus HA is either cleaved by subtilisin proteases in the secretory pathway or by secreted proteases present in the lung lumen. However, subsequent binding to the receptor determinant sialic acid does not trigger membrane fusion but internalization into endosomal vesicles, where fusion is triggered by low pH (Eckert and Kim 2001). Thus, infectious entry of influenza viruses is pH-dependent and is facilitated by fusion of the viral membrane with endosomal membranes (Eckert and Kim 2001).

The SARS-S protein employs a mixture of the entry strategies described above. At present, there is no evidence for appreciable cleavage of SARS-S produced in infected cells (Xiao et al. 2003; Yang et al. 2004; Simmons et al. 2004; Yao et al. 2004; Hofmann et al. 2004b), with the exception of a single report (Wu et al. 2004). It has been documented that the presence of furin can augment SARS-S activity and that a furin inhibitor blocks SARS-CoV infection (Bergeron et al. 2005; Follis et al. 2006). However, cleavage of SARS-S has not been detected under these conditions (Bergeron et al. 2005; Follis et al. 2006). Instead, a seminal study by Simmons and colleagues showed that SARS-S is activated by the endosomal, pH-dependent protease cathepsin L upon uptake into target cells, and that cathepsin L activity is essential for infectious entry (Simmons et al. 2005). Cathepsin B can also contribute to SARS-S activation but seems to be of minor importance compared to cathepsin L (Simmons et al. 2005). Importantly, appropriate SARS-S cleavage by cathepsin L seems to require a modest conformational rearrangement of SARS-S (Simmons et al. 2005), which is induced upon binding to ACE2 (Beniac et al. 2007). Thus, SARS-S-driven entry is pH-dependent and relies on fusion of viral and endosomal membranes (Yang et al. 2004; Simmons et al. 2004, 2005; Hofmann et al. 2004b). However, acidic conditions are required for cathepsin activity and have no triggering effect on SARS-S (Simmons et al. 2005).

The cathepsin L cleavage site in SARS-S was mapped to T678, when recombinant proteins were employed (Bosch et al. 2008), but evidence that T678 is important for SARS-CoV entry is lacking and cathepsin L-mediated cleavage of virion-associated SARS-S in target cells remains to be demonstrated. It is also unclear if cellular proteases other than cathepsin B and L can allow SARS-S-driven entry into certain target cells. An activating function of factor Xa has recently been suggested (Du et al. 2007) but the results await confirmation. Finally, it is noteworthy that engineered cleavage of SARS-S in virus-producing cells can ablate the need for cathepsin activity in target cells (Watanabe et al. 2008). This finding highlights the need to analyze if SARS-S is cleaved in primary lung cells and to determine if cathepsin activity is indeed required for viral spread in vivo – information pivotal to efforts aiming at the development of cathepsin inhibitors for antiviral therapy.

1.6 Membrane Fusion is Driven by Conserved Elements Located in the S2 Subunit of the SARS Spike Protein

The functional organization of SARS-S2 resembles that of the TMs of other class I fusion proteins and SARS-S-driven membrane fusion reaction follows the principles previously established for other class I fusion proteins (Hofmann and Pöhlmann 2004): membrane fusion commences by insertion of the fusion peptide into the target cell membrane (Fig. 1.3). In this context, it is worth noting that SARS-S, in contrast to, for example, HIV Env and influenza HA, contains an "internal" fusion peptide, which does not constitute the N terminus of S2 but may comprise amino acids 770–788 (Sainz et al. 2005). Upon fusion peptide insertion, the S2 subunit is connected with the viral and the target cell membrane. Subsequently, the C-terminal HR (termed HR2) folds back onto the N-terminal HR (termed HR1), forming an energetically stable six-helix bundle structure, in which HR1 and HR2 are oriented in an antiparallel fashion (Bosch et al. 2003; Tripet et al. 2004; Liu et al. 2004; Supekar et al. 2004; Ingallinella et al. 2004; Xu et al. 2004; Hsu et al. 2004). Thereby, viral and target cell membranes are pulled into close proximity, allowing the membranes to merge (Fig. 1.3). Peptides derived from HR2, which bind to HR1 and block the formation of the six-helix bundle, are used for therapy of HIV infection (Este and Telenti 2007). A similar approach was successful for blockade of SARS-CoV spread in cell culture (Bosch et al. 2003; Liu et al. 2004; Zhu et al. 2004; Yuan et al. 2004; Ni et al. 2005), but the inhibitors developed were not as potent as those used to treat HIV infection. One reason

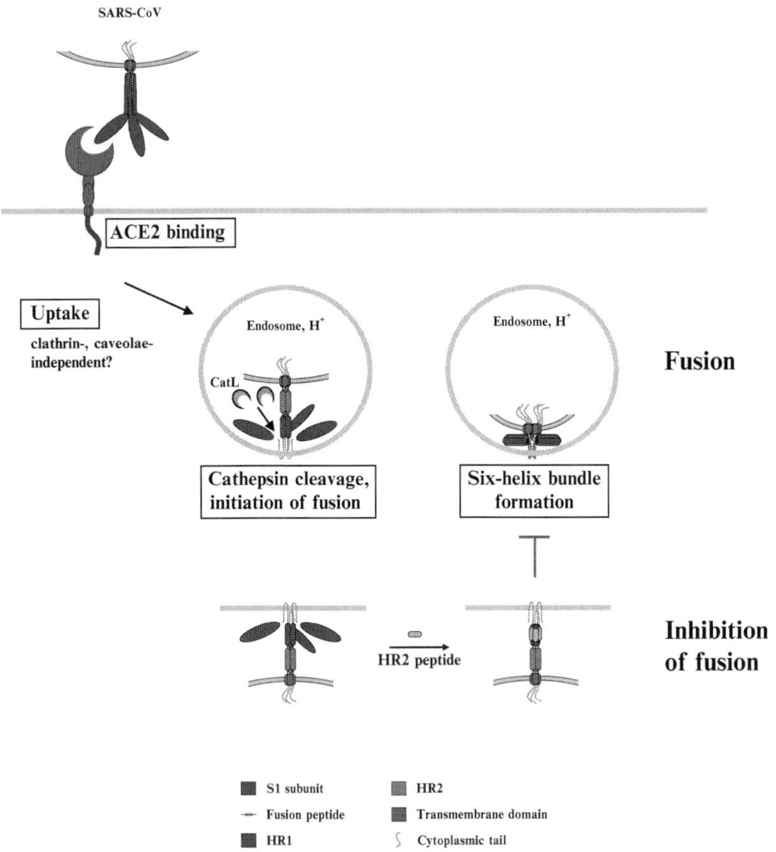

Fig. 1.3 Cellular entry of SARS-CoV and its inhibition (adapted from Hofmann and Pöhlmann 2004). The cellular entry of SARS-CoV commences by binding of the S protein to its receptor ACE2. Bound virus is then taken up into target cells, possibly by a clathrin- and caveolae-independent mechanism (Wang et al. 2008). The S protein is cleaved by the pH-dependent cellular protease cathepsin L in endosomes, and cathepsin L activity is essential for infectious entry. The membrane fusion reaction starts with the insertion of the fusion peptide into the target cell membrane. Formation of the stable six-helix bundle structure brings the viral and the target cell membrane into close proximity and is intimately associated with membrane fusion. The fusion reaction can be inhibited by HR2-derived peptides, which bind into a groove on HR1 and thereby prevent back-folding of HR2 onto HR1 and thus the formation of the six-helix bundle structure

for the decreased potency might be inherent to the cellular location of the membrane fusion reaction: the HIV Env protein drives fusion with the plasma cell membrane, and the target of the inhibitory peptides is readily accessible. In contrast, SARS-CoV fuses with endosomal membranes, and inhibitors must be taken up into endosomes to efficiently block the fusion reaction. Potentially, this could present a significant hurdle to the development of fusion inhibitors for therapy of SARS-CoV infection (Watanabe et al. 2008).

1.7 Conclusions

The cellular entry of SARS-CoV is a multistep process which involves the formation of several transient intermediates. All structures participating in the entry cascade are potential targets for inhibitors and the feasibility of several approaches to prevent entry has already been demonstrated. The first step, SARS-S engagement of ACE2, is an attractive target for both preventive and therapeutic approaches. Thus, the immunization with the RBD has been shown to elicit neutralizing antibodies (He et al. 2004a, 2005), and monoclonal RBD-specific antibodies which exhibit potent antiviral effects in animal models have been identified (Sui et al. 2004, 2005; Rockx et al. 2007; Zhu et al. 2007). In addition, SARS-S binding to ACE2 can be inhibited by nonpeptidic molecules targeting the receptor (Huentelman et al. 2004), and soluble ACE2 was shown to block SARS-CoV infection (Hofmann et al. 2004a) and to protect against ARDS (Kuba et al. 2005), making this approach particularly promising. The S protein can also be targeted by lectins, which bind glycans on the S protein and thereby block viral entry (van der Meer et al. 2007; Keyaerts et al. 2007), albeit issues concerning potential toxicity and antigenicity remain to be addressed. After binding to ACE2 and uptake into target cells, the S protein must be activated by cathepsin L, and potent cathepsin L inhibitors are available (Simmons et al. 2005). However, the role of cathepsin L in viral spread in vivo remains to be assessed, and knock-out mice (Reinheckel et al. 2001) might be suitable tools for these endeavours. Finally, inhibitors of the fusion reaction have been described (Bosch et al. 2003; Liu et al. 2004; Zhu et al. 2004; Yuan et al. 2004; Ni et al. 2005), but optimization of available compounds and generation of nonpeptidic compounds is desirable. In summary, the approaches described above, particularly combinations thereof, should allow development of compounds suitable for effectively preventing or combating future outbreaks of SARS-CoV.

Acknowledgments We thank T.F. Schulz for support and BMBF for funding.

References

Babcock GJ, Esshaki DJ, Thomas WD Jr, Ambrosino DM (2004) Amino acids 270 to 510 of the severe acute respiratory syndrome coronavirus spike protein are required for interaction with receptor. J Virol 78:4552–4560

Bashirova AA, Geijtenbeek TB, van Duijnhoven GC, van Vliet SJ, Eilering JB, Martin MP, Wu L, Martin TD, Viebig N, Knolle PA, Kewalramani VN, van Kooyk Y, Carrington M (2001) A dendritic cell-specific intercellular adhesion molecule 3-grabbing nonintegrin (DC-SIGN)-related protein is highly expressed on human liver sinusoidal endothelial cells and promotes HIV-1 infection. J Exp Med 193:671–678

Beniac DR, deVarennes SL, Andonov A, He R, Booth TF (2007) Conformational reorganization of the SARS coronavirus spike following receptor binding: implications for membrane fusion. PLoS ONE 2:e1082

Bergeron E, Vincent MJ, Wickham L, Hamelin J, Basak A, Nichol ST, Chretien M, Seidah NG (2005) Implication of proprotein convertases in the processing and spread of severe acute respiratory syndrome coronavirus. Biochem Biophys Res Commun 326:554–563

Bosch BJ, van der Zee R, de Haan CA, Rottier PJ (2003) The coronavirus spike protein is a class I virus fusion protein: structural and functional characterization of the fusion core complex. J Virol 77:8801–8811

Bosch BJ, Bartelink W, Rottier PJ (2008) Cathepsin L functionally cleaves the severe acute respiratory syndrome coronavirus class I fusion protein upstream of rather than adjacent to the fusion peptide. J Virol 82:8887–8890

Chan VS, Chan KY, Chen Y, Poon LL, Cheung AN, Zheng B, Chan KH, Mak W, Ngan HY, Xu X, Screaton G, Tam PK, Austyn JM, Chan LC, Yip SP, Peiris M, Khoo US, Lin CL (2006) Homozygous L-SIGN (CLEC4M) plays a protective role in SARS coronavirus infection. Nat Genet 38:38–46

de Haan CA, Stadler K, Godeke GJ, Bosch BJ, Rottier PJ (2004) Cleavage inhibition of the murine coronavirus spike protein by a furin-like enzyme affects cell–cell but not virus–cell fusion. J Virol 78:6048–6054

Ding Y, He L, Zhang Q, Huang Z, Che X, Hou J, Wang H, Shen H, Qiu L, Li Z, Geng J, Cai J, Han H, Li X, Kang W, Weng D, Liang P, Jiang S (2004) Organ distribution of severe acute respiratory syndrome (SARS) associated coronavirus (SARS-CoV) in SARS patients: implications for pathogenesis and virus transmission pathways. J Pathol 203:622–630

Drosten C, Gunther S, Preiser W, van der WS, Brodt HR, Becker S, Rabenau H, Panning M, Kolesnikova L, Fouchier RA, Berger A, Burguiere AM, Cinatl J, Eickmann M, Escriou N, Grywna K, Kramme S, Manuguerra JC, Muller S, Rickerts V, Sturmer M, Vieth S, Klenk HD, Osterhaus AD, Schmitz H, Doerr HW (2003) Identification of a novel coronavirus in patients with severe acute respiratory syndrome. N Engl J Med 348:1967–1976

Du L, Kao RY, Zhou Y, He Y, Zhao G, Wong C, Jiang S, Yuen KY, Jin DY, Zheng BJ (2007) Cleavage of spike protein of SARS coronavirus by protease factor Xa is associated with viral infectivity. Biochem Biophys Res Commun 359:174–179

Eckert DM, Kim PS (2001) Mechanisms of viral membrane fusion and its inhibition. Annu Rev Biochem 70:777–810

Este JA, Telenti A (2007) HIV entry inhibitors. Lancet 370:81–88

Follis KE, York J, Nunberg JH (2006) Furin cleavage of the SARS coronavirus spike glycoprotein enhances cell–cell fusion but does not affect virion entry. Virology 350:358–369

Fouchier RA, Hartwig NG, Bestebroer TM, Niemeyer B, de Jong JC, Simon JH, Osterhaus AD (2004) A previously undescribed coronavirus associated with respiratory disease in humans. Proc Natl Acad Sci USA 101:6212–6216

Geijtenbeek TB, Kwon DS, Torensma R, van Vliet SJ, van Duijnhoven GC, Middel J, Cornelissen IL, Nottet HS, Kewalramani VN, Littman DR, Figdor CG, van Kooyk Y (2000) DC-SIGN, a dendritic cell-specific HIV-1-binding protein that enhances trans-infection of T cells. Cell 100:587–597

Gramberg T, Hofmann H, Moller P, Lalor PF, Marzi A, Geier M, Krumbiegel M, Winkler T, Kirchhoff F, Adams DH, Becker S, Münch J, Pöhlmann S (2005) LSECtin interacts with filovirus glycoproteins and the spike protein of SARS coronavirus. Virology 340:224–236

Gu J, Korteweg C (2007) Pathology and pathogenesis of severe acute respiratory syndrome. Am J Pathol 170:1136–1147

Gu J, Gong E, Zhang B, Zheng J, Gao Z, Zhong Y, Zou W, Zhan J, Wang S, Xie Z, Zhuang H, Wu B, Zhong H, Shao H, Fang W, Gao D, Pei F, Li X, He Z, Xu D, Shi X, Anderson VM, Leong AS (2005) Multiple organ infection and the pathogenesis of SARS. J Exp Med 202:415–424

Guan Y, Zheng BJ, He YQ, Liu XL, Zhuang ZX, Cheung CL, Luo SW, Li PH, Zhang LJ, Guan YJ, Butt KM, Wong KL, Chan KW, Lim W, Shortridge KF, Yuen KY, Peiris JS, Poon LL (2003) Isolation and characterization of viruses related to the SARS coronavirus from animals in southern China. Science 302:276–278

Haga S, Yamamoto N, Nakai-Murakami C, Osawa Y, Tokunaga K, Sata T, Yamamoto N, Sasazuki T, Ishizaka Y (2008) Modulation of TNF-alpha-converting enzyme by the spike protein of SARS-CoV and ACE2 induces TNF-alpha production and facilitates viral entry. Proc Natl Acad Sci USA 105:7809–7814

Hamming I, Timens W, Bulthuis ML, Lely AT, Navis GJ, van Goor H (2004) Tissue distribution of ACE2 protein, the functional receptor for SARS coronavirus. A first step in understanding SARS pathogenesis. J Pathol 203:631–637

Han DP, Lohani M, Cho MW (2007) Specific asparagine-linked glycosylation sites are critical for DC-SIGN- and L-SIGN-mediated severe acute respiratory syndrome coronavirus entry. J Virol 81:12029–12039

He Y, Zhou Y, Liu S, Kou Z, Li W, Farzan M, Jiang S (2004a) Receptor-binding domain of SARS-CoV spike protein induces highly potent neutralizing antibodies: implication for developing subunit vaccine. Biochem Biophys Res Commun 324:773–781

He Y, Zhou Y, Wu H, Luo B, Chen J, Li W, Jiang S (2004b) Identification of immunodominant sites on the spike protein of severe acute respiratory syndrome (SARS) coronavirus: implication for developing SARS diagnostics and vaccines. J Immunol 173:4050–4057

He Y, Lu H, Siddiqui P, Zhou Y, Jiang S (2005) Receptor-binding domain of severe acute respiratory syndrome coronavirus spike protein contains multiple conformation-dependent epitopes that induce highly potent neutralizing antibodies. J Immunol 174:4908–4915

Hofmann H, Pöhlmann S (2004) Cellular entry of the SARS coronavirus. Trends Microbiol 12:466–472

Hofmann H, Geier M, Marzi A, Krumbiegel M, Peipp M, Fey GH, Gramberg T, Pöhlmann S (2004a) Susceptibility to SARS coronavirus S protein-driven infection correlates with expression of angiotensin converting enzyme 2 and infection can be blocked by soluble receptor. Biochem Biophys Res Commun 319:1216–1221

Hofmann H, Hattermann K, Marzi A, Gramberg T, Geier M, Krumbiegel M, Kuate S, Uberla K, Niedrig M, Pöhlmann S (2004b) S protein of severe acute respiratory syndrome-associated coronavirus mediates entry into hepatoma cell lines and is targeted by neutralizing antibodies in infected patients. J Virol 78:6134–6142

Hofmann H, Pyrc K, van der HL, Geier M, Berkhout B, Pöhlmann S (2005) Human coronavirus NL63 employs the severe acute respiratory syndrome coronavirus receptor for cellular entry. Proc Natl Acad Sci USA 102:7988–7993

Hofmann H, Simmons G, Rennekamp AJ, Chaipan C, Gramberg T, Heck E, Geier M, Wegele A, Marzi A, Bates P, Pöhlmann S (2006) Highly conserved regions within the spike proteins of human coronaviruses 229E and NL63 determine recognition of their respective cellular receptors. J Virol 80:8639–8652

Hsieh PK, Chang SC, Huang CC, Lee TT, Hsiao CW, Kou YH, Chen IY, Chang CK, Huang TH, Chang MF (2005) Assembly of severe acute respiratory syndrome coronavirus RNA packaging signal into virus-like particles is nucleocapsid dependent. J Virol 79:13848–13855

Hsu CH, Ko TP, Yu HM, Tang TK, Chen ST, Wang AH (2004) Immunological, structural, and preliminary X-ray diffraction characterizations of the fusion core of the SARS-coronavirus spike protein. Biochem Biophys Res Commun 324:761–767

Huang Y, Yang ZY, Kong WP, Nabel GJ (2004) Generation of synthetic severe acute respiratory syndrome coronavirus pseudoparticles: implications for assembly and vaccine production. J Virol 78:12557–12565

Huang IC, Bosch BJ, Li F, Li W, Lee KH, Ghiran S, Vasilieva N, Dermody TS, Harrison SC, Dormitzer PR, Farzan M, Rottier PJ, Choe H (2006) SARS coronavirus, but not human coronavirus NL63, utilizes cathepsin L to infect ACE2-expressing cells. J Biol Chem 281:3198–3203

Huentelman MJ, Zubcevic J, Hernandez Prada JA, Xiao X, Dimitrov DS, Raizada MK, Ostrov DA (2004) Structure-based discovery of a novel angiotensin-converting enzyme 2 inhibitor. Hypertension 44:903–906

Imai Y, Kuba K, Rao S, Huan Y, Guo F, Guan B, Yang P, Sarao R, Wada T, Leong-Poi H, Crackower MA, Fukamizu A, Hui CC, Hein L, Uhlig S, Slutsky AS, Jiang C, Penninger JM (2005) Angiotensin-converting enzyme 2 protects from severe acute lung failure. Nature 436:112–116

Imai Y, Kuba K, Penninger JM (2008) The discovery of angiotensin-converting enzyme 2 and its role in acute lung injury in mice. Exp Physiol 93:543–548

Ingallinella P, Bianchi E, Finotto M, Cantoni G, Eckert DM, Supekar VM, Bruckmann C, Carfi A, Pessi A (2004) Structural characterization of the fusion-active complex of severe acute respiratory syndrome (SARS) coronavirus. Proc Natl Acad Sci USA 101:8709–8714

Jeffers SA, Tusell SM, Gillim-Ross L, Hemmila EM, Achenbach JE, Babcock GJ, Thomas WD Jr, Thackray LB, Young MD, Mason RJ, Ambrosino DM, Wentworth DE, Demartini JC, Holmes KV (2004) CD209L (L-SIGN) is a receptor for severe acute respiratory syndrome coronavirus. Proc Natl Acad Sci USA 101:15748–15753

Keyaerts E, Vijgen L, Pannecouque C, Van Damme E, Peumans W, Egberink H, Balzarini J, Van Ranst M (2007) Plant lectins are potent inhibitors of coronaviruses by interfering with two targets in the viral replication cycle. Antivir Res 75:179–187

Khoo US, Chan KY, Chan VS, Lin CL (2008) DC-SIGN and L-SIGN: the SIGNs for infection. J Mol Med 86:861–874

Kielian M (2006) Class II virus membrane fusion proteins. Virology 344:38–47

Kielian M, Rey FA (2006) Virus membrane-fusion proteins: more than one way to make a hairpin. Nat Rev Microbiol 4:67–76

Ksiazek TG, Erdman D, Goldsmith CS, Zaki SR, Peret T, Emery S, Tong S, Urbani C, Comer JA, Lim W, Rollin PE, Dowell SF, Ling AE, Humphrey CD, Shieh WJ, Guarner J, Paddock CD, Rota P, Fields B, DeRisi J, Yang JY, Cox N, Hughes JM, LeDuc JW, Bellini WJ, Anderson LJ (2003) A novel coronavirus associated with severe acute respiratory syndrome. N Engl J Med 348:1953–1966

Kuba K, Imai Y, Rao S, Gao H, Guo F, Guan B, Huan Y, Yang P, Zhang Y, Deng W, Bao L, Zhang B, Liu G, Wang Z, Chappell M, Liu Y, Zheng D, Leibbrandt A, Wada T, Slutsky AS, Liu D, Qin C, Jiang C, Penninger JM (2005) A crucial role of angiotensin converting enzyme 2 (ACE2) in SARS coronavirus-induced lung injury. Nat Med 11:875–879

Kuba K, Imai Y, Rao S, Jiang C, Penninger JM (2006) Lessons from SARS: control of acute lung failure by the SARS receptor ACE2. J Mol Med 84:814–820

Lanoue A, Clatworthy MR, Smith P, Green S, Townsend MJ, Jolin HE, Smith KG, Fallon PG, McKenzie AN (2004) SIGN-R1 contributes to protection against lethal pneumococcal infection in mice. J Exp Med 200:1383–1393

Lau SK, Woo PC, Li KS, Huang Y, Tsoi HW, Wong BH, Wong SS, Leung SY, Chan KH, Yuen KY (2005) Severe acute respiratory syndrome coronavirus-like virus in Chinese horseshoe bats. Proc Natl Acad Sci USA 102:14040–14045

Li F (2008) Structural analysis of major species barriers between humans and palm civets for severe acute respiratory syndrome coronavirus infections. J Virol 82:6984–6991

Li W, Moore MJ, Vasilieva N, Sui J, Wong SK, Berne MA, Somasundaran M, Sullivan JL, Luzuriaga K, Greenough TC, Choe H, Farzan M (2003) Angiotensin-converting enzyme 2 is a functional receptor for the SARS coronavirus. Nature 426:450–454

Li W, Greenough TC, Moore MJ, Vasilieva N, Somasundaran M, Sullivan JL, Farzan M, Choe H (2004) Efficient replication of severe acute respiratory syndrome coronavirus in mouse cells is limited by murine angiotensin-converting enzyme 2. J Virol 78:11429–11433

Li F, Li W, Farzan M, Harrison SC (2005a) Structure of SARS coronavirus spike receptor-binding domain complexed with receptor I2005. Science 309:1864–1868

Li W, Shi Z, Yu M, Ren W, Smith C, Epstein JH, Wang H, Crameri G, Hu Z, Zhang H, Zhang J, McEachern J, Field H, Daszak P, Eaton BT, Zhang S, Wang LF (2005b) Bats are natural reservoirs of SARS-like coronaviruses. Science 310:676–679

Li W, Zhang C, Sui J, Kuhn JH, Moore MJ, Luo S, Wong SK, Huang IC, Xu K, Vasilieva N, Murakami A, He Y, Marasco WA, Guan Y, Choe H, Farzan M (2005c) Receptor and viral determinants of SARS-coronavirus adaptation to human ACE2. EMBO J 24:1634–1643

Li W, Sui J, Huang IC, Kuhn JH, Radoshitzky SR, Marasco WA, Choe H, Farzan M (2007) The S proteins of human coronavirus NL63 and severe acute respiratory syndrome coronavirus bind overlapping regions of ACE2. Virology 367:367–374

Lin HX, Feng Y, Wong G, Wang L, Li B, Zhao X, Li Y, Smaill F, Zhang C (2008) Identification of residues in the receptor-binding domain (RBD) of the spike protein of human coronavirus NL63 that are critical for the RBD-ACE2 receptor interaction. J Gen Virol 89:1015–1024

Liu S, Xiao G, Chen Y, He Y, Niu J, Escalante CR, Xiong H, Farmar J, Debnath AK, Tien P, Jiang S (2004) Interaction between heptad repeat 1 and 2 regions in spike protein of SARS-associated coronavirus: implications for virus fusogenic mechanism and identification of fusion inhibitors. Lancet 363:938–947

Marzi A, Gramberg T, Simmons G, Moller P, Rennekamp AJ, Krumbiegel M, Geier M, Eisemann J, Turza N, Saunier B, Steinkasserer A, Becker S, Bates P, Hofmann H, Pöhlmann S (2004) DC-SIGN and DC-SIGNR interact with the glycoprotein of Marburg virus and the S protein of severe acute respiratory syndrome Coronavirus. J Virol 78:12090–12095

Mathewson AC, Bishop A, Yao Y, Kemp F, Ren J, Chen H, Xu X, Berkhout B, van der HL, Jones IM (2008) Interaction of severe acute respiratory syndrome-coronavirus and NL63 coronavirus spike proteins with angiotensin converting enzyme-2. J Gen Virol 89:2741–2745

McBride CE, Li J, Machamer CE (2007) The cytoplasmic tail of the severe acute respiratory syndrome coronavirus spike protein contains a novel endoplasmic reticulum retrieval signal that binds COPI and promotes interaction with membrane protein. J Virol 81:2418–2428

Mossel EC, Huang C, Narayanan K, Makino S, Tesh RB, Peters CJ (2005) Exogenous ACE2 expression allows refractory cell lines to support severe acute respiratory syndrome coronavirus replication. J Virol 79:3846–3850

Mossel EC, Wang J, Jeffers S, Edeen KE, Wang S, Cosgrove GP, Funk CJ, Manzer R, Miura TA, Pearson LD, Holmes KV, Mason RJ (2008) SARS-CoV replicates in primary human alveolar type II cell cultures but not in type I-like cells. Virology 372:127–135

Nal B, Chan C, Kien F, Siu L, Tse J, Chu K, Kam J, Staropoli I, Crescenzo-Chaigne B, Escriou N, van der WS, Yuen KY, Altmeyer R (2005) Differential maturation and subcellular localization of severe acute respiratory syndrome coronavirus surface proteins S, M and E. J Gen Virol 86:1423–1434

Nash TC, Buchmeier MJ (1997) Entry of mouse hepatitis virus into cells by endosomal and nonendosomal pathways. Virology 233:1–8

Ni L, Zhu J, Zhang J, Yan M, Gao GF, Tien P (2005) Design of recombinant protein-based SARS-CoV entry inhibitors targeting the heptad-repeat regions of the spike protein S2 domain. Biochem Biophys Res Commun 330:39–45

Nie Y, Wang P, Shi X, Wang G, Chen J, Zheng A, Wang W, Wang Z, Qu X, Luo M, Tan L, Song X, Yin X, Chen J, Ding M, Deng H (2004) Highly infectious SARS-CoV pseudotyped virus reveals the cell tropism and its correlation with receptor expression. Biochem Biophys Res Commun 321:994–1000

Peiris JS, Lai ST, Poon LL, Guan Y, Yam LY, Lim W, Nicholls J, Yee WK, Yan WW, Cheung MT, Cheng VC, Chan KH, Tsang DN, Yung RW, Ng TK, Yuen KY (2003a) Coronavirus as a possible cause of severe acute respiratory syndrome. Lancet 361:1319–1325

Peiris JS, Yuen KY, Osterhaus AD, Stohr K (2003b) The severe acute respiratory syndrome. N Engl J Med 349:2431–2441

Peiris JS, Guan Y, Yuen KY (2004) Severe acute respiratory syndrome. Nat Med 10:S88–S97

Penninger J, Imai Y, Kuba K (2008) The discovery of ACE2 and its role in acute lung injury. Exp Physiol 93:543–548

Pöhlmann S, Soilleux EJ, Baribaud F, Leslie GJ, Morris LS, Trowsdale J, Lee B, Coleman N, Doms RW (2001) DC-SIGNR, a DC-SIGN homologue expressed in endothelial cells, binds to human and simian immunodeficiency viruses and activates infection in trans. Proc Natl Acad Sci USA 98:2670–2675

Pyrc K, Jebbink MF, Berkhout B, van der HL (2004) Genome structure and transcriptional regulation of human coronavirus NL63. Virol J 1:7

Pyrc K, Berkhout B, van der HL (2007) Identification of new human coronaviruses. Expert Rev Anti Infect Ther 5:245–253

Qiu Z, Hingley ST, Simmons G, Yu C, Das SJ, Bates P, Weiss SR (2006) Endosomal proteolysis by cathepsins is necessary for murine coronavirus mouse hepatitis virus type 2 spike-mediated entry. J Virol 80:5768–5776

Reinheckel T, Deussing J, Roth W, Peters C (2001) Towards specific functions of lysosomal cysteine peptidases: phenotypes of mice deficient for cathepsin B or cathepsin L. Biol Chem 382:735–741

Ren W, Qu X, Li W, Han Z, Yu M, Zhou P, Zhang SY, Wang LF, Deng H, Shi Z (2008) Difference in receptor usage between severe acute respiratory syndrome (SARS) coronavirus and SARS-like coronavirus of bat origin. J Virol 82:1899–1907

Rockx B, Sheahan T, Donaldson E, Harkema J, Sims A, Heise M, Pickles R, Cameron M, Kelvin D, Baric R (2007) Synthetic reconstruction of zoonotic and early human severe acute respiratory syndrome coronavirus isolates that produce fatal disease in aged mice. J Virol 81:7410–7423

Sainz B Jr, Rausch JM, Gallaher WR, Garry RF, Wimley WC (2005) Identification and characterization of the putative fusion peptide of the severe acute respiratory syndrome-associated coronavirus spike protein. J Virol 79:7195–7206

Schaefer M, Reiling N, Fessler C, Stephani J, Taniuchi I, Hatam F, Yildirim AO, Fehrenbach H, Walter K, Ruland J, Wagner H, Ehlers S, Sparwasser T (2008) Decreased pathology and prolonged survival of human DC-SIGN transgenic mice during mycobacterial infection. J Immunol 180:6836–6845

Shih YP, Chen CY, Liu SJ, Chen KH, Lee YM, Chao YC, Chen YM (2006) Identifying epitopes responsible for neutralizing antibody and DC-SIGN binding on the spike glycoprotein of the severe acute respiratory syndrome coronavirus. J Virol 80:10315–10324

Simmons G, Reeves JD, Rennekamp AJ, Amberg SM, Piefer AJ, Bates P (2004) Characterization of severe acute respiratory syndrome-associated coronavirus (SARS-CoV) spike glycoprotein-mediated viral entry. Proc Natl Acad Sci USA 101:4240–4245

Simmons G, Gosalia DN, Rennekamp AJ, Reeves JD, Diamond SL, Bates P (2005) Inhibitors of cathepsin L prevent severe acute respiratory syndrome coronavirus entry. Proc Natl Acad Sci USA 102:11876–11881

Sims AC, Burkett SE, Yount B, Pickles RJ (2008) SARS-CoV replication and pathogenesis in an in vitro model of the human conducting airway epithelium. Virus Res 133:33–44

Siu YL, Teoh KT, Lo J, Chan CM, Kien F, Escriou N, Tsao SW, Nicholls JM, Altmeyer R, Peiris JS, Bruzzone R, Nal B (2008) The M, E, and N structural proteins of the severe acute respiratory syndrome coronavirus are required for efficient assembly, trafficking, and release of virus-like particles. J Virol 82:11318–11330

Song HD, Tu CC, Zhang GW, Wang SY, Zheng K, Lei LC, Chen QX, Gao YW, Zhou HQ, Xiang H, Zheng HJ, Chern SW, Cheng F, Pan CM, Xuan H, Chen SJ, Luo HM, Zhou DH, Liu YF, He JF, Qin PZ, Li LH, Ren YQ, Liang WJ, Yu YD, Anderson L, Wang M, Xu RH, Wu XW, Zheng HY, Chen JD, Liang G, Gao Y, Liao M, Fang L, Jiang LY, Li H, Chen F, Di B, He LJ, Lin JY, Tong S, Kong X, Du L, Hao P, Tang H, Bernini A, Yu XJ, Spiga O, Guo ZM, Pan HY, He WZ, Manuguerra JC, Fontanet A, Danchin A, Niccolai N, Li YX, Wu CI, Zhao GP (2005) Cross-host evolution of severe acute respiratory syndrome coronavirus in palm civet and human. Proc Natl Acad Sci USA 102:2430–2435

Stadler K, Rappuoli R (2005) SARS: understanding the virus and development of rational therapy. Curr Mol Med 5:677–697

Stadler K, Masignani V, Eickmann M, Becker S, Abrignani S, Klenk HD, Rappuoli R (2003) SARS–beginning to understand a new virus. Nat Rev Microbiol 1:209–218

Stertz S, Reichelt M, Spiegel M, Kuri T, Martinez-Sobrido L, Garcia-Sastre A, Weber F, Kochs G (2007) The intracellular sites of early replication and budding of SARS-coronavirus. Virology 361:304–315

Sui J, Li W, Murakami A, Tamin A, Matthews LJ, Wong SK, Moore MJ, Tallarico AS, Olurinde M, Choe H, Anderson LJ, Bellini WJ, Farzan M, Marasco WA (2004) Potent neutralization of severe acute respiratory syndrome (SARS) coronavirus by a human mAb to S1 protein that blocks receptor association. Proc Natl Acad Sci USA 101:2536–2541

Sui J, Li W, Roberts A, Matthews LJ, Murakami A, Vogel L, Wong SK, Subbarao K, Farzan M, Marasco WA (2005) Evaluation of human monoclonal antibody 80R for immunoprophylaxis of severe acute respiratory syndrome by an animal study, epitope mapping, and analysis of spike variants. J Virol 79:5900–5906

Supekar VM, Bruckmann C, Ingallinella P, Bianchi E, Pessi A, Carfi A (2004) Structure of a proteolytically resistant core from the severe acute respiratory syndrome coronavirus S2 fusion protein. Proc Natl Acad Sci USA 101:17958–17963

Tang NL, Chan PK, HDS To KF, Zhang W, Chan FK, Sung JJ, Lo YM (2007) Lack of support for an association between CLEC4M homozygosity and protection against SARS coronavirus infection. Nat Genet 39:691–692

To KF, Lo AW (2004) Exploring the pathogenesis of severe acute respiratory syndrome (SARS): the tissue distribution of the coronavirus (SARS-CoV) and its putative receptor, angiotensin-converting enzyme 2 (ACE2). J Pathol 203:740–743

Towler P, Staker B, Prasad SG, Menon S, Tang J, Parsons T, Ryan D, Fisher M, Williams D, Dales NA, Patane MA, Pantoliano MW (2004) ACE2 X-ray structures reveal a large hinge-bending motion important for inhibitor binding and catalysis. J Biol Chem 279:17996–18007

Tripet B, Howard MW, Jobling M, Holmes RK, Holmes KV, Hodges RS (2004) Structural characterization of the SARS-coronavirus spike S fusion protein core. J Biol Chem 279: 20836–20849

van der Hoek L, Pyrc K, Jebbink MF, Vermeulen-Oost W, Berkhout RJ, Wolthers KC, Wertheim-van Dillen PM, Kaandorp J, Spaargaren J, Berkhout B (2004) Identification of a new human coronavirus. Nat Med 10:368–373

van der Meer FJ, de Haan CA, Schuurman NM, Haijema BJ, Peumans WJ, Van Damme EJ, Delputte PL, Balzarini J, Egberink HF (2007) Antiviral activity of carbohydrate-binding agents against Nidovirales in cell culture. Antivir Res 76:21–29

Wang P, Chen J, Zheng A, Nie Y, Shi X, Wang W, Wang G, Luo M, Liu H, Tan L, Song X, Wang Z, Yin X, Qu X, Wang X, Qing T, Ding M, Deng H (2004) Expression cloning of functional receptor used by SARS coronavirus. Biochem Biophys Res Commun 315:439–444

Wang H, Yang P, Liu K, Guo F, Zhang Y, Zhang G, Jiang C (2008) SARS coronavirus entry into host cells through a novel clathrin- and caveolae-independent endocytic pathway. Cell Res 18: 290–301

Watanabe R, Matsuyama S, Shirato K, Maejima M, Fukushi S, Morikawa S, Taguchi F (2008) Entry from the cell surface of severe acute respiratory syndrome coronavirus with cleaved s protein as revealed by pseudotype virus bearing cleaved s protein. J Virol 82:11985–11991

Williams RK, Jiang GS, Holmes KV (1991) Receptor for mouse hepatitis virus is a member of the carcinoembryonic antigen family of glycoproteins. Proc Natl Acad Sci USA 88:5533–5536

Wong SK, Li W, Moore MJ, Choe H, Farzan M (2004) A 193-amino acid fragment of the SARS coronavirus S protein efficiently binds angiotensin-converting enzyme 2. J Biol Chem 279:3197–3201

Wu XD, Shang B, Yang RF, Yu H, Ma ZH, Shen X, Ji YY, Lin Y, Wu YD, Lin GM, Tian L, Gan XQ, Yang S, Jiang WH, Dai EH, Wang XY, Jiang HL, Xie YH, Zhu XL, Pei G, Li L, Wu JR, Sun B (2004) The spike protein of severe acute respiratory syndrome (SARS) is cleaved in virus infected Vero-E6 cells. Cell Res 14:400–406

Xiao X, Chakraborti S, Dimitrov AS, Gramatikoff K, Dimitrov DS (2003) The SARS-CoV S glycoprotein: expression and functional characterization. Biochem Biophys Res Commun 312:1159–1164

Xu Y, Lou Z, Liu Y, Pang H, Tien P, Gao GF, Rao Z (2004) Crystal structure of severe acute respiratory syndrome coronavirus spike protein fusion core. J Biol Chem 279:49414–49419

Yang ZY, Huang Y, Ganesh L, Leung K, Kong WP, Schwartz O, Subbarao K, Nabel GJ (2004) pH-Dependent entry of severe acute respiratory syndrome coronavirus is mediated by the spike glycoprotein and enhanced by dendritic cell transfer through DC-SIGN. J Virol 78:5642–5650

Yao YX, Ren J, Heinen P, Zambon M, Jones IM (2004) Cleavage and serum reactivity of the severe acute respiratory syndrome coronavirus spike protein. J Infect Dis 190:91–98

Yen YT, Liao F, Hsiao CH, Kao CL, Chen YC, Wu-Hsieh BA (2006) Modeling the early events of severe acute respiratory syndrome coronavirus infection in vitro. J Virol 80:2684–2693

Yuan K, Yi L, Chen J, Qu X, Qing T, Rao X, Jiang P, Hu J, Xiong Z, Nie Y, Shi X, Wang W, Ling C, Yin X, Fan K, Lai L, Ding M, Deng H (2004) Suppression of SARS-CoV entry by peptides corresponding to heptad regions on spike glycoprotein. Biochem Biophys Res Commun 319:746–752

Zhi L, Zhou G, Zhang H, Zhai Y, Yang H, Zhang F, Wang S, Wei M, He F (2007) Lack of support for an association between CLEC4M homozygosity and protection against SARS coronavirus infection. Nat Genet 39:692–694

Zhu J, Xiao G, Xu Y, Yuan F, Zheng C, Liu Y, Yan H, Cole DK, Bell JI, Rao Z, Tien P, Gao GF (2004) Following the rule: formation of the 6-helix bundle of the fusion core from severe acute respiratory syndrome coronavirus spike protein and identification of potent peptide inhibitors. Biochem Biophys Res Commun 319:283–288

Zhu Z, Chakraborti S, He Y, Roberts A, Sheahan T, Xiao X, Hensley LE, Prabakaran P, Rockx B, Sidorov IA, Corti D, Vogel L, Feng Y, Kim JO, Wang LF, Baric R, Lanzavecchia A, Curtis KM, Nabel GJ, Subbarao K, Jiang S, Dimitrov DS (2007) Potent cross-reactive neutralization of SARS coronavirus isolates by human monoclonal antibodies. Proc Natl Acad Sci USA 104:12123–12128

Chapter 2
The Cell Biology of the SARS Coronavirus Receptor, Angiotensin-Converting Enzyme 2

Daniel W. Lambert

Abstract The identification of angiotensin-converting enzyme 2 (ACE2) as a cellular receptor for the SARS coronavirus (SARS-CoV) rejuvenated research into what was regarded by some as a minor player in the renin–angiotensin system. The discovery of its double life led to breathtaking advances in the understanding of virtually all aspects of its biology, including its structure, physiological and patho-physiological roles and cell biology. ACE2, like its well-known homologue, ACE, is a metallopeptidase which resides on the cell surface of the epithelial, and sometimes endothelial, cells of the heart, kidney, testes, lung and gastrointestinal tract. It is a type I transmembrane protein with a large catalytic extracellular domain which acts as both a peptidase and a viral receptor. This extracellular domain can be cleaved from the cell surface by other peptidases, modulating its activity. The levels of the enzyme on the cell surface are also thought to be regulated by internalisation on S-protein binding and by clustering in membrane microdomains known as lipid rafts. This chapter summarises the current understanding of how the cell biology of ACE2 is regulated and may influence and determine its function, and concludes by discussing the future challenges and opportunities for studies of this increasingly important enzyme.

2.1 Introduction

Angiotensin-converting enzyme 2 (ACE2) was first identified in 2000 simultaneously by two groups using distinct methodologies (Donoghue et al. 2000; Tipnis et al. 2000). Its close mammalian homologue, angiotensin-converting enzyme (ACE), is a well-characterised angiotensinase and prominent therapeutic target in

D.W. Lambert
Department of Oral and Maxillofacial Pathology, Faculty of Medicine, Dentistry and Health, University of Sheffield, UK S10 2TA
e-mail: d.w.lambert@sheffield.ac.uk

S.K. Lal (ed.), *Molecular Biology of the SARS-Coronavirus*,
DOI 10.1007/978-3-642-03683-5_2,

hypertension, leading to a concentration of early studies of ACE2 on its substrate specificities and role in the renin–angiotensin system (RAS). Like ACE, ACE2 is a zinc metallopeptidase which is able to hydrolyse a wide variety of substrates. Of these, the best studied in the context of ACE2 are angiotensin I (Ang I) and angiotensin II (Ang II), peptides involved in regulating blood pressure and tissue fibrosis. Although able to cleave both peptides, it has become clear that the mitogenic and hypertensive peptide Ang II is the predominant physiological substrate of ACE2, being cleaved to the vasodilatory peptide angiotensin-(1–7). This suggested that ACE2 is therefore likely to have a beneficial role in cardiovascular disease, a finding which slowed research efforts due to its unsuitability as a target for conventional pharmacological intervention. The discovery of a role for ACE2 as a receptor for the SARS coronavirus (SARS-CoV) (Li et al. 2005), however, led to a reinvigoration and diversification of research effort toward understanding the tissue distribution and cell biology of ACE2.

2.2 Clues from Homologous Proteins

ACE2 is an 805-amino-acid glycoprotein bearing significant sequence homology in its N-terminal domain to somatic ACE and in its cytoplasmic, C-terminal domain to collectrin, also known as Tmem27 (Fig. 2.1). Analysis of the amino acid sequence of ACE2 reveals a putative signal peptide and transmembrane domain, indicating that it, like its homologue ACE, is expressed as a type I (N-terminal domain extracellular) transmembrane protein. The extracellular domain shares significant sequence identity with the equivalent region of ACE, but unlike somatic ACE, contains only a single HEMGH zinc-binding catalytic motif, as is the case with the germinal isoform of ACE (Fig. 2.1). The intracellular, carboxy-terminal region of ACE2, however, shares no homology with ACE but instead closely resembles that

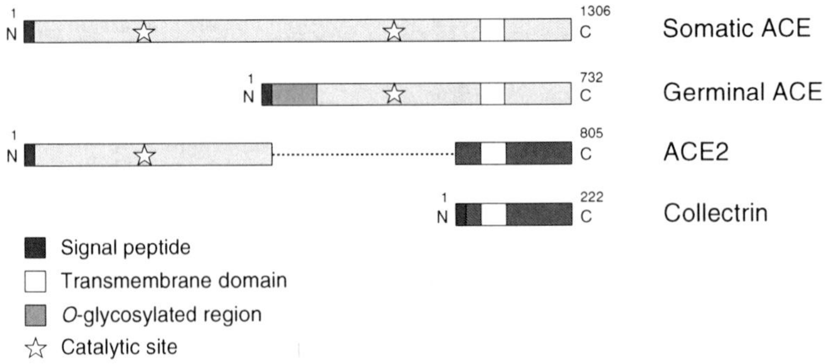

Fig. 2.1 Alignment of ACE2 sequence with homologous proteins. Regions of homology are indicated with *shading*. All four proteins contain signal peptides and transmembrane regions, but collectrin contains no catalytic residues. ACE2 is homologous with both the N terminus of somatic ACE and the C terminus of collectrin

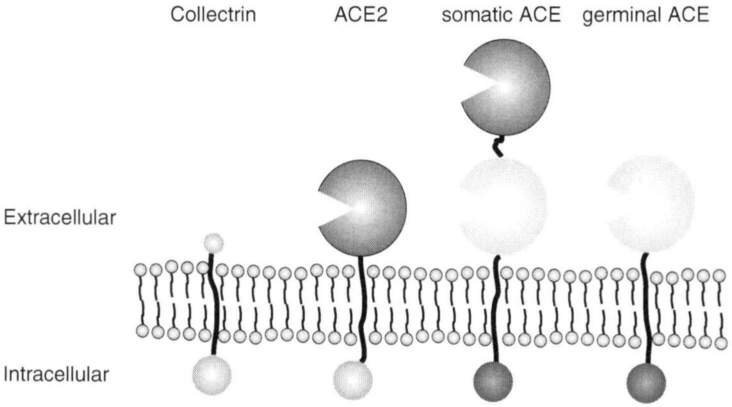

Fig. 2.2 Orientation of ACE2 and its homologues in the plasma membrane. ACE2 and its homologues are type I membrane proteins, with an extracellular amino-terminal domain and an intracellular carboxy-terminal domain. ACE2, somatic ACE and germinal ACE contain catalytic sites (represented as "*Pacman*" *shapes*) in the extracellular domain; collectrin does not

of collectrin, a non-catalytic protein with a small extracellular domain expressed in the kidney (Zhang et al. 2001) and pancreas (Fukui et al. 2005). This structure suggests the possibility that ACE2 may represent a gene fusion product between ACE and collectrin. The regions of homology between the four proteins are further illustrated in Fig. 2.2, which illustrates the orientation of the proteins in the plasma membrane.

The membrane localisation of ACE2 and ACE is in keeping with their roles in the RAS, allowing their extracellular catalytic sites to cleave circulating angiotensin (and other) peptides. In polarised epithelial cells in culture, ACE2 is trafficked predominantly to the apical membrane, with little detectable in the basolateral compartment (Warner et al. 2005). Interestingly, ACE displays a different localisation, being equally distributed between apical and basolateral membrane compartments. While the mechanisms responsible for this difference have yet to be identified, it is likely that distinct targeting motifs may reside in the disparate cytoplasmic domains of the two proteins. This suggestion is reinforced by the primarily apical expression of collectrin (Zhang et al. 2001), which shares homology in its cytoplasmic domain with ACE2 but not ACE, in collecting duct epithelial cells in the kidney.

In vivo, ACE2 is expressed predominantly in the heart, kidneys and testes (Tipnis et al. 2000), and to a lesser extent the lung and gastrointestinal tract, with low levels detectable in most tissues (Hamming et al. 2004). In the heart, ACE2 is expressed predominantly in cardiac myofibroblasts (Guy et al. 2008), cardiac myocytes and endothelial cells (Burrell et al. 2005), although this distribution is reported to vary between species. In the kidney, ACE2 is expressed in proximal and distal tubular epithelial cells, with low levels detectable in the glomeruli. Immunohistochemical analysis demonstrates a predominantly membranous expression pattern for ACE2 in these cells, with immunoreactivity strongest in the apical brush border (Brosnihan et al. 2003). These findings are in keeping with the observed localisation in polarised kidney epithelial cells in culture (Warner et al. 2005). In the lung ACE2 is primarily confined to the epithelium, with cell surface expression detected in Clara and type II cells, but is also found in smooth muscle and endothelial cells (Wiener et al. 2007). In lung epithelial cells grown in culture, ACE2 is expressed predominantly in the apical membrane compartment (Ren et al. 2006), in keeping with its role as a receptor for SARS-CoV.

2.3 Regulation of ACE2 Expression on the Cell Surface

The levels and function of cell-surface proteins may be controlled in a number of ways, including modulation of gene expression, shedding of the protein from the cell surface, internalisation and clustering in lipid microdomains within the plasma membrane. This chapter will concentrate on the mechanisms regulating the levels of the mature ACE2 protein on the cell surface.

2.3.1 Proteolytic Cleavage Secretion

Many membrane proteins, particularly type I transmembrane proteins, undergo a proteolytic cleavage secretion event, more commonly referred to as "shedding," in which the ectodomain of the protein is cleaved by a proteinase, often a member of the matrix metalloproteinase (MMP) or a disintegrin and metalloproteinase (ADAM) families, and released into the extracellular milieu (illustrated in Fig. 2.3) (Huovila et al. 2005). This process may serve to release a ligand, allowing it to bind to its receptor (e.g., cytokines such as TNF-α), or simply to downregulate

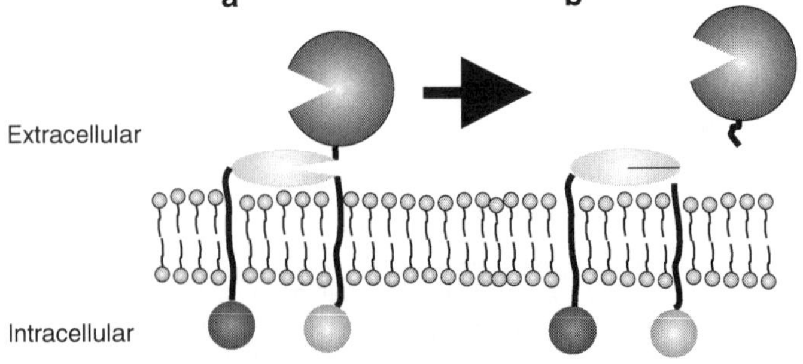

Fig. 2.3 Ectodomain shedding. Many transmembrane proteins, particularly those with an extracellular amino-terminal domain, are subject to a "shedding" event in which an intramembrane proteinase cleaves the juxtamembrane region of the target protein (**a**), releasing its ectodomain into the extracellular milieu (**b**)

the levels or activity of a protein on the cell surface. ACE2 (along with its homologue ACE) is subject to such an ectodomain shedding event, releasing a catalytically active ectodomain, a process regulated by protein kinase C activation and involving a member of the ADAM family, TACE (TNF-α converting enzyme) (Lambert et al. 2005, 2008). While the physiological significance of this shedding event is not clear, increased levels of circulating ACE2 have been detected in cardiovascular disease (Shaltout et al. 2008), and the ability of cleaved (soluble) ACE2 to reduce SARS-CoV infectivity is well established (Li et al. 2003). Intriguingly, however, siRNA-mediated TACE downregulation reduces the ability of SARS to infect Huh7 cells (Haga et al. 2008), suggesting the role of ACE2 shedding in SARS infection is more complex than is readily apparent.

Commonly, the transmembrane regions of shed proteins are subsequently subject to further intramembrane cleavage, generating a short carboxy-terminal fragment, a process termed regulated intramembrane proteolysis (RIP) (Medina and Dotti 2003). It has been demonstrated for a number of proteins, most notably notch and Alzheimer's precursor protein (APP) but also the ACE2 homologue, ACE (Fleming 2006), that this carboxy-terminal fragment is able to trigger signalling events leading to changes in the expression of target genes. Whether such a signalling mechanism occurs following ectodomain shedding of ACE2 remains to be established. The cytoplasmic domain of ACE2 is known, however, to have a regulatory role, both in terms of ectodomain shedding (Lambert et al. 2008) and SARS infectivity (Haga et al. 2008). Association of the cytoplasmic tail with a ubiquitous calcium-binding protein, calmodulin, reduces the release of its ectodomain suggesting a role for calmodulin in regulating ACE2 expression on the cell surface. The role of the cytoplasmic domain on SARS infection is controversial; Haga et al. (2008) recently reported that entry of SARS-CoV is dependent on the presence of the cytoplasmic domain of ACE2, a finding in direct contrast to those of Pohlmann et al. (2006) and Inoue et al. (2007) who suggest that entry is not dependent on the presence of this domain. These differences remain to be resolved but are likely due to the different experimental systems used.

2.3.2 *The Role of Membrane Microdomains*

It is thought that within the plane of plasma membranes, clusters of lipids such as sphingolipids and cholesterol form microdomains often termed lipid rafts. Although still somewhat controversial, a large body of evidence indicates that lipid rafts influence signalling and protein–protein interactions by partitioning and clustering proteins. Much of this evidence comes from studies in which cellular cholesterol is depleted using agents such as methyl-β-cyclodextrin. Cholesterol depletion alters the ability of a number of viruses to infect mammalian cells, including SARS-CoV. Studies by Glende et al. (2008) have revealed cholesterol dependence for SARS-CoV entry into cells on the presence of lipid rafts, possibly due to clustering of ACE2 into these microdomains. Furthermore, it has been demonstrated that virus entry is mediated by internalisation of ACE2 upon S-protein binding into endosomes by a clathrin- and caveolin-independent mechanism involving lipid rafts (Wang et al. 2008). A degree of controversy remains about the role of membrane microdomains in regulating SARS-CoV entry, however, as others have failed to detect ACE2 in lipid raft preparations (Warner et al. 2005). The reasons for these discrepancies remain unclear, but are likely to be due to the use of heterologously- or endogenously-expressed ACE2 and/or differences in lipid raft preparation methodologies.

2.4 Conclusions and Future Perspectives

The serendipitous discovery of ACE2 as the cellular receptor for SARS-CoV rejuvenated studies analysing the cell biology of a protein previously thought by some only to be a minor player in the RAS. This reinvigoration of research not only led to important discoveries regarding the mechanisms regulating the expression of ACE2 at the cell surface, impacting on its function as the SARS-CoV receptor, but also helped stimulate studies which revealed an unexpectedly significant role for ACE2 in the RAS. Further work is required to fully elucidate the mechanisms regulating the cell surface function of ACE2; it is likely to interact with as-yet-unidentified proteins and may turn out to have intracellular signalling functions which influence its function and the function of other proteins. At present, most of the cell biological studies of ACE2 have been directed at analysing post-transcriptional events regulating its function. Changes in the levels of ACE2, however, have been identified in a wide variety of pathologies, suggesting that transcriptional and post-transcriptional regulatory mechanisms may also have an important role. Indeed, recent studies have indicated a number of pathways which may regulate ACE2 at the molecular level. Whatever the focus of future studies turns out to be, however, it seems unlikely that ACE2 has given up all its secrets yet.

References

Brosnihan KB, Neves LA, Joyner J, Averill DB, Chappell MC, Sarao R, Penninger J, Ferrario CM (2003) Enhanced renal immunocytochemical expression of ANG-(1–7) and ACE2 during pregnancy. Hypertension 42:749–753

Burrell LM, Risvanis J, Kubota E, Dean RG, MacDonald PS, Lu S, Tikellis C, Grant SL, Lew RA, Smith AI, Cooper ME, Johnston CI (2005) Myocardial infarction increases ACE2 expression in rat and humans. Eur Heart J 26:369–375; discussion 322–364

Donoghue M, Hsieh F, Baronas E, Godbout K, Gosselin M, Stagliano N, Donovan M, Woolf B, Robison K, Jeyaseelan R, Breitbart RE, Acton S (2000) A novel angiotensin-converting enzyme-related carboxypeptidase (ACE2) converts angiotensin I to angiotensin 1–9. Circ Res 87:E1–E9

Fleming I (2006) Signaling by the angiotensin-converting enzyme. Circ Res 98:887–896

Fukui K, Yang Q, Cao Y, Takahashi N, Hatakeyama H, Wang H, Wada J, Zhang Y, Marselli L, Nammo T, Yoneda K, Onishi M, Higashiyama S, Matsuzawa Y, Gonzalez FJ, Weir GC, Kasai H, Shimomura I, Miyagawa J, Wollheim CB, Yamagata K (2005) The HNF-1 target collectrin controls insulin exocytosis by SNARE complex formation. Cell Metab 2:373–384

Glende J, Schwegmann-Wessels C, Al-Falah M, Pfefferle S, Qu X, Deng H, Drosten C, Naim HY, Herrler G (2008) Importance of cholesterol-rich membrane microdomains in the interaction of the S protein of SARS-coronavirus with the cellular receptor angiotensin-converting enzyme 2. Virology 381:215–221

Guy JL, Lambert DW, Turner AJ, Porter KE (2008) Functional angiotensin-converting enzyme 2 is expressed in human cardiac myofibroblasts. Exp Physiol 93:579–588

Haga S, Yamamoto N, Nakai-Murakami C, Osawa Y, Tokunaga K, Sata T, Sasazuki T, Ishizaka Y (2008) Modulation of TNF-alpha-converting enzyme by the spike protein of SARS-CoV and ACE2 induces TNF-alpha production and facilitates viral entry. Proc Natl Acad Sci USA 105:7809–7814

Hamming I, Timens W, Bulthuis ML, Lely AT, Navis GJ, van Goor H (2004) Tissue distribution of ACE2 protein, the functional receptor for SARS coronavirus. A first step in understanding SARS pathogenesis. J Pathol 203:631–637

Huovila AP, Turner AJ, Pelto-Huikko M, Karkkainen I, Ortiz RM (2005) Shedding light on ADAM metalloproteinases. Trends Biochem Sci 30:413–422

Inoue Y, Tanaka N, Tanaka Y, Inoue S, Morita K, Zhuang M, Hattori T, Sugamura K (2007) Clathrin-dependent entry of severe acute respiratory syndrome coronavirus into target cells expressing ACE2 with the cytoplasmic tail deleted. J Virol 81:8722–8729

Lambert DW, Yarski M, Warner FJ, Thornhill P, Parkin ET, Smith AI, Hooper NM, Turner AJ (2005) Tumor necrosis factor-alpha convertase (ADAM17) mediates regulated ectodomain shedding of the severe-acute respiratory syndrome-coronavirus (SARS-CoV) receptor, angiotensin-converting enzyme-2 (ACE2). J Biol Chem 280:30113–30119

Lambert DW, Clarke NE, Hooper NM, Turner AJ (2008) Calmodulin interacts with angiotensin-converting enzyme-2 (ACE2) and inhibits shedding of its ectodomain. FEBS Lett 582:385–390

Li W, Moore MJ, Vasilieva N, Sui J, Wong SK, Berne MA, Somasundaran M, Sullivan JL, Luzuriaga K, Greenough TC, Choe H, Farzan M (2003) Angiotensin-converting enzyme 2 is a functional receptor for the SARS coronavirus. Nature 426:450–454

Li W, Zhang C, Sui J, Kuhn JH, Moore MJ, Luo S, Wong SK, Huang IC, Xu K, Vasilieva N, Murakami A, He Y, Marasco WA, Guan Y, Choe H, Farzan M (2005) Receptor and viral determinants of SARS-coronavirus adaptation to human ACE2. EMBO J 24:1634–1643

Medina M, Dotti CG (2003) RIPped out by presenilin-dependent gamma-secretase. Cell Signal 15:829–841

Pohlmann S, Gramberg T, Wegele A, Pyrc K, van der Hoek L, Berkhout B, Hofmann H (2006) Interaction between the spike protein of human coronavirus NL63 and its cellular receptor ACE2. Adv Exp Med Biol 581:281–284

Ren X, Glende J, Al-Falah M, de Vries V, Schwegmann-Wessels C, Qu X, Tan L, Tschernig T, Deng H, Naim HY, Herrler G (2006) Analysis of ACE2 in polarized epithelial cells: surface expression and function as receptor for severe acute respiratory syndrome-associated coronavirus. J Gen Virol 87:1691–1695

Shaltout HA, Figueroa JP, Rose JC, Diz DI, Chappell MC (2008) Alterations in circulatory and renal angiotensin-converting enzyme and angiotensin-converting enzyme 2 in fetal programmed hypertension. Hypertension 53:404–408

Tipnis SR, Hooper NM, Hyde R, Karran E, Christie G, Turner AJ (2000) A human homolog of angiotensin-converting enzyme. Cloning and functional expression as a captopril-insensitive carboxypeptidase. J Biol Chem 275:33238–33243

Wang H, Yang P, Liu K, Guo F, Zhang Y, Zhang G, Jiang C (2008) SARS coronavirus entry into host cells through a novel clathrin- and caveolae-independent endocytic pathway. Cell Res 18:290–301

Warner FJ, Lew RA, Smith AI, Lambert DW, Hooper NM, Turner AJ (2005) Angiotensin-converting enzyme 2 (ACE2), but not ACE, is preferentially localized to the apical surface of polarized kidney cells. J Biol Chem 280:39353–39362

Wiener RS, Cao YX, Hinds A, Ramirez MI, Williams MC (2007) Angiotensin converting enzyme 2 is primarily epithelial and is developmentally regulated in the mouse lung. J Cell Biochem 101:1278–1291

Zhang H, Wada J, Hida K, Tsuchiyama Y, Hiragushi K, Shikata K, Wang H, Lin S, Kanwar YS, Makino H (2001) Collectrin, a collecting duct-specific transmembrane glycoprotein, is a novel homolog of ACE2 and is developmentally regulated in embryonic kidneys. J Biol Chem 276:17132–17139

Chapter 3
Structural Molecular Insights into SARS Coronavirus Cellular Attachment, Entry and Morphogenesis

Daniel R. Beniac and Timothy F. Booth

Abstract Coronavirus spikes have the largest mass of any known viral spike molecule. The spike is a type 1 viral fusion protein, a class of trimeric surface glycoprotein proteins from diverse viral families that share many common structural and functional characteristics. Fusion proteins are mainly responsible for host cell receptor recognition and subsequent membrane fusion, and may perform other roles such as virus assembly and release via budding. The conformational changes that occur in the spike of intact SARS coronavirus (SARS-CoV) when it binds to the viral receptor, angiotensin-converting enzyme 2 (ACE2) are described. Clues to the structural/functional relationships of membrane fusion have been made possible by the development of viral purification and inactivation methods, along with cryo-electron microscopy (cryo-EM) and three-dimensional (3D) image processing of many different images containing multiple views of the spikes. These methods have allowed study of the spikes while still attached to virions that are noninfectious, but fusionally competent. The receptor-binding and fusion core domains within the SARS-CoV spike have been precisely localized within the spike. Receptor binding results in structural changes that have been observed in the spike molecule, and these appear to be the initial step in viral membrane fusion. A working model for the stepwise process of receptor binding, and subsequent membrane fusion in SARS-CoV is presented. Uniquely, the large size of the SARS-CoV spike allows structural changes to be observed by cryo-EM in the native state. This provides a useful model for studying the basic process of membrane fusion in general, which forms an essential part of the function of many cellular processes.

T.F. Booth (✉)
Viral Diseases Division, National Microbiology Laboratory, Public Health Agency of Canada, 1015 Arlington Street, Winnipeg, Manitoba R3E 3R2, Canada
e-mail: Tim_Booth@phac-aspc.gc.ca

3.1 Structure of SARS Coronavirus (SARS-CoV)

The earliest coronavirus isolates were identified from mammalian and avian sources in the 1930s and 1940s (Beaudette and Hudson 1937; Doyle and Hutchings 1946; Cheevers and Daniels 1949). At this time, electron microscopy revealed prominent large spikes on the surface of virions, whose resulting crown-like appearance gave rise to the name coronavirus (Fig. 3.1). Coronaviruses have 4–5 structural proteins including the spike protein (S), envelope protein (E), membrane protein (M), nucleocapsid protein (N), and members of the coronavirus phylogenetic subgroup 2a have a shorter S protein called hemagglutinin esterase (HE). The SARS-CoV genome is unusually large at ~29.7 kb, and encodes 14 open reading frames for several proteins, (Marra et al. 2003; Rota et al. 2003). On the interior, the lipid envelope of SARS-CoV appears to have a gap observed as a low density in three-dimensional (3D) structures (Fig. 3.2) and then a more dense layer, which is presumably the surface of the nucleocapsid comprising mainly the N protein. Details of how this nucleocapsid is organized are not clear; however, the N protein is presumably anchored to the cytoplasmic side of the virion envelope via the

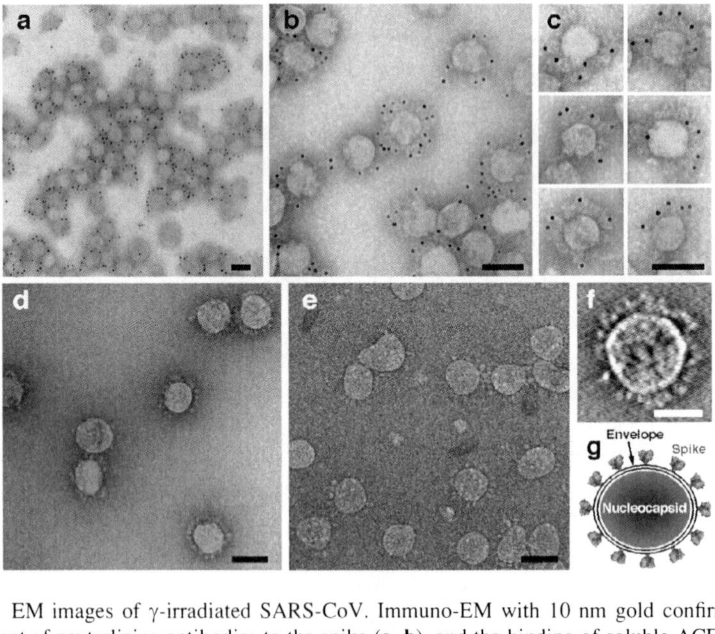

Fig. 3.1 EM images of γ-irradiated SARS-CoV. Immuno-EM with 10 nm gold confirmed the attachment of neutralizing antibodies to the spike (**a**, **b**), and the binding of soluble ACE2 to the spike (**c**). (**d**) SARS-CoV negative-stained with methylamine tungstate shows the virions to be spherical/pleomorphic, with the spikes clearly visible from the side perspective. (**e**) Cryo-EM provides additional details including the end-on perspective views of the spikes. (**f**) The central section of an electron tomogram of a negative-stained SARS-CoV clearly shows the viral envelope with the spikes attached. (**g**) Schematic model of SARS-CoV. Scale bars: (*black*) 1,000 Å, (*white*) 500 Å

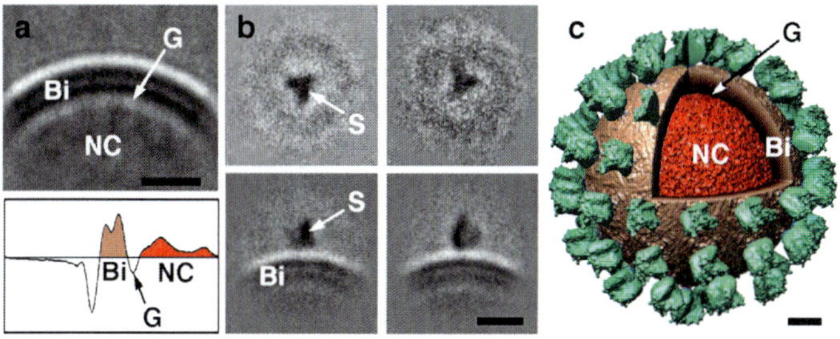

Fig. 3.2 Image analysis was employed to investigate the spike of SARS-CoV. (**a**) Image average of the virion clearly shows the lipid bilayer (Bi), a prominent gap (G), and the nucleocapsid (NC). (**b**) 2D class averages of the spike (S) presenting end-on and side-view perspectives. (**c**) 3D model of SARS-CoV, with a wedge cut out of it to reveal the nucleocapsid. Legend: (S) spike, *green*; (Bi) lipid bilayer, *beige*; (G) gap; (NC) nucleocapsid, *red*. Scale bar: 150 Å

M protein, which is a low abundance protein and may account for the "gap" of low density between the envelope and the nucleocapsid. Electron microscopic examination shows no evidence for supercoiled RNA such as that seen in paramyxoviruses and the interior appears amorphous by cryo-electron microscopy (cryo-EM); hence the arrangement whereby the RNA may be bound to the nucleocapsid protein is not clear.

A striking feature of the SARS-CoV spike is its huge mass (~500 kD per trimer). However, despite the size differences, the SARS-CoV spike performs the same fundamental task in viral entry to the host cell as other smaller type 1 viral fusion proteins, such as the influenza hemagglutinin (HA) (~220 kD per trimer). The SARS-CoV spike can be subdivided into four structural domains (from N to C terminus); two large external domains S1 and S2 are largely responsible for receptor binding and membrane fusion, respectively. In most type 1 viral fusion proteins the analogous peptides are generated by proteolysis of the spike precursor during the maturation process in the host cell, yielding two peptides with the fusion peptide on the N terminus of S2. In SARS-CoV the S1/S2 assignment is given based on sequence homology to other viral fusion proteins, although there appears to be no peptide cleavage. The final two small domains are comprised of a transmembrane domain, and a carboxyterminal cytoplasmic domain. The cell-surface molecule angiotensin-converting enzyme 2 (ACE2) is the receptor for the SARS-CoV S protein (Li et al. 2003) which is a relatively large macromolecule with a diameter of 70 Å. By comparison, the receptor for influenza HA, sialic acid, is much smaller with a 10 Å diameter.

One of the challenges of achieving structural molecular studies with native SARS-CoV is that it is classified as a biological safety level 3 organism, requiring handling in containment. However, once it was shown that specimens could be γ-irradiated with a sufficient dose (2 Mrad) for viral inactivation, while still preserving protein structure, it was possible to carry out cryo-EM of intact virions and to obtain the 3D structure of the native, unfixed virions and the spike, using single particle image processing and averaging from multiple images containing many different viewing angles of the molecule (Booth et al. 2005; Beniac et al. 2006). Immunolabeling showed that these virions were intact antigenically as well as structurally (Fig. 3.1) and still able to bind to the SARS-CoV receptor, ACE2.

3.2 Structure of the Coronavirus Spike

Cryo-EM coupled with 3D single-particle image analysis has been used to determine the structure of the SARS-CoV spike, and positioning of the binding of ACE2 to the spike (Beniac et al. 2006, 2007). Spikes on the surface of virus particles are readily imaged by cryo-EM in the frozen-hydrated native state (Fig. 3.1e). 3D image processing was carried out on selected spikes using single-particle image processing (Penczek et al. 1994; Frank et al. 1996; Beniac et al. 2006). The structures of both the spike and the spike–ACE2 complex have been solved to 18.5 Å resolution (Figs. 3.2 and 3.3; Beniac et al. 2006, 2007). The spike shows a striking structure, being about 180 Å in diameter and with three distinct lobes or domains 50 Å thick on each subunit of the trimer (similar in appearance to the blades of a propeller), and a thin stalk connecting the spike to the viral envelope. The blades are twisted at an angle of ~30° to the axis of symmetry, and are almost certainly composed of the spike S1 domain.

Fig. 3.3 3D reconstructions presented as *shaded surfaces* are shown from the side (*upper*) and end-on (*lower*) perspectives. Cryo-EM reconstruction of the SARS-CoV spike (**a**), and SARS-CoV spike–ACE2 complex (**b**). (**c**) The atomic resolution structures were docked within the SARS-CoV spike–ACE2 3D reconstruction; PDB ID code: 2AJF (ACE2, *blue*; receptor-binding domain, *red*), and 2FXP (*yellow*). The *arrow* points to the C terminus of ACE2. Color scheme: ACE2, *violet*; spike, *green*; stalk, *blue*; envelope, *beige*; nucleocapsid, *red*. Scale bar: 100 Å

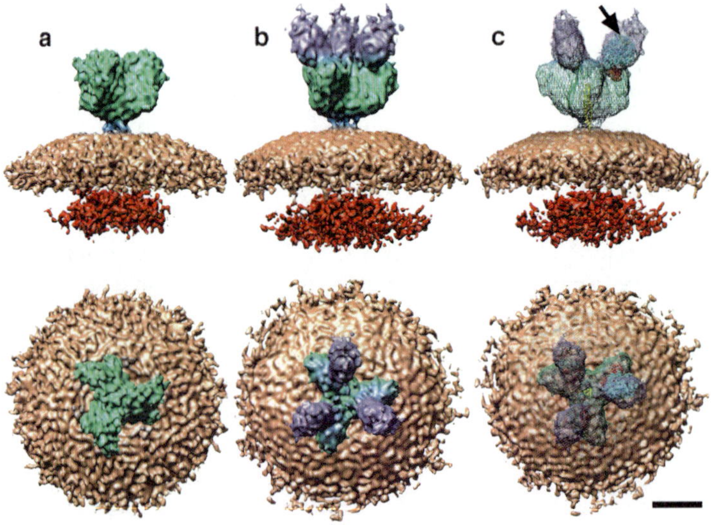

Fig. 3.4 The cryo-EM reconstruction of the SARS-CoV spike (**a,f**) was subtracted from the SARS-CoV spike–ACE2 complex (**e,j**). The positive component attributed to the SARS-CoV spike in (**b,g**) indicates a rearrangement in the S2 core, and the positive component attributed the SARS-CoV–ACE2 complex shows the addition of ACE2 and an exterior rearrangement in S1 (**d,i**). The net difference map is presented in (**c,h**). The structures are presented from a side perspective (**a–e**) and end-on perspective (**f–j**). The *arrows* in (**f**) and (**j**) illustrate the mass reorganization that occurs in the central axis of the spike where one small central blob splits into three nubs. In (**a,f**) the region on the spike adjacent to ACE2 which corresponds to the receptor-binding domain has been highlighted with a *dotted line* and is colored *purple*. The color scheme is the same as in Fig. 3.3. Scale bar: 100 Å

ACE2 binding does not result in a fundamental structural unfolding of the spike. However, the overall height of the spike was reduced from 160 Å to 150 Å following binding. When viewed end-on, the spike undergoes a rotation of ~5° following binding, and the mass at the center of the axis of symmetry on the distal end of the spike redistributes itself. These redistributions of mass were further identified in difference maps between the two reconstructions (Fig. 3.4). Upon ACE2 binding the spike undergoes a decondensation of mass around the central axis (Fig. 3.4b,g; blue). This region is the putative location of the S2 domain. The difference map for the bound spike (Fig. 3.4d,i) shows changes in both the ACE2 component (purple) and the outer edges of the three "blades" of the S1 domain (green).

The precise location of ACE2 binding on the distal end of the spike is centered at 70 Å from the central axis of the spike, with a 30 Å gap between the axis of symmetry and ACE2. One ACE2 molecule can bind to each of the three propeller-like blades of the spike, making a structure 220 Å high (Fig. 3.3b). Binding of more than one ACE2 to each spike (on one or both of the other two propeller blades of each trimer) is possible, hence binding of one ACE2 molecule does not stearically hinder binding of additional ACE2 molecules.

The cryo-EM 3D structures of the spike and the spike–ACE2 complex, when combined with the atomic resolution structures of the SARS-CoV spike receptor-binding domain – ACE2 complex (Li et al. 2005a) and the heptad repeat pre- and postfusion cores (Supekar et al. 2004; Hakansson-McReynolds et al. 2006), show that the receptor-binding domain docks to the distal end of the spike with ACE2

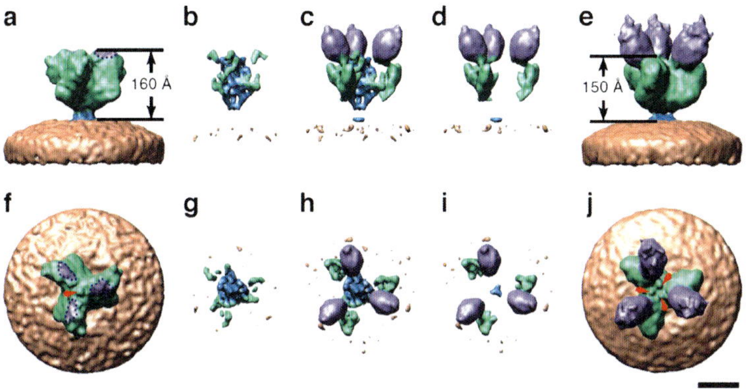

filling the extra mass on the spike (shown by the color violet in Fig. 3.3). The empty upper region of the mass appears to be components of the second ACE2 and the Fc component of the chimeric protein, and the location of the C-terminus of the docked ACE2 was consistent with this interpretation.

3.3 Viral Membrane Fusion in SARS-CoV

Viral membrane fusion proteins are responsible both for binding to cellular receptors, and the subsequent fusion of viral and cellular membranes. The paradigm for type 1 fusion proteins consists of two heptad repeat regions, and a hydrophobic fusion peptide (Dutch et al. 2000). This motif is present in SARS-CoV (Hakansson-McReynolds et al. 2006) and other coronaviruses (Xu et al. 2004), as well as the hemagglutinin (HA) of influenza (Skehel and Wiley 2000), gp21 of human T-cell leukemia virus type 1 (Kobe et al. 1999), gp41 of HIV(Weissenhorn et al. 1997), GP2 of Ebola (Weissenhorn et al. 1998; Malashkevich et al. 1999), and the fusion protein of paramyxovirus (Zhao et al. 2000; Chen et al. 2001). Type 1 viral fusion proteins can also be divided into two subtypes: those whose fusion mechanism is low pH-dependent such as influenza HA, and those that are pH-independent like the retroviral fusion proteins. In retroviruses, receptor binding itself can trigger fusion, with temperature and redox conditions also influencing the fusion mechanism (Hernandez et al. 1997; Damico et al. 1998). The SARS-CoV spike appears to be insensitive to redox conditions (Fenouillet et al. 2007). Although the factors which trigger fusion (endocytosis, pH sensitivity, single receptor vs. primary and coreceptor binding, redox change) differ amongst diverse virus families, all viral fusion proteins are thought to share the same basic fusion mechanism (Baker et al. 1999; Skehel and Wiley 2000; Dutch et al. 2000; Colman and Lawrence 2003; Dimitrov 2004; Hofmann and Pohlmann 2004).

The precise mechanisms by which type 1 viral fusion proteins gain access to the host cell remain unknown. The hypothetical entry process includes several steps that take place in sequence: receptor binding, fusion core rearrangement, fusion peptide insertion in host cell membrane, refolding of heptad repeats, membrane fusion, and finally viral nucleocapsid transfer (Earp et al. 2005).

3.4 Cellular Attachment and Entry of SARS-CoV

In most proposed models of membrane fusion it is postulated that the S1 domain or analogous receptor-binding domains dissociate from the spike during the membrane fusion process. This dynamic process was demonstrated for influenza HA by Kemble et al. (1992) in their investigation where they engineered intermonomer disulfide bonds between the HA S1 subunits. The result of this was that fusion activity was impaired; however it could be restored under reducing conditions. It is probable that the SARS-CoV spike shares a similar mechanism, with the structural changes detected by cryo-EM representing the initial step in this process.

By analogy with other type 1 viral fusion proteins, the fusion core of the SARS-CoV spike is thought to undergo similar structural rearrangements during fusion. The receptor-binding domain is localized in a position on the distal end of the molecule, closer to the 3-fold axis than anticipated, yet still in a position that would not impede these structural rearrangements. Putative mechanisms by which type 1 viral fusion proteins achieve membrane fusion have been proposed (Baker et al. 1999; Skehel and Wiley 2000; Dutch et al. 2000; Colman and Lawrence 2003; Dimitrov 2004), but complete structural evidence for the role of intermediate structures in these mechanisms has yet to be obtained. The structural biology of this process has been best characterized for the influenza HA, and paramyxovirus fusion protein (F) for which the prefusion and membrane fusion pH structures have been determined by X-ray crystallography (Sauter et al. 1992; Bullough et al. 1994; Skehel and Wiley 2000; Yin et al. 2005, 2006) (Fig. 3.5, inset). All of the subsequent models for type 1 viral fusion proteins are based on the structural data of these two fusion proteins. A drawback in all of these models is that they are based on recombinant ectodomains that are not proven to exist as a component in the complete molecule, and they lack both membrane-interacting residues and lipids (Skehel and Wiley 2000). The cryo-EM structures of intact SARS-CoV spike bound to native virion lipid envelopes are very instructive when atomic resolution fragments are docked within the overall molecule, especially as the entire SARS-CoV spike has proven to be a difficult subject for X-ray crystallography, and atomic resolution data exist for only a few fragments of the SARS-CoV spike. This structural data has been modeled into a scheme to propose a mechanism for SARS-CoV spike-mediated membrane fusion (Figs. 3.5 and 3.6). In the initial step the receptor-binding domain of the spike attaches to its human receptor ACE2. At this point the fusion core is in the prefusion configuration with the three heptad 2 repeats (HR2) forming a coiled-coil symmetric trimer at the center of the stalk of the spike (Hakansson-McReynolds et al. 2006).

During the next step of the membrane fusion process the virus is internalized in the cell by endocytosis and is exposed to a low pH environment, and may undergo proteolytic cleavage between the S1 and S2 domains (Simmons et al. 2005). The next step is fusion core rearrangement, so that the fusion peptide (FP) inserts into the host cell membrane. In Fig. 3.5 this initial process is illustrated with models M0–M3, based on the atomic structure of the HR2 prefusion core, which begins to collapse upon itself in model M3 in a manner similar to that which occurs with influenza HA (Skehel and Wiley 2000). The inset in Fig. 3.5 illustrates this process in HA by coloring segments of S2 to illustrate the rearrangement from M0 to M3 that takes place. During this process we propose that the receptor-binding domain still holds on to ACE2 so that the fully extended fusion peptide will be positioned to penetrate through the host membrane. Our cryo-EM results show that it is possible for the spike to attach to three ACE2 receptors at once; this may serve to hold on to the host membrane like a tripod so as to accurately orientate the fusion core (Fig. 3.7). In addition the 30 Å gap between the axis of symmetry and ACE2

Fig. 3.5 Seven models (M0–M6) are presented which show the hypothetical rearrangement of the SARS-CoV fusion core which takes place during membrane fusion. For simplicity we show only one of the three HR1/HR2 structures for each cylindrical model. Models were constructed based on the structures for influenza HA, presented in the *inset* at the same scale. The cryo-EM, docked ACE2–receptor-binding domain, and prefusion core (M0) and postfusion core (M6) structures provided start and end points for modeling spike rearrangement. Five intermediate models illustrate the "jack-knife" mechanism of the fusion core. The following color scheme was used: cryo-EM surface: same as in Fig. 3.3. Ribbon structures: ACE2, *white* (C terminus *blue*; spike receptor-binding domain, *red*; HR1, *pink*; HR2, *yellow*. Cylindrical models: FP, *red*; HR1, *pink*; HR2, *yellow*. Scale bar: 100 Å

provides sufficient space for fusion core rearrangement. Damico et al. (1998) demonstrated that the kinetics of binding of the Rous sarcoma virus envelope protein ectodomain to liposomes was not linear with respect to receptor concentration. This suggested that activation of the trimeric ectodomain favored binding to multiple receptor monomers. One can therefore infer that other structurally homologous viral envelope proteins can also bind multiple receptors, which may be a general adaptation that provides the correct temporal and spatial arrangement to bring about membrane fusion. The observation that the SARS-CoV spike could bind three soluble ACE2 receptors provides three possible binding states with one, two or three membrane-bound receptors attached to the spike. In Fig. 3.7 we present these three states; with only one receptor bound the spike and virus have a wide range of movement possible, whereas with two receptors bound the movement is

Fig. 3.6 A schematic of the SARS-CoV spike protein with the location of the known atomic structures is presented in (**a**). The following abbreviations are used: RBD, receptor-binding domain; FP: fusion peptide; HR1: heptad repeat 1; HR2: heptad repeat 2; TM: transmembrane; CY: cytoplasmic tail. There are several steps involved in viral entry; they can be broken down into at least six components: (**b**) receptor targeting, (**c**) viral attachment, (**d**) fusion core rearrangement and fusion peptide insertion, (**e**) fusion core refolding, (**f**) membrane fusion, and (**g**) nucleocapsid transfer. In (**b–g**) the host membrane is represented using a *shaded blue line*. The color scheme used is the same as in Figs. 3.3 and 3.5. Scale bar: 100 Å

greatly restricted to motion in one plane only. Only in the case of three bound receptors will the spike and its fusion core be arranged perpendicular to the cell surface with minimal movement possible. At present it has not been demonstrated that membrane fusion requires the fusion core to be oriented perpendicular to the host cell membrane to function. However, one can hypothesize this based on the orientation of ACE2 on the distal end of the SARS-CoV spike. It is interesting to note that binding to three receptor molecules is the minimum number of binding events required to achieve this perpendicular orientation in 3D space. This observation matches up with the conserved trimeric structures of type 1 fusion proteins which are common amongst enveloped viruses, thus indicating that a possible conserved structural–functional relationship may exist.

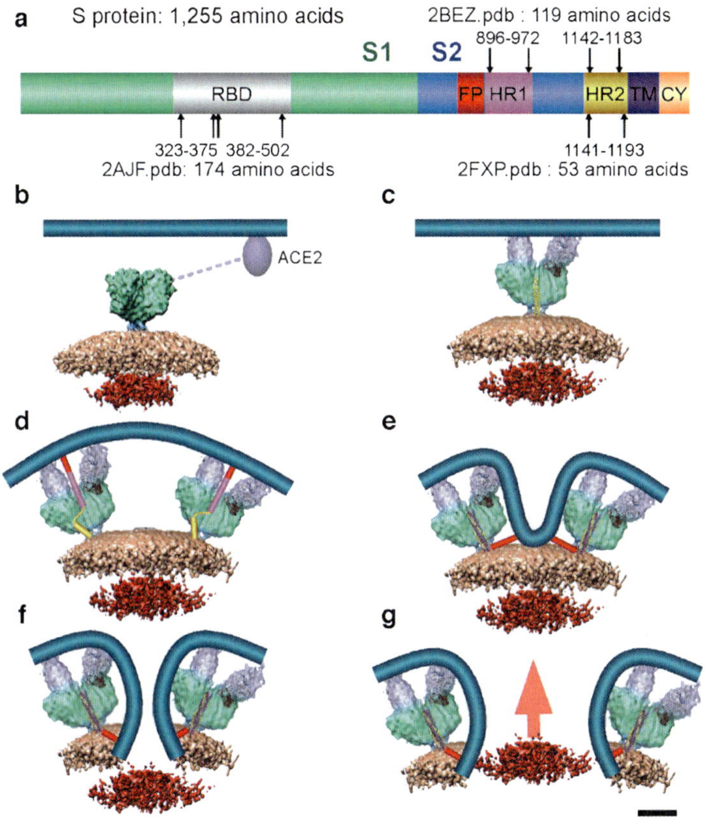

Fig. 3.7 The binding of the SARS-CoV spike to multiple receptors is presented from three 90° orthographic views (X, Y, and Z axes). When one receptor is bound there is a wide range of motion possible (**a**). By binding two receptors the freedom of movement is greatly reduced to motion in one plane (**b**). When three receptors are bound, the spike is positioned in such a way that movement is restrained and the fusion core is perpendicular to the host cell membrane (**c**). The host membrane is represented using a *shaded blue surface* (same as in Fig. 3.6), and the color scheme used for the SARS-CoV spike is the same as in Fig. 3.3. Scale bar: 100 Å

The next step in the membrane fusion process involves the refolding of the fusion core back upon itself to adopt the postfusion configuration (Supekar et al. 2004) (Fig. 3.5; models M4–M6). In this configuration three heptad 1 repeats (HR1) form a parallel coiled-coil trimer, and the three HR2 pack in an antiparallel fashion in the hydrophobic grooves of the HR1 trimer. The result of this is that both the FP and transmembrane regions of the spike are brought in close proximity to each other, resulting in the host and viral membranes being in close contact. Each SARS-CoV virion has an average of 65 spikes (Beniac et al. 2006). At this density, several spikes would be close enough together to act in concert to disrupt the plasma membrane and induce pore formation between viral and cellular membranes. Opening of these pores would allow the SARS-CoV nucleocapsid to enter the host cell cytoplasm, as shown in our model in Fig. 3.6. Multiple SARS-CoV spike trimers may be involved in formation of the fusion pore, similar to the situation in influenza, where it has been suggested that between three and six HA trimers may be involved in the production of each fusion pore (Skehel and Wiley 2000). In Figs. 3.5 and 3.6 we have presented the ACE2–SARS:S1 domains as they

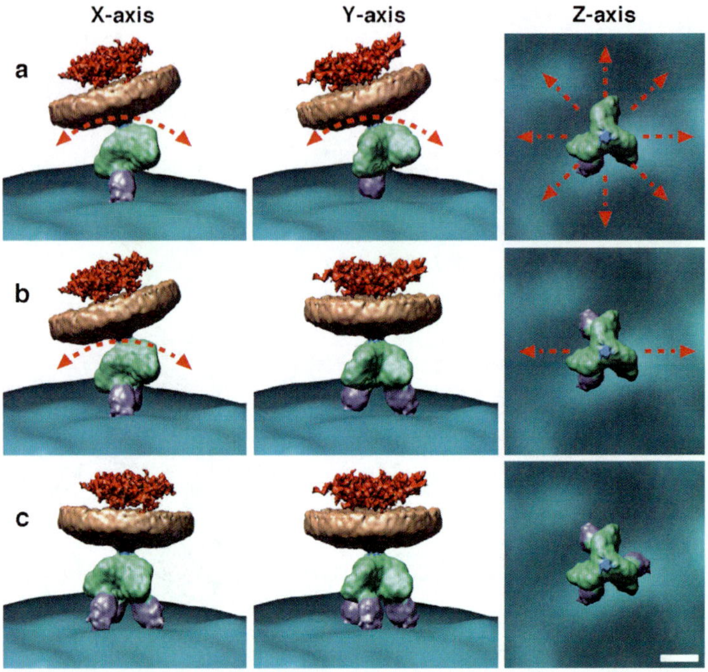

were solved in this cryo-EM investigation, for the model of membrane fusion that we present. For other fusion proteins like influenza HA1 and HIV GP120 it has been modeled that the rearrangements upon membrane fusion are dramatic involving a shedding of the above-mentioned domains. The cryo-EM investigations have detected structural movement of S1 upon ACE2 binding, which could represent the initial phase of this dramatic process that is postulated to occur in the course of membrane fusion.

An analysis of the structure of the spike–receptor complex demonstrates how SARS-CoV can adapt to utilize receptors from different species and how they may evolve to gain specificity for new receptor types, in that there is redundancy and a great deal of protein mass that can accommodate evolutionary changes. RNA viruses have a high rate of mutation and recombination (Moya et al. 2004). In SARS-CoV the spike is able to retain specific binding affinity for the ACE2 of more than one host species, and rapid evolution to gain specificity for novel ACE2 species has been demonstrated (Li et al. 2005b, 2006). The large size of the spike of coronaviruses may be related to the use of large host cell-surface molecules such as ACE2 as specific receptors. Amongst the coronavirus family, specific cell-surface receptors for the S protein are all in the range of 60–110 kD (Wentworth and Holmes 2001). These large host receptor molecules are of course functionally constrained and, in turn, relatively well conserved across species barriers. In utilizing binding to a large receptor molecule, the spike S1 domain also acts as a "spacer arm" holding the receptor far enough away from the threefold axis of symmetry of the spike S2 domain to permit fusion core rearrangement and subsequent membrane fusion. Such a property necessitates having a large spike molecule. Moreover, multiple receptor binding can have functional significance, enhancing the binding and entry of viruses. Cross-linking of adjacent host receptor molecules could increase the affinity of the virus for its target cell, as well as improving the kinetics of fusion. The SARS-CoV spike is a useful model system for the investigation of type 1 viral fusion protein dynamics. Utilizing this system for further research may lead to the possibility of developing broad-spectrum antivirals that target conserved cell fusion mechanisms shared by diverse virus families.

References

Baker KA, Dutch RE, Lamb RA, Jardetzky TS (1999) Structural basis for paramyxovirus-mediated membrane fusion. Mol Cell 3:309–319

Beaudette FW, Hudson CB (1937) Cultivatiion of the virus of infectious bronchitis. J Am Vet Med Assoc 90:51–60

Beniac DR, Andonov A, Grudeski E, Booth TF (2006) Architecture of the SARS coronavirus prefusion spike. Nat Struct Mol Biol 13:751–752

Beniac DR, deVarennes SL, Andonov A, He R, Booth TF (2007) Conformational reorganization of the SARS Coronavirus spike following receptor binding: implications for membrane fusion. PLoS ONE 2(10):e1082. doi:10.1371/journal.pone.0001082

Booth TF, Kournikakis B, Bastien N, Ho J, Kobasa D, Stadnyk L, Li Y, Spence M, Paton S, Henry B, Mederski B, White D, Low DE, McGeer A, Simor A, Vearncombe M, Downey J, Jamieson FB, Tang P, Plummer F (2005) Detection of airborne severe acute respiratory syndrome (SARS) coronavirus and environmental contamination in SARS outbreak units. J Infect Dis 191:1472–1477

Bullough PA, Hughson FM, Skehel JJ, Wiley DC (1994) Structure of influenza haemagglutinin at the pH of membrane fusion. Nature 371:37–43

Cheevers FS, Daniels JB (1949) A murine virus (JHM) causing disseminated encephalomyelitis with extensive destruction of myelin. J Exp Med 90:181–210

Chen L, Gorman JJ, McKimm-Breschkin J, Lawrence LJ, Tulloch PA, Smith BJ, Colman PM, Lawrence MC (2001) The structure of the fusion glycoprotein of Newcastle disease virus suggests a novel paradigm for the molecular mechanism of membrane fusion. Structure 9:255–266

Colman PM, Lawrence MC (2003) The structural biology of type I viral membrane fusion. Nat Rev Mol Cell Biol 4:309–319

Damico RL, Crane J, Bates P (1998) Receptor-triggered membrane association of a model retroviral glycoprotein. Proc Natl Acad Sci USA 95:2580–2585

Dimitrov DS (2004) Virus entry: molecular mechanisms and biomedical applications. Nat Rev Microbiol 2:109–122

Doyle LP, Hutchings LM (1946) A transmissible gastroenteritis in pigs. J Am Vet Med Assoc 108:257–259

Dutch RE, Jardetzky TS, Lamb RA (2000) Virus membrane fusion proteins: biological machines that undergo a metamorphosis. Biosci Rep 20:597–612

Earp LJ, Delos SE, Park HE, White JM (2005) The many mechanisms of viral membrane fusion proteins. Curr Top Microbiol Immunol 285:25–66

Fenouillet E, Barbouche R, Jones IM (2007) Cell entry by enveloped viruses: redox considerations for HIV and SARS-Coronavirus. Antioxid Redox Signal 9:1009–1034

Frank J, Radermacher M, Penczek P, Zhu J, Li Y, Ladjadj M, Leith A (1996) SPIDER and WEB: processing and visualization of images in 3D electron microscopy and related fields. J Struct Biol 116:190–199

Hakansson-McReynolds S, Jiang S, Rong L, Caffrey M (2006) Solution structure of the severe acute respiratory syndrome-coronavirus heptad repeat 2 domain in the prefusion state. J Biol Chem 281:11965–11971

Hernandez LD, Peters RJ, Delos SE, Young JA, Agard DA, White JM (1997) Activation of a retroviral membrane fusion protein: soluble receptor-induced liposome binding of the ALSV envelope glycoprotein. J Cell Biol 139:1455–1464

Hofmann H, Pohlmann S (2004) Cellular entry of the SARS coronavirus. Trends Microbiol 12:466–472

Kemble GW, Bodian DL, Rose J, Wilson IA, White JM (1992) Intermonomer disulfide bonds impair the fusion activity of influenza virus hemagglutinin. J Virol 66:4940–4950

Kobe B, Center RJ, Kemp BE, Poumbourios P (1999) Crystal structure of human T cell leukemia virus type 1 gp21 ectodomain crystallized as a maltose-binding protein chimera reveals structural evolution of retroviral transmembrane proteins. Proc Natl Acad Sci USA 96:4319–4324

Li W, Moore MJ, Vasilieva N, Sui J, Wong SK, Berne MA, Somasundaran M, Sullivan JL, Luzuriaga K, Greenough TC, Choe H, Farzan M (2003) Angiotensin-converting enzyme 2 is a functional receptor for the SARS coronavirus. Nature 426:450–454

Li F, Li W, Farzan M, Harrison SC (2005a) Structure of SARS coronavirus spike receptor-binding domain complexed with receptor. Science 309:1864–1868

Li W, Zhang C, Sui J, Kuhn JH, Moore MJ, Luo S, Wong SK, Huang IC, Xu K, Vasilieva N, Murakami A, He Y, Marasco WA, Guan Y, Choe H, Farzan M (2005b) Receptor and viral determinants of SARS-coronavirus adaptation to human ACE2. EMBO J 24:1634–1643

Li W, Wong SK, Li F, Kuhn JH, Huang IC, Choe H, Farzan M (2006) Animal origins of the severe acute respiratory syndrome coronavirus: insight from ACE2-S-protein interactions. J Virol 80:4211–4219

Malashkevich VN, Schneider BJ, McNally ML, Milhollen MA, Pang JX, Kim PS (1999) Core structure of the envelope glycoprotein GP2 from Ebola virus at 1.9-A resolution. Proc Natl Acad Sci USA 96:2662–2667

Marra MA, Jones SJ, Astell CR, Holt RA, Brooks-Wilson A, Butterfield YS, Khattra J, Asano JK, Barber SA, Chan SY, Cloutier A, Coughlin SM, Freeman D, Girn N, Griffith OL, Leach SR, Mayo M, McDonald H, Montgomery SB, Pandoh PK, Petrescu AS, Robertson AG, Schein JE, Siddiqui A, Smailus DE, Stott JM, Yang GS, Plummer F, Andonov A, Artsob H, Bastien N, Bernard K, Booth TF, Bowness D, Czub M, Drebot M, Fernando L, Flick R, Garbutt M, Gray M, Grolla A, Jones S, Feldmann H, Meyers A, Kabani A, Li Y, Normand S, Stroher U, Tipples GA, Tyler S, Vogrig R, Ward D, Watson B, Brunham RC, Krajden M, Petric M, Skowronski DM, Upton C, Roper RL (2003) The Genome sequence of the SARS-associated coronavirus. Science 300:1399–1404

Moya A, Holmes EC, Gonzalez-Candelas F (2004) The population genetics and evolutionary epidemiology of RNA viruses. Nat Rev Microbiol 2:279–288

Penczek PA, Grassucci RA, Frank J (1994) The ribosome at improved resolution: new techniques for merging and orientation refinement in 3D cryo-electron microscopy of biological particles. Ultramicroscopy 53:251–270

Rota PA, Oberste MS, Monroe SS, Nix WA, Campagnoli R, Icenogle JP, Penaranda S, Bankamp B, Maher K, Chen MH, Tong S, Tamin A, Lowe L, Frace M, DeRisi JL, Chen Q, Wang D, Erdman DD, Peret TC, Burns C, Ksiazek TG, Rollin PE, Sanchez A, Liffick S, Holloway B, Limor J, McCaustland K, Olsen-Rasmussen M, Fouchier R, Gunther S, Osterhaus AD, Drosten C, Pallansch MA, Anderson LJ, Bellini WJ (2003) Characterization of a novel coronavirus associated with severe acute respiratory syndrome. Science 300:1394–1399

Sauter NK, Hanson JE, Glick GD, Brown JH, Crowther RL, Park SJ, Skehel JJ, Wiley DC (1992) Binding of influenza virus hemagglutinin to analogs of its cell-surface receptor, sialic acid: analysis by proton nuclear magnetic resonance spectroscopy and X-ray crystallography. Biochemistry 31:9609–9621

Simmons G, Gosalia DN, Rennekamp AJ, Reeves JD, Diamond SL, Bates P (2005) Inhibitors of cathepsin L prevent severe acute respiratory syndrome coronavirus entry. Proc Natl Acad Sci USA 102:11876–11881

Skehel JJ, Wiley DC (2000) Receptor binding and membrane fusion in virus entry: the influenza hemagglutinin. Annu Rev Biochem 69:531–569

Supekar VM, Bruckmann C, Ingallinella P, Bianchi E, Pessi A, Carfi A (2004) Structure of a proteolytically resistant core from the severe acute respiratory syndrome coronavirus S2 fusion protein. Proc Natl Acad Sci USA 101:17958–17963

Weissenhorn W, Dessen A, Harrison SC, Skehel JJ, Wiley DC (1997) Atomic structure of the ectodomain from HIV-1 gp41. Nature 387:426–430

Weissenhorn W, CarfiA LKH, Skehel JJ, Wiley DC (1998) Crystal structure of the Ebola virus membrane fusion subunit, GP2, from the envelope glycoprotein ectodomain. Mol Cell 2:605–616

Wentworth DE, Holmes KV (2001) Molecular determinants of species specificity in the coronavirus receptor aminopeptidase N (CD13): influence of N-linked glycosylation. J Virol 75: 9741–9752

Xu Y, Lou Z, Liu Y, Pang H, Tien P, Gao GF, Rao Z (2004) Crystal structure of severe acute respiratory syndrome coronavirus spike protein fusion core. J Biol Chem 279:49414–49419

Yin HS, Paterson RG, Wen X, Lamb RA, Jardetzky TS (2005) Structure of the uncleaved ectodomain of the paramyxovirus (hPIV3) fusion protein. Proc Natl Acad Sci USA 102: 9288–9293

Yin HS, Wen X, Paterson RG, Lamb RA, Jardetzky TS (2006) Structure of the parainfluenza virus 5 F protein in its metastable, prefusion conformation. Nature 439:38–44

Zhao X, Singh M, Malashkevich VN, Kim PS (2000) Structural characterization of the human respiratory syncytial virus fusion protein core. Proc Natl Acad Sci USA 97:14172–14177

Part II
Structures Involved in Viral Replication and Gene Expression

Chapter 4
RNA Higher-Order Structures Within the Coronavirus 5′ and 3′ Untranslated Regions and Their Roles in Viral Replication

Pinghua Liu and Julian Leibowitz

Abstract The 5′ and 3′ untranslated regions (UTRs) of all coronaviruses contain RNA higher-order structures which play essential roles in viral transcription and replication. In this chapter we present our current knowledge of how those *cis*-acting elements were defined and their functional roles in viral transcription and replication. Cellular proteins which have been shown binding to those *cis*-acting elements and potentially support the RNA discontinuous synthesis model are also discussed. A conserved RNA structure model for the 5′ and 3′ UTRs of group 2 coronaviruses is presented with the known cellular protein binding sites.

4.1 Introduction

Coronaviruses are single-stranded, positive-sense, nonsegmented enveloped RNA viruses belonging to the family *Coronaviridae*, one of the three families in the order *Nidovirales*. They are the largest known RNA viruses with 27–31 kb genomes. Coronaviruses are classified as group 1, 2, and 3 based on serologic relatedness, genome organization and sequence similarity. Extensive phylogenetic comparisons placed the SARS coronavirus (SARS-CoV) as an early branch of the group 2 coronaviruses (Snijder et al. 2003). For all coronaviruses the 5′ two-thirds of the genome comprise the replicase gene, and the 3′ genes encode structural proteins and nonessential accessory proteins.

Coronaviruses infect cells by binding to specific receptors and enter cells by direct membrane fusion at the plasma membrane or by an endocytotic mechanism (Nash and Buchmeier 1997; Wang et al. 2008). SARS-CoV uses angiotension-

J. Leibowitz (✉)
Department of Microbial and Molecular Pathogenesis, Texas A&M University System College of Medicine, 407 Reynolds Medical Science Building, 1114 TAMUS, College Station, TX 77843-1114, USA
e-mail: jleibowitz@tamu.edu

S.K. Lal (ed.), *Molecular Biology of the SARS-Coronavirus*,
DOI 10.1007/978-3-642-03683-5_4,

converting enzyme 2 (ACE2) as its functional receptor (Li et al. 2003) and enters cells through pH- and receptor-dependent endocytosis (Wang et al. 2008). Upon entering the cytoplasm the virus particle is uncoated, releasing the RNA genome. The viral genome directs the synthesis of two large polypeptides, pp1a and pp1ab, via a frameshifting mechanism involving a pseudoknot structure (Brierley et al. 1987). The resulting polypeptide contains a conserved array of functional domains, which upon proteolytic processing results in 15–16 nonstructural proteins (nsp), many of which are likely to be involved in either RNA synthesis or proteolytic processing of the polyprotein precursors of nsp1–16 (Snijder et al. 2003). The 3′ one-third of the genome contains the genes for viral structural proteins and accessory proteins. These genes are expressed by transcription of a 3′ coterminal nested set of 7–9 mRNAs that also contain a common ~70–90 nucleotide (nt) 5′ leader identical in sequence to the 5′ end of the genome (Lai et al. 1983, 1984; Spaan et al. 1982). The 3′ end of the leader sequence contains a short (6–8 nt) sequence, the transcriptional regulatory sequence (TRS) also present in the genome just 5′ to the coding sequence for each mRNA (Budzilowicz et al. 1985).

Subgenomic negative-sense RNAs that correspond to each subgenomic RNA are found in infected cells (Sethna et al. 1989), as are replication intermediates containing subgenome-length negative strands (Sawicki and Sawicki 1990). In the currently accepted model, subgenomic mRNAs are transcribed from a complementary set of subgenome-size minus-strand RNAs, produced by discontinuous minus-strand synthesis. Molecular genetic studies with viruses containing mutations in the TRS support a model where leader-body joining takes place during synthesis of subgenomic negative-sense RNAs (Zuniga et al. 2004; Pasternak et al. 2001; van Marle et al. 1999). Sense–antisense base-pairing interactions between short conserved sequences play a key regulatory role in this process.

4.2 *cis*-Acting RNA Elements in Coronavirus Replication

The 5′ and 3′ untranslated regions (UTRs) of all coronavirus genomes contain *cis*-acting sequences required for viral transcription and replication (Chang et al. 1994; Dalton et al. 2001; Izeta et al. 1999; Kim et al. 1993). Additional *cis*-acting sequences such as packaging signals needed for assembly have been identified and mapped to internal positions in the genome. Many of these *cis*-acting sequences have been defined by studying defective interfering (DI) RNAs. These DI RNAs are extensively deleted, retain their 5′ and 3′ UTRs plus some additional genomic RNA, and are replication competent in the presence of helper virus able to provide replicase components in *trans*. Thus they retain *cis*-acting sequences needed for genome replication. DI RNAs have also been used to study the *cis*-acting signals needed for transcription (subgenomic mRNA synthesis) and for virion assembly. Recently reverse genetic systems for a number of coronaviruses have been developed, enabling the study of coronavirus *cis*-acting sequences in the context of the viral genome.

4.2.1 The Transcription Regulatory Sequence

Coronavirus RNA transcription occurs in the cytoplasm. All the coronavirus mRNAs have a common leader sequence at their 5' ends (Spaan et al. 1982). The leader sequence contains a transcription regulatory sequence (TRS-L) at its 3' end. This sequence motif constitutes part of the signal for subgenomic mRNA transcription. Preceding every transcription unit on the viral genomic RNA are additional transcription regulatory elements, named body transcription regulatory sequence (TRS-B) (Budzilowicz et al. 1985). All coronavirus TRSs can be divided into three sequence blocks, the core 6–8 nt sequence (CS), plus 5' and 3' flanking sequences (Sola et al. 2005). The most frequently used CS for group 1 coronaviruses is a hexamer (5'-CUAAAC-3'). For group 2 coronaviruses a heptameric sequence, 5'-UCUAAAC-3' is the consensus sequence; it is almost identical to the group I CS. Interestingly, SARS-CoV has a CS (5'-ACGAAC-3') which differs from other group 2 coronaviruses (Marra et al. 2003; Rota et al. 2003). The CS for group 3 coronaviruses is a divergent octamer, 5'-CUUAACAA-3' (Alonso et al. 2002). The related arterivirus CS is 5'-UCAACU-3' and partially resembles the infectious bronchitis virus (IBV) CS (van Marle et al. 1999).

Mutational analysis in a DI system found that the sequence flanking the CS-B affected the efficiency of subgenomic DI RNA transcription and that CS-B was necessary but not sufficient for the synthesis of the subgenomic DI RNA (Makino et al. 1991). Further analysis of MHV subgenomic mRNA transcription revealed that the 5' leader sequence of MHV serves as a *cis*-acting element required for the transcription of subgenomic mRNAs (Liao and Lai 1994). Analysis of transmissible gastroenteritis virus (TGEV) mRNA synthesis in a minigenome system showed that the CS is essential for mediating a 100- to 1,000-fold increase in mRNA synthesis (Alonso et al. 2002). However, the CS flanking sequences also influenced transcription levels.

The functional importance of the TRS-L and TRS-B in the synthesis of subgenomic mRNA was shown by a mutagenesis study in equine arteritis virus (EAV), a member of the related arterivirus genus, utilizing a reverse genetic system (van Marle et al. 1999). Mutagenesis of the RNA 7 TRS-B significantly reduced its transcription. In contrast, mutagenesis of TRS-L affected all subgenomic mRNA transcription, and compensatory mutations in both TRS-L and RNA7 TRS-B restored RNA 7 transcription. This evidence strongly supports the mechanism of discontinuous minus-strand transcription. An additional comprehensive covariation mutagenesis study of several EAV TRSs demonstrated that discontinuous RNA synthesis depends not only on base-pairing between sense TRS-L and antisense TRS-B, but also upon the primary sequence of the TRS-B (Pasternak et al. 2001). While the TRS-L merely plays a targeting role for strand transfer, the TRS-B fulfills multiple functions. The sequences of mRNA leader-body junctions of TRS mutants strongly suggested the discontinuous step occurs during minus-strand synthesis. The development of reverse genetic systems for several coronaviruses has allowed a similar molecular genetic approach to investigating the role of the coronavirus

TRS rather than using DI replicons. For TGEV, analysis of the role of TRS demonstrated that the canonical CS-B was not absolutely required for the generation of subgenomic mRNAs, but its presence led to transcription levels at least 1,000-fold higher than those in its absence (Zuniga et al. 2004). A recent study in SARS-CoV rewired the TRS circuit (Yount et al. 2006). Recombinant genomes were constructed that contained mixtures of the wild-type and mutant regulatory TRS. Viable viruses were isolated from wild-type and recombinant genomes containing homogeneous transcription circuits; chimeras that contained mixed regulatory networks were invariantly lethal. In the lethal mutants the mixed TRS circuits promoted inefficient subgenomic transcription from inappropriate start sites, resulting in truncated open reading frames (ORFs) and minimized viral structural protein expression (Yount et al. 2006). This experiment provides further evidence for the discontinuous synthesis of minus-strand RNAs and the key role of TRS-L and TRS-B sequences in regulating subgenomic RNA synthesis.

In TGEV, there is a good correlation between subgenomic mRNA levels and the free energy of TRS-L and TRS-B duplex formation except for subgenomic mRNA N (Sola et al. 2005), leading the Enjuanes laboratory to seek additional regulatory sequences. Recently, a 9-nt transcriptional enhancer sequence was demonstrated 449 nts upstream of the TGEV N gene TRS core sequence (CS-N) (Moreno et al. 2008). This enhancer sequence interacts with a complementary sequence just upstream of CS-N, specifically increasing the accumulation of subgenomic mRNA N. This interaction is exclusively conserved in group 1a coronaviruses.

4.2.2 The 5′ cis*-Acting RNA Elements*

5′ *cis*-acting elements have been defined for several coronaviruses using DI deletion mapping analysis (Dalton et al. 2001; Chang et al. 1994; Izeta et al. 1999). A series of studies with MHV DIs demonstrated that as little as 466 nts at the 5′ terminus were needed for DI RNA replication (Kim et al. 1993; Kim and Makino 1995; Luytjes et al. 1996). Similar analyses with a BCoV synthetic DI RNA indicated that only the 5′ 498 nts were needed for DI RNA replication. Currently, no SARS-CoV 5′ *cis*-acting elements have been defined by a deletion analysis.

The DI RNA deletion analyses cited above defined the minimal *cis*-acting elements required for DI replication without further dissecting these sequences. Chang et al. (1994) demonstrated that the BCoV 68 nt leader is a necessary part of the 498 nt *cis*-acting signal for DI RNA replication. RNA higher-order structures contained in group 2 coronaviruses 5′ UTR were first predicted for BCoV using the Mfold algorithm (Chang et al. 1994). Subsequently, enzymatic probing and functional mutational analysis (Chang et al. 1996; Raman et al. 2003; Raman and Brian 2005) defined four stem-loops, denoted I, II, III, and IV within the first 210 nts of BCoV. The predicted stem-loop I (nts 11–42) contains three contiguous Waston–Crick base-pairs and a large 16 nt loop and is not conserved among group 2 coronaviruses (Chang et al. 1994). The predicted stem-loop II (nts 51–84) is a A–U base-pair-rich hairpin with a low free energy that folds the TRS into the terminal loop. A poorly conserved stem-loop II homolog has been predicted in other coronaviruses, and in EAV (van der Born et al. 2004; Raman and Brian 2005). Stem-loop III (nts 97–116) is phylogenetically conserved and appears to have homologs in coronavirus groups 1 and 3; enzymatic probing and mutational analysis in DI RNA replication assays supports its existence (Raman et al. 2003). Stem-loop IV (nts 186–215), a bulged stem-loop, is also conserved amongst group 2 coronaviruses and may have a homolog in group 1 and 3. However, the predicted stem-loop IV homolog in SARS-CoV appears to be group 1-like (Raman and Brian 2005). Stem-loop IV exists as a higher-order structure based on enzymatic probing and it is required for DI RNA replication. Recently, two stem-loops, SLV (nts 239–310) and SLVI (nts 311–340), extending into the nsp1 coding region were demonstrated by RNase structure probing and sequence covariation among closely related group 2 coronaviruses. SLVI is required for DI RNA replication (Brown et al. 2007).

The recent establishment of reverse genetic systems for coronaviruses representing all of the coronavirus subgroups has facilitated the functional analysis of *cis*-acting elements in the context of the whole genome. Recently, we proposed a consensus secondary structural model of the 5′ 140 nts of the 5′ UTR based on nine representative coronaviruses (including SARS-CoV) from all three coronavirus groups (Kang et al. 2006; Liu et al. 2007). The 5′ ~140 nts of the nine coronaviruses genomes were predicted to fold into three major stem-loops, denoted SL1, SL2, and SL4 (see Fig. 4.1). Some sequences were predicted to contain a fourth stem-loop, SL3, which folds the TRS-L into a hairpin loop. SL1, SL2, and SL4 were structurally conserved amongst all coronaviruses examined. SL3 is only predicted to be stable for human coronavirus OC43 (HCoV-OC43) and SARS-CoV. It should be noted that SL1 and SL2 differ from the structures studied in BCoV pDrep 1 RNA by the Brian group (Chang et al. 1996; Raman et al. 2003; Raman and Brian 2005). However, SLIII in Brian's model is almost identical with our SL4b (Liu et al. 2007; Raman et al. 2003).

Fig. 4.1 Conserved RNA higher-order structural model within the 5′ and 3′ UTRs of group 2 coronaviruses. Proteins binding to the positive-strand RNA are shown as *solid symbols*; proteins binding to the minus-strand RNA are shown as *open symbols*

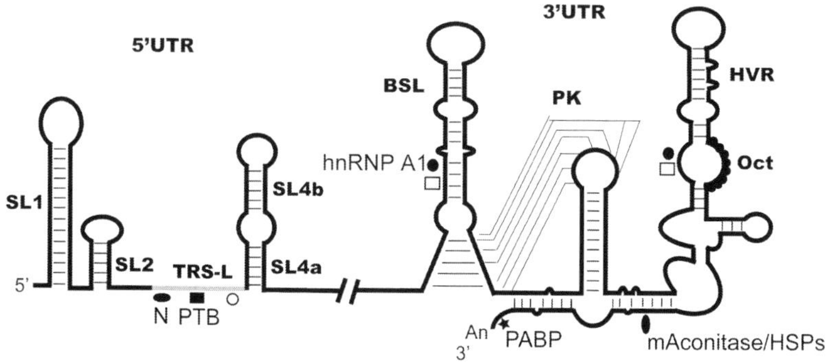

Although the full MHV and SARS-CoV 5′ UTRs are significantly different in terms of sequences and predicted secondary structures, the SARS-CoV SL1, SL2, and SL4 can functionally replace their MHV counterparts in the MHV genome and produce viable chimeric viruses (Kang et al. 2006). However, MHV chimeras containing the complete SARS-CoV 5′ UTR or the SARS-CoV SL3 were not viable. Replacing the SARS-CoV TRS with the MHV TRS in the MHV/5′ UTR SARS-CoV chimera permitted the synthesis of minus-strand genomic RNA but did not support the production of positive- or minus-strand subgenomic RNA7. This study supports the idea that SL1, SL2, and SL4 are conserved and interchangeable within the same group without affecting viral viability.

A detailed mutational and biophysical study of MHV SL1 revealed that this stem-loop is functionally and structurally bipartite. SL1 contains one or more noncanonical base-pairs in the central portion of the stem. In MHV, two pyrimidine–pyrimidine base-pairs are present in the middle of SL1, as demonstrated by NMR studies (Liu et al. 2007). These noncanonical base-pairs divide the SL1 helical stem into upper and lower segments. The upper region of SL1 is required to be base-paired; mutations that disrupted base-pairing of this region were not viable or severely impaired (Li et al. 2008). Combining both sets of mutations in the upper region of SL1 restored the base-pairing and yielded a viable virus comparable to the wild-type virus in its growth phenotype. In contrast, mutations in the lower region of SL1 that destroyed base-pairing were viable, and genomes with compensatory mutations predicted to restore base-pairing were nonviable. Deletion of a bulged or extruded A in the lower portion of the stem (mutation ΔA35), a mutation that increased the thermal stability of the lower portion of the SL1 helix, was strongly selected against. ΔA35-containing viruses were rapidly replaced by viruses containing destabilizing second-site mutations near ΔA35. Additionally, mutations that increased the stability of the lower portion of SL1 were lethal, suggesting that structural lability in the lower portion of the SL1 stem was required. Thermal denaturation and imino proton exchange experiments further demonstrated that the lower half of SL1 is unstable. SL1 second-site mutants also contained an additional second-site mutation, A29G or A78G, in their 3′ UTR, providing genetic evidence for an interaction between the 5′ and 3′ UTRs. Thus we hypothesized that the base of SL1 has an optimized lability required to mediate a physical interaction between the 5′ UTR and the 3′ UTR (Li et al. 2008). These data, plus the observed defects in subgenomic RNA synthesis in our nonviable SL1 mutants, are consistent with the genome circularization model for coronavirus transcription put forward by Zuniga et al. (2004) and suggest that replication complexes and transcription complexes have different structural requirements in the 5′ UTR.

SL2 is the most conserved secondary structure in the 5′ UTRs of all coronaviruses examined (Liu et al. 2007). Except for the core TRS leader sequence, the (C/U)UUG(U/C) sequence encompassing the predicted SL2 loop is the most conserved contiguous run of nucleotides in the entire 5′ UTR and contains features of a canonical U-turn motif, in which the middle 3 nts of the loop, UNR ($U_0 \bullet N_{+1} \bullet R_{+2}$), form a triloop that stacks on a Y:Y, Y:A, or G:A noncanonical base-pair. The basic structural feature of the canonical U-turn is a sharp turn in the phosphate backbone between U_0 and N_{+1}, with U_0 engaged in two critical hydrogen bonds: the U_0 imino proton donates a hydrogen bond to the nonbridging phosphate oxygen following R_{+2}, and the U_0 2'-OH proton donates a hydrogen bond to the N7 of R2. NMR studies of SL2 indicated that the U_0 imino proton donates a hydrogen bond in SL2, consistent with a U-turn structure. However, there was no evidence for the predicted noncanonical pyrimidine–pyrimidine base-pair between positions 47 and 51. Additional NMR studies indicate that U51 was extruded from the loop and that the Watson–Crick faces of C47 and G50 were in apposition (Li and Giedroc, unpublished). Formation of the stem was required for virus viability, although the sequence of the stem was unimportant (Liu et al. 2007). Replacing U48 with either cytosine or adenosine was lethal, consistent with a UNR loop structure for SL2 (Liu et al. 2007). However, viruses containing a U48G mutation were viable and replicated almost as well as wild-type virus. NMR studies indicated that a guanine at position 48 engaged in a hydrogen bond, similar to that observed for U48. Mutagenesis of U49 and U51 demonstrated that any nucleotide can function in these positions, whereas the G at position 50 is required (Liu et al. 2007; Liu et al. 2009). RT-PCR analyses of cells electroporated with genomes containing lethal mutations in SL2 demonstrate that SL2 is required for subgenomic RNA synthesis, as was SL1. Taken together the functional and structural data suggests that SL2 more closely resembles a YNMG-like tetraloop than a U-turn. Additional NMR studies should provide an atomic resolution structure to determine the preise geometry of the loop.

4.2.3 The 3' cis-Acting RNA Elements

Experiments to dissect the *cis*-acting elements in the coronavirus 3' UTR have generated a comprehensive view of *cis*-acting elements in this region. Initial DI deletion analyses found that the minimal 3' terminus sequence required to support MHV DI RNA replication is 436 nts, a region containing part of the upstream N gene and the entire 301 nt 3' UTR (Lin and Lai 1993; Luytjes et al. 1996). For TGEV and IBV, the minimal sequence requirements were 492 nts and 338 nts respectively, and did not include any part of the N gene (Mendez et al. 1996; Dalton et al. 2001). It was later confirmed this was also true for MHV, as a recombinant virus containing the N gene translocated into an upstream genomic position was viable (Goebel et al. 2004a). For MHV, the differing conclusions resulting from DI assays and intact virus may reflect the fact that DI assays are inherently competition assays with wild-type genomes and thus may be more sensitive at detecting minor decreases in relative fitness then assays with infectious viruses that focus on recovering viable viruses. This was clearly true in experiments in which DI RNAs carrying mutations at the 3' end failed to replicate at detectable levels, but recombinant viruses with these same mutations were viable with only modestly impaired replication phenotypes (Johnson et al. 2005). A deletion analysis utilizing a DI RNA replicon defined the minimum sequence needed for minus-strand RNA synthesis as the 3′-most 55 nts plus the poly(A) tail (Lin et al. 1994). Spangnolo and Hogue demonstrated that the poly(A) tail was required for DI replication, although as little as five As would suffice to initiate replication (Spagnolo and Hogue 2000).

Genetic and enzymatic probing of MHV and BCoV 3′ UTR secondary structure demonstrated the presence of three RNA secondary structures (Fig. 4.1) (Hsue and Masters 1997). The 5′-most of these, a 68 nt bulged stem-loop immediately downstream of the N gene stop codon, was predicted to be absolutely conserved in MHV, BCoV, HCoV-OC43, and bovine enteric coronavirus. This stem-loop was further characterized biochemically, and mutagenesis demonstrated that it was essential for DI RNA and for viral replication (Hsue et al. 2000). A 54 nt hairpin-type pseudoknot 3′ to the 68 nt bulged stem-loop was first found to be required for DI RNA replication in the 3′ UTR of BCoV (Williams et al. 1999). This pseudoknot is phylogenetically conserved among coronaviruses, including the SARS-CoV (Goebel et al. 2004b), both in location and in shape but only partially in nucleotide sequence. In a later study with MHV (Goebel et al. 2004a), this pseudoknot was demonstrated to partially overlap with the bulged stem-loop, such that the last part of the bulged stem-loop overlaps with stem 1 of the pseudoknot; thus these two structures cannot be formed simultaneously. This finding led to the proposal that the bulged stem-loop and the pseudoknot are components of a molecular switch that regulate viral RNA synthesis (Goebel et al. 2004a).

The third RNA secondary structure, a complex multiple stem-loop structure, is further downstream in the MHV 3′ UTR. This structure was predicted by computer-assisted analysis of the last 166 nts of the genome 3′ to the pseudoknot using the Mfold algorithm (Liu et al. 2001). Enzymatic probing of RNA secondary structure supported the existence of the predicted long bulged stem-loop encompassing nts 143–68 and with a second stem-loop from nts 67 to 52. Within the long stem-loop, a conserved bulged-stem structure (nts 142–132 and nts 79–68) also present in BCoV was identified by covariation analysis. Site-directed mutagenesis and DI RNA replication assays indicated that the long bulged-stem loop between nts 143 and 68 plays an important role in DI RNA replication. Similar assays to examine the shorter stem-loop between nts 67 and 52 failed to provide evidence for a role in DI replication (Liu et al. 2001). The long bulged stem-loop contains an octanucleotide sequence, 5′-GGAAGAGC-3′ (nts 81–74 in MHV), that is conserved in the 3′ UTR of coronaviruses from all three groups, and thus might have important biological functions.

Further analysis of the long bulged stem-loop revealed that although the octanucleotide sequence is almost universally conserved in coronaviruses, the remainder of this complex stem-loop resides in a hypervariable region (HVR) of the 3′ UTR that is poorly conserved in group 2 coronaviruses. An extensive mutational analysis of the HVR was carried out by deletion, rearrangements, and point mutations (Goebel et al. 2007). All these mutations have only modest effects on viral replication, indicating that the HVR is not essential for viral RNA synthesis. This result differs from the results obtained in DI systems by Liu et al. (2001) and by Lin et al. (1994). Since the most extensive HVR mutant deleted nts 30–170, it is clear that not all 55 nts are required for minus-strand replication. A possible explanation for these discrepancies is the inherent competitive nature of DI replication assays greatly increasing the effects of mutations that are only moderately deleterious in the context of the intact genome. The HVR deletion mutant was highly attenuated in mice, suggesting that the HVR might play a significant role in viral pathogenesis (Goebel et al. 2007). However, it should be kept in mind that the HVR deletion virus grew to a titer 2–3-fold less than that of wild-type virus in cell culture, making the interpretation of its effect on pathogenesis difficult.

Most recently, multiple second-site revertants of the pseudoknot were recovered by characterizing an unstable mutant Alb391, with a 6 nt insertion of AACAAG in loop 1 of the pseudoknot of MHV 3′ UTR. These second-site suppressor mutations were localized to two separate regions of the genome: one group of mutations was mapped to nsp8 and nsp9 and the second group mapped to the extreme 3′ end of the genome. These observations led the authors to point out that coronavirus replicase gene products might interact with the 3′ end of the genome, and that the loop 1 of the pseudoknot has the potential to base-pair with the extreme end of the genome (Zust et al. 2008). This observation is supported by structural predictions, phylogenetic conservation of the interaction amongst all known group 2 coronaviruses, and the ability of a drastically minimized truncation mutant ΔHVR3 in which all sequences between nts 29 and 171 were replaced by a tetraloop capping the helix downstream of pseudoknot stem 2 (Zust et al. 2008). However, this base-pairing interaction has not yet been demonstrated biochemically or by mutagenesis.

Interestingly, the 3′ UTR stem-loop structures of the group 2 coronaviruses seem to be different from both group 1 and group 3 coronaviruses. All the group 1 coronaviruses contain a highly conserved pseudoknot (Williams et al. 1999), but there is no bulged stem-loop structure in the 3′ UTR. The group 3 coronaviruses have a highly conserved and functionally essential stem-loop (Dalton et al. 2001), but only a poor candidate for the pseudoknot structure can be found nearby (Williams et al. 1999). Only the group 2 coronaviruses have both the pseudoknot and the bulged stem-loop in close proximity and they all overlap in the same fashion. Despite their primary sequence divergence among the 3′ UTRs of group 2 coronaviruses, the secondary structures are all highly conserved and functionally equivalent, as shown by the replication of a BCoV DI RNA in the presence of various group 2 helper viruses (Wu et al. 2003), and by the isolation of chimeras in which the 3′ UTRs of BCoV and SARS-CoV, both group 2 coronaviruses, replaced their MHV counterpart without affecting viral viability (Hsue and Masters 1997; Goebel et al. 2004b). However, the MHV 3′ UTR cannot be replaced with either the group 1 TGEV 3′ UTR or the group 3 IBV 3′ UTR (Hsue and Masters 1997; Goebel et al. 2004b).

4.2.4 Proteins Binding to the 5′ and 3′ cis-Acting Elements

Although exactly how a coronavirus synthesizes its RNAs is still unclear, there is increasing evidence that coronavirus discontinuous synthesis of subgenomic mRNAs is directed by *cis*-acting sequences present on the viral RNAs with the help of *trans*-acting factors encoded by the virus, as well as cellular proteins. Although it has been demonstrated that continuous protein synthesis is required for viral RNA synthesis (Sawicki and Sawicki 1986), little is known as to which viral and cellular proteins are involved in viral RNA transcription and replication.

The current discontinuous RNA synthesis model proposes that the TRS-L sequence is brought into close proximity to sequences located at the 3' end of the genomic RNA through RNA–RNA, or RNA–protein and protein–protein interactions. Indeed, several cellular proteins have been shown to interact with the 5' and 3' ends of the coronavirus genome (see Fig. 4.1). At the 3' end of the coronavirus genome, the 73 kDa cytoplasmic poly(A) binding protein (PABP) binds to the poly (A) tail (Spagnolo and Hogue 2000). A series of host proteins were found to bind to the MHV 3'-most 42 nt RNA probe using RNase protection/gel mobility shift and UV cross-linking assays (Yu and Leibowitz 1995a, 1995b). Further analysis revealed that these proteins include mitochondrial aconitase and the chaperones mitochondrial HSP70, HSP60, and HSP40 (Nanda et al. 2004; Nanda and Leibowitz 2001). DI replication assays suggested that proteins binding to both the poly(A) tail and the last 42 nts of the MHV genome had a role in replication. Mutations in the 3'(+)42 host protein binding element had a deleterious effect on the accumulation of DI RNA, and when the same mutations were introduced into the MHV genome, one mutant was found to be nonviable. This mutant had a defect in subgenomic mRNA synthesis which points to a potential role for sequences at the extreme 3' end of the MHV genome in subgenomic RNA synthesis (Johnson et al. 2005), a finding consistent with the model proposed by the Enjuanes group (Zuniga et al. 2004). Polypyrimidine tract-binding (PTB) protein was shown to bind to two regions of the MHV 3' UTR, a strong PTB binding site was mapped to nts 53–149, and another weak binding site was mapped to nts 270–307 on the complementary strand of the 3' UTR (Huang and Lai 1999). Since a number of these binding sites are deleted in the replication competent HVR mutant virus discussed above, it is unlikely that most of these proteins are required for viral replication (Goebel et al. 2007).

The viral protein N binds to the TRS-L specifically and with high affinity (Nelson et al. 2000). It has been suggested that N protein binding to TRS-L favors translation of viral RNAs (Tahara et al. 1998) and may also play a role in MHV RNA synthesis (Li et al. 1999). Recently, BCoV NSP1 was shown to bind three 5' UTR and one 3' UTR-located *cis*-replication stem-loops and may function to regulate viral genome translation or replication (Gustin et al. 2009). Another cellular protein, the heterogeneous nuclear ribonucleoprotein (hnRNP) A1 was demonstrated to bind to the MHV minus-strand leader and TRS-B complementary sequences through immunoprecipitation after UV cross-linking and by in vitro binding assays with recombinant protein (Li et al. 1997). hnRNP A1 was shown to have two binding sites at the MHV 3' end and these binding sites are complementary to the sites on the minus-strand RNA that bind PTB (Huang and Lai 2001). Mutations that affect PTB binding to the negative strand of the 3' UTR also inhibited hnRNP A1 binding on the positive strand, indicating a possible relationship between these two proteins. Furthermore, both hnRNP A1 and PTB bind to the complementary strands at the 5′ end of MHV RNA. Based on these observations, it was proposed that hnRNP A1–PTB interactions provide a molecular mechanism for potential 5′–3′ cross-talk in MHV RNA, which may be important for RNA replication and transcription. However, the role of hnRNP A1 in viral replication is controversial. Paul Masters' group (Shen and Masters 2001) tested the role of hnRNP A1 in viral transcription and replication by inserting a high-affinity hnRNP A1 binding site in place of, or adjacent to, an intergenic sequence in the MHV genome. This inserted hnRNP A1 binding site was not able to functionally replace or enhance transcription from the intergenic sequence. Additionally, MHV was able to replicate normally and synthesize normal levels of genome and subgenomic RNAs in cells lacking hnRNP A1, suggesting that hnRNP A1 is not required for MHV discontinuous transcription or genome replication. However, it was subsequently shown that other members of the hnRNP family can substitute for hnRNAP A1 (Choi et al. 2004).

A recent study also showed that a cytoplasmic host factor is indispensable for SARS-CoV in vitro RNA synthesis, although this host factor has not yet been identified (van Hemert et al. 2008). How these and other potential cellular proteins interact with the coronavirus 5′ and 3′ cis-acting elements to initiate and support genomic and subgenomic RNA synthesis is a long unanswered interesting question.

4.3 Future Directions

There is a paucity of data demonstrating viral proteins binding to specific *cis*-acting sequences in the coronavirus genome, with the exception of the N protein to TRS-L discussed above. Although there is genetic evidence for nsp8 and nsp9 interacting with the 3′ UTR (Zust et al. 2008), there is no data defining precisely where these proteins bind, nor is there any evidence as to how any other replicase components bind to the genome. We anticipate that research over the next few years will answer these questions and clarify how the various virus and host proteins function in viral replication.

References

Alonso S, Izeta A, Sola I, Enjuanes L (2002) Transcription regulatory sequences and mRNA expression levels in the coronavirus transmissible gastroenteritis virus. J Virol 76(3): 1293–1308

Brierley I, Boursnell ME, Binns MM, Bilimoria B, Blok VC, Brown TD, Inglis SC (1987) An efficient ribosomal frame-shifting signal in the polymerase-encoding region of the coronavirus IBV. EMBO J 6:3779–3785

Brown CG, Nixon KS, Senanayake SD, Brian DA (2007) An RNA stem-loop within the bovine coronavirus nsp1 coding region is a *cis*-acting element in defective interfering RNA replication. J Virol 81:7716–7724

Budzilowicz CJ, Wilczynski SP, Weiss SR (1985) Three intergenic regions of coronavirus mouse hepatitis virus strain A59 genome RNA contain a common nucleotide sequence that is homologous to the 3' end of the viral mRNA leader sequence. J Virol 53:834–840

Chang RY, Hofmann MA, Sethana PB, Brian DA (1994) A cis-acting function for the coronavirus leader in defective interfering RNA replication. J Virol 68:8223–8231

Chang RY, Krishnan R, Brian DA (1996) The UCUAAAC promoter motif is not required for high-frequency leader recombination in bovine coronavirus defective interfering RNA. J Virol 70(5):2720–2729

Choi KS, Mizutani A, Lai MM (2004) SYNCRIP, a member of the heterogeneous nuclear ribonucleoprotein family, is involved in mouse hepatitis virus RNA synthesis. J Virol 78(23):13153–13162

Dalton K, Casais R, Shaw K, Stirrups K, Evans S, Britton P, Brown TDK, Cavanagh, D. (2001) cis-Acting sequences required for coronavirus infectious bronchitis virus defective-RNA replication and packaging. J Virol 75:125–133

Goebel SJ, Hsue B, Dombrowski TF, Masters PS (2004a) Characterization of the RNA components of a putative molecular switch in the 3' untranslated region of the murine coronavirus genome. J Virol 78(2):669–682

Goebel SJ, Taylor J, Masters PS (2004b) The 3' cis-acting genomic replication element of the severe acute respiratory syndrome coronavirus can function in the murine coronavirus genome. J Virol 78(14):7846–7851

Goebel SJ, Miller TB, Bennett CJ, Bernard KA, Masters PS (2007) A hypervariable region within the 3' cis-acting element of the murine Coronavirus genome is nonessential for RNA synthesis but affects pathogenesis. J Virol 81(3):1274–1287

Gustin KM, Guan B, Dziduszka A, Brian DA (2009) Bovine coronavirus nonstructural protein 1 (p28) is an RNA binding protein that binds terminal genomic cis-replication elements. J Virol 83(12):6087–6097

Hsue B, Masters PS (1997) A bulged stem-loop structure in the 3' untranslated region of the genome of the coronavirus mouse hepatitis virus is essential for replication. J Virol 71(10):7567–7578

Hsue B, Hartshorne T, Masters PS (2000) Characterization of an essential RNA secondary structure in the 3' untranslated region of the murine coronavirus genome. J Virol 74(15):6911–6921

Huang P, Lai MM (1999) Polypyrimidine tract-binding protein binds to the complementary strand of the mouse hepatitis virus 3' untranslated region, thereby altering RNA conformation. J Virol 73(11):9110–9116

Huang P, Lai MMC (2001) Heterogeneous nuclear ribonucleoprotein A1 binds to the 3'-untranslated region and mediates potential 5'-3'-end cross talks of mouse Hepatitis Virus RNA. J Virol 75(11):5009–5017

Izeta A, Smerdou C, Alonso S, Penzes Z, Mendez A, Plana-Duran J, Enjuanes L. (1999) Replication and packaging of transmissible gastroenteritis coronavirus-derived synthetic minigenomes. J Virol 73:1535–1545

Johnson RF, Feng M, Liu P, Millership JJ, Yount B, Baric RS, Leibowitz JL (2005) The effect of mutations in the mouse hepatitis virus 3'(+)42 protein binding element on RNA replication. J Virol 79(23):14570–14585

Kang H, Feng M, Schroeder ME, Giedroc DP, Leibowitz JL (2006) Putative cis-acting stem-loops in the 5' untranslated region of the severe acute respiratory syndrome coronavirus can substitute for their mouse hepatitis virus counterparts. J Virol 80(21):10600–10614

Kim Y-N, Makino S (1995) Characterization of a murine coronavirus defective interfering RNA internal cis-acting replication signal. J Virol 69:4963–4971

Kim Y-N, Jeong YS, Makino S (1993) Analysis of cis-acting sequences essential for coronavirus defective interfering RNA replication. Virology 197:53–63

Lai MM, Patton CD, Baric RS, Stohlman SA (1983) Presence of leader in the mRNA of mouse hepatitis virus. J Virol 46(3):1027–1033

Lai MMC, Baric RS, Brayton PR, Stohlman SA (1984) Characterization of leader RNA sequences on the virion and mRNAs of mouse hepatitis virus, a cytoplasmic RNA virus. Proc Natl Acad Sci USA 81:3626–3630

Li HP, Zhang X, Duncan R, Comai L, Lai MM (1997) Heterogeneous nuclear ribonucleoprotein A1 binds to the transcription- regulatory region of mouse hepatitis virus RNA. Proc Natl Acad Sci USA 94(18):9544–9549

Li HP, Huang P, Park S, Lai MM (1999) Polypyrimidine tract-binding protein binds to the leader RNA of mouse hepatitis virus and serves as a regulator of viral transcription. J Virol 73(1): 772–777

Li W, Moore MJ, Vasilieva N, Sui J, Wong SK, Berne MA, Somasundaran M, Sullivan JL, Luzuriaga K, Greenough TC, Choe H, Farzan M (2003) Angiotensin-converting enzyme 2 is a functional receptor for the SARS coronavirus. Nature 426(6965):450–454

Li L, Kang H, Liu P, Makkinje N, Williams ST, Leibowitz JL, Giedroc DP (2008) Structural lability in stem-loop 1 drives a 5′ UTR-3′ UTR interaction in coronavirus replication. J Mol Biol 377:790–803

Liao CL, Lai MMC (1994) Requirement of the 5′-end genomic sequence as an upstream *cis*-acting element for coronavirus subgenomic mRNA transcription. J Virol 68(8):4727–4737

Lin Y-J, Lai MMC (1993) Deletion mapping of a mouse hepatitis virus defective interfering RNA reveals the requirement of an internal and discontinous sequence for replication. J Virol 67:6110–6118

Lin Y-J, Liao CL, Lai MMC (1994) Identification of the *cis*-acting signal for minus-strand RNA synthesis of a murine coronavirus: implications for the role of minus-strand RNA in RNA replication and transcription. J Virol 68:8131–8140

Liu Q, Johnson RF, Leibowitz JL (2001) Secondary structural elements within the 3′ untranslated region of mouse Hepatitis Virus strain JHM genomic RNA. J Virol 75(24):12105–12113

Liu P, Li L, Keane SC, Yang D, Leibowitz JL, Giedroc DP (2009) Mouse hepatitis virus stem-loop 2 adopts a uYNMG(U)a-like tetraloop structure that is highly functionally tolerant of base substitutions. J Virol, In press 2009, doi: 10.1128/JVI.00915-09

Liu P, Li L, Millership JJ, Kang H, Leibowitz JL, Giedroc DP (2007) A U-turn motif-containing stem-loop in the coronavirus 5′ untranslated region plays a functional role in replication. RNA 13:763–780

Luytjes W, Gerritsma H, Spaan WJM (1996) Replication of synthetic defective interfering RNAs derived from coronavirus mouse hepatitis virus-A59. Virology 216:174–183

Makino S, Joo M, Makino JK (1991) A system for study of coronavirus mRNA synthesis: a regulated, expressed subgenomic defective interfering RNA results form intergenic site insertion. J Virol 65(11):6031–6041

Marra MA, Jones SJ, Astell CR, Holt RA, Brooks-Wilson A, Butterfield YS, Khattra J, Asano JK, Barber SA, Chan SY, Cloutier A, Coughlin SM, Freeman D, Girn N, Griffith OL, Leach SR, Mayo M, McDonald H, Montgomery SB, Pandoh PK, Petrescu AS, Robertson AG, Schein JE, Siddiqui A, Smailus DE, Stott JM, Yang GS, Plummer F, Andonov A, Artsob H, Bastien N, Bernard K, Booth TF, Bowness D, Czub M, Drebot M, Fernando L, Flick R, Garbutt M, Gray M, Grolla A, Jones S, Feldmann H, Meyers A, Kabani A, Li Y, Normand S, Stroher U, Tipples GA, Tyler S, Vogrig R, Ward D, Watson B, Brunham RC, Krajden M, Petric M, Skowronski DM, Upton C, Roper RL (2003) The genome sequence of the SARS-associated coronavirus. Science 300(5624):1399–1404

Mendez A, Smerdou C, Izeta A, Gebauer F, Enjuanes L (1996) Molecular characterization of transmissible gastroenteritis coronavirus defective interfering genomes: packaging and heterogeneity. Virology 217:495–507

Moreno JL, Zuniga S, Enjuanes L, Sola I (2008) Identification of a coronavirus transcription enhancer. J Virol 82(8):3882–3893

Nanda SK, Leibowitz JL (2001) Mitochondrial aconitase binds to the 3′-untranslated region of the mouse hepatitis virus genome. J Virol 75:3352–3362

Nanda SK, Johnson RF, Liu Q, Leibowitz JL (2004) Mitochondrial HSP70, HSP40, and HSP60 bind to the 3′ untranslated region of the murine hepatitis virus genome. Arch Virol 149(1):93–111

Nash TC, Buchmeier MJ (1997) Entry of mouse hepatitis virus into cells by endosomal and nonendosomal pathways. Virology 23(1):1–8

Nelson GW, Stohlman SA, Tahara SM (2000) High affinity interaction between nucleocapsid protein and leader/intergenic sequence of mouse hepatitis virus RNA. J Gen Virol 81 (Pt 1):181–188

Pasternak A, van der Born E, Spaan WJM, Snijder EJ (2001) Sequence requirements for RNA strand transfer during nidovirus discontinuous subgenomic RNA synthesis. EMBO J 20(24): 7220–7228

Raman S, Brian DA (2005) Stem-loop IV in the 5′ untranslated region is a *cis*-acting element in bovine coronavirus defective interfering RNA replication. J Virol 79(19):12434–12446

Raman S, Bouma P, Williams GD, Brian DA (2003) Stem-loop III in the 5′ untranslated region is a *cis*-acting element in bovine coronavirus defective interfering RNA replication. J Virol 77(12):6720–6730

Rota PA, Oberste MS, Monroe SS, Nix WA, Campagnoli R, Icenogle JP, Penaranda S, Bankamp B, Maher K, Chen MH, Tong S, Tamin A, Lowe L, Frace M, DeRisi JL, Chen Q, Wang D, Erdman DD, Peret TC, Burns C, Ksiazek TG, Rollin PE, Sanchez A, Liffick S, Holloway B, Limor J, McCaustland K, Olsen-Rasmussen M, Fouchier R, Gunther S, Osterhaus AD, Drosten C, Pallansch MA, Anderson LJ, Bellini WJ (2003) Characterization of a novel coronavirus associated with severe acute respiratory syndrome. Science 300(5624):1394–1399

Sawicki SG, Sawicki DL (1986) Coronavirus minus-strand RNA synthesis and effect of cycloheximide on coronavirus RNA synthesis. J Virol 57:328–334

Sawicki SG, Sawicki DL (1990) Coronavirus transcription: subgenomic mouse hepatitis virus replicative intermediates function in RNA synthesis. J Virol 64:1050–1056

Sethna PB, Hung S-L, Brian DA (1989) Coronavirus subgenomic minus-strand RNAs and the potential for mRNA replicons. Proc Natl Acad Sci USA 86:5626–5630

Shen X, Masters PS (2001) Evaluation of the role of heterogeneous nuclear ribonucleoprotein A1 as a host factor in murine coronavirus discontinuous transcription and genome replication. Proc Natl Acad Sci USA 98(5):2717–2722

Snijder EJ, Bredenbeek PJ, Dobbe JC, Thiel V, Ziebuhr J, Poon LL, Guan Y, Rozanov M, Spaan WJ, Gorbalenya AE (2003) Unique and conserved features of genome and proteome of SARS-coronavirus, an early split-off from the Coronavirus group 2 lineage. J Mol Biol 331(5): 991–1004

Sola I, Moreno JL, Zuniga S, Alonso S, Enjuanes L (2005) Role of nucleotides immediately flanking the transcription-regulating sequence core in coronavirus subgenomic mRNA synthesis. J Virol 79(4):2506–2516

Spaan WJM, Rottier PJM, Horzinek MC, van der Zeijst BAM (1982) Sequence relationships between the genome and the intracellular RNA species 1, 3, 6, and 7 of mouse hepatitis virus strain A59. J Virol 42:432–439

Spagnolo JF, Hogue BG (2000) Host protein interactions with the 3′ end of bovine coronavirus RNA and the requirement of the poly(A) tail for coronavirus defective genome replication. J Virol 74(11):5053–5065

Tahara SM, Dietlin TA, Nelson GW, Stohlman SA, Manno DJ (1998) Mouse hepatitis virus nucleocapsid protein as a translational effector of viral mRNAs. Adv Exp Med Biol 440: 313–318

van der Born E, Gultyaev AP, Snijder EJ (2004) Secondary structure and function of the 5′-proximal region of the equine arteritis virus RNA genome. RNA 10:424–437

van Hemert MJ, Van den Worm SHE, Knoops K, Mommaas AM, Gorbalenya AE, Snijder EJ (2008) SARS-Coronavirus replication/transcription complexes are membrane-protected and need a host factor for activity in vitro. PLos Pathog 4(5):1–10 van Marle G, Dobbe JC, Gultyaev AP, Luytjes W, Spaan WJM, Snijder EJ (1999) Arterivirus discontinuous mRNA transcription is guided by base pairing between sense and antisense transcription-regulating sequences. Proc Natl Acad Sci USA 96:3501–3506

Wang H, Yang P, Liu K, Guo F, Zhang Y (2008) SARS coronavirus entry into host cells through a novel clathrin- and caveolae-independent endocytic pathway. Cell Res 18:290–301

Williams GD, Chang RY, Brian DA (1999) A phylogenetically conserved hairpin-type 3′ untranslated region pseudoknot functions in coronavirus RNA replication. J Virol 73(10):8349–8355

Wu HY, Guy JS, Yoo D, Vlasak R, Urbach E, Brian DA (2003) Common RNA replication signals exist among group 2 coronaviruses: evidence for in vivo recombination between animal and human coronavirus molecules. Virology 315(1):174–183

Yount B, Roberts RS, Lindesmith L, Baric RS (2006) Rewiring the severe acute respiratory syndrome coronavirus (SARS-CoV) transcription circuit: engineering a recombination-resistant genome. Proc Natl Acad Sci USA 103(33):12546–12551

Yu W, Leibowitz JL (1995a) A conserved motif at the 3′ end of mouse hepatitis virus genomic RNA required for host protein binding and viral RNA replication. Virology 214:128–138

Yu W, Leibowitz JL (1995b) Specific binding of host cellular proteins to multiple sites within the 3′ end of mouse hepatitis virus genomic RNA. J Virol 69:5033–5038

Zuniga S, Sola I, Alonso S, Enjuanes L (2004) Sequence motifs involved in the regulation of discontinuous coronavirus subgenomic RNA synthesis. J Virol 78(2):980–994

Zust R, Miller TB, Goebel SJ, Thiel V, Masters PS (2008) Genetic interactions between an essential 3′ *cis*-acting RNA pseudoknot, replicase gene products, and the extreme 3′ end of the mouse coronavirus genome. J Virol 82(3):1214–1228

Chapter 5
Programmed −1 Ribosomal Frameshifting in SARS Coronavirus

Jonathan D. Dinman

Abstract In coronaviruses such as the SARS coronavirus (SARS-CoV), programmed −1 ribosomal frameshifting (−1 PRF) is used to direct the synthesis of immediate early proteins, e.g., RNA-dependent RNA polymerase (RDRP) and proteases, that are thought to prepare the infected cell for takeover by the virus. Unlike other RNA viruses which make their structural proteins first, this class of proteins is synthesized after −1 PRF, from subgenomic mRNAs produced subsequent to production of RDRP. Also unique among the coronaviruses is the inclusion of mRNA structural elements that do not appear to be essential for frameshifting. Understanding the differences between –1 PRF signals from coronaviruses and other viruses will enhance our understanding of –1 PRF in general, and will be instructive in designing new classes of antiviral therapeutics. In this chapter we summarize current knowledge and add additional insight to the function of the programmed –1 ribosomal frameshift signal present in the SARS-associated coronavirus.

5.1 Introduction

The emergence of a new infectious disease, known as severe acute respiratory syndrome (SARS), became a global concern following an outbreak in southern China in late 2002. The World Health Organization released a global alert on the 12 March 2003 and at the close of the epidemic more than 8,000 people had been infected in 29 countries (reviewed in Lai 2003; Stadler et al. 2003). The mortality rate was approximately 10%, and morbidity was significantly greater (Stadler et al. 2003 and references within).

J.D. Dinman
Department of Cell Biology and Molecular Genetics, University of Maryland, College Park, Maryland 20742, USA
e-mail: dinman@umd.edu

S.K. Lal (ed.), *Molecular Biology of the SARS-Coronavirus*,
DOI 10.1007/978-3-642-03683-5_5, © Springer-Verlag Berlin Heidelberg 2010

An international effort resulted in the rapid sequencing of many isolates, the first two published on May 1, 2003 (Marra et al. 2003; Rota et al. 2003). Using various portions of those sequences, phylogenetic analyses were performed with coronaviruses from the three previously described groups. The initial unrooted trees suggested that the SARS-CoV represented a new group (Holmes 2003; Lai 2003). Subsequent analyses using rooted trees indicated that the SARS-CoV is most likely an early split from the group 2 lineage (Stadler et al. 2003; Lio and Goldman 2004; Snijder et al. 2003). Although a body of research has been performed on some of the related viruses that impact economically important industries, and comparisons have been made based on the relationships, actual analysis of the SARS-CoV is more limited because of its recent emergence. Although there are no approved antiviral drugs that are highly effective against SARS-CoV, this virus has many unique steps in its replication that could be targeted (Holmes 2003). Various options for preventing additional outbreaks of the disease are addressed in other chapters in this book: these include methods of prevention, limiting spread, and targeting the virus after infection. Post-infection targets have been effective in limiting replication of other RNA viruses including HIV; for example, protease inhibitors are designed to target an early part of the viral lifecycle. In both HIV and SARS-CoV, polyproteins are synthesized as a consequence of programmed −1 ribosomal frameshift (−1 PRF) events, and the polyproteins are subsequently autocatalytically cleaved by their encoded proteases.

5.2 Programmed −1 Ribosomal Frameshifting

Programmed −1 ribosomal frameshifting is a mechanism in which *cis*-acting elements in the mRNA direct elongating ribosomes to shift reading frame by 1 base in the 5′ direction. The use of a −1 PRF mechanism for expression of a viral gene was first published in 1985 for the Rous sarcoma virus (Jacks and Varmus 1985) and subsequently for other retroviruses (Jacks et al. 1988b). The first complete coronavirus sequence was published in 1987 (Boursnell et al. 1987) and later that year Brierley and co-workers used an in vitro translation system to demonstrate that a −1 PRF mechanism was used to translate ORF1ab (Brierley et al. 1987). The IBV frameshift signal was further analyzed by Brierley and co-workers and in the following years became one of the best characterized −1 PRF signals.

The −1 PRF signal can be broken down into three discrete parts: the "slippery site," a linker region, and a downstream stimulatory region of mRNA secondary structure, typically an mRNA pseudoknot (reviewed in Baranov et al. 2002; Brierley 1995; Dinman and Berry 2006). Mutagenesis studies from many different laboratories have demonstrated that the primary sequence of the slippery site and its placement in relation to the incoming translational reading frame is critical: it must be X XXY YYZ (codons are shown in the incoming or 0-frame) where X is a stretch of three identical nucleotides, Y is either AAA or UUU, and Z is A, C, or U. Although less is known about the linker region, whose length and base composition

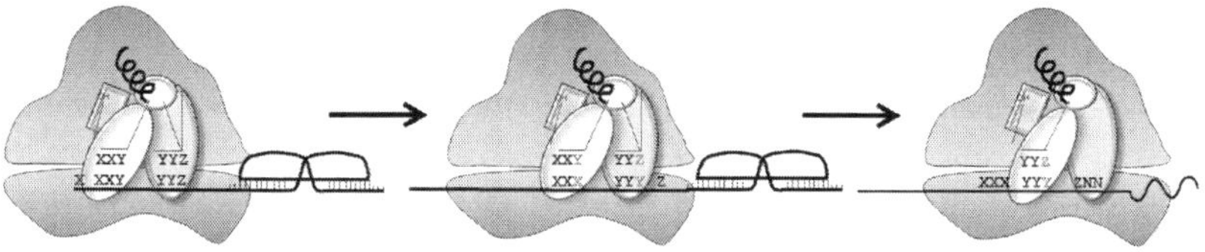

Fig. 5.1 Model of programmed –1 ribosomal frameshifting. An elongating ribosome is forced to pause by a strong mRNA secondary structure such as a pseudoknot. The length of the linker is such that the ribosomal A- and P-site bound aminoacyl- (aa-) and peptidyl-tRNAs are positioned over the slippery site. The sequence of the slippery site allows for re-pairing of the tRNAs to the –1 frame codons after they "simultaneously slip" by one base in the 5′ direction along the mRNA. Subsequent denaturation of the downstream mRNA secondary structure allows the ribosome to continue elongation of the nascent polypeptide in the new translational reading frame

varies, these parameters are thought to be important for determining the extent of –1 PRF in a virus-specific manner. The function of the downstream secondary structure is to induce elongating ribosomes to pause, a critical step for efficient –1 PRF to occur. The generally accepted mechanism of –1 PRF is as follows: (1) the mRNA secondary structure forces elongating ribosomes to pause, and the length of the linker is such that the ribosomal A- and P-site bound aminoacyl- (aa-) and peptidyl-tRNAs are positioned over the slippery site; (2) the sequence of the slippery site allows for re-pairing of the tRNAs to the –1 frame codons after they "simultaneously slip" by one base in the 5′ direction along the mRNA; (3) subsequent melting of the downstream mRNA secondary structure allows the ribosome to continue elongation of the nascent polypeptide in the new translational reading frame. This is diagrammed in Fig. 5.1.

5.3 Programmed Frameshifting Rates and Virus Propagation

In the best characterized examples of RNA viruses that utilize programmed ribosomal frameshifting (e.g., most retroviruses, totiviruses, and Ty elements), the open reading frame (ORF) encoding the major viral nucleocapsid proteins (e.g., Gag) is located at the 5′ end of the mRNA whereas the ORFs encoding proteins with enzymatic functions (typically Pro and Pol) are located 3′ of, and out of frame with, the Gag ORF. The mRNAs transcribed from these viral templates contain two overlapping ORFs. The enzymatic proteins are only translated as a result of a programmed ribosomal frameshift event that occurs with an efficiency of 1–40% depending on the specific virus and assay system employed (reviewed in Brierley 1995). Thus, the majority of translational events result in the production of the Gag protein, while a minority of frameshifts yield viral enzymatic proteins. The ratio of Gag to Gag–pol synthesized in viruses as a consequence of programmed frameshifting varies between a narrow window of 20:1 to 60:1 (reviewed in Farabaugh 1997).

The importance of maintaining this precise ratio on viral propagation has been demonstrated using two endogenous viruses of the yeast *Saccharomyces cerevisiae* and with two retroviruses (reviewed in Dinman et al. 1998). Small alterations in programmed frameshifting efficiencies promote rapid loss of the yeast dsRNA L–A "killer" virus, and in inhibition of HIV-1 replication (reviewed in Dinman and Berry 2006). Similarly, increasing or decreasing the efficiency of the +1 ribosomal frameshift in the Ty*1* retrotransposable element of yeast results in reduced retrotranspostion frequencies (reviewed in Dinman 1995). In L–A, Gag–pol dimerization nucleates formation of the virus particle (reviewed in Wickner 1996). Increasing the amount of Gag–pol protein synthesized may cause too many particles to initiate nonproductively while producing too little may prevent efficient dimerization (Dinman and Wickner 1992). Proteolytic processing of the TyA–TyB (Gag–pol equivalent) polyprotein of Ty*1* is more akin to the situation observed in retroviruses. In Ty*1*, increasing the amount of Gag–pol protein synthesized inhibited proteolytic processing of the polyprotein (Kawakami et al. 1993). As a consequence, formation of the mature forms of RNase H, integrase and reverse transcriptase is blocked (Kawakami et al. 1993). Similarly, changing the ratio of Gag to Gag–pol proteins in retroviruses like HIV or Moloney murine leukemia virus interferes with virus particle formation (reviewed in Baranov et al. 2002; Brierley 1995; Dinman and Berry 2006). In these cases, over-expression of the Gag–pol protein results in inefficient processing of the polyprotein and inhibition of virus production. To summarize, viral PRF efficiencies have been finely tuned to deliver the precise ratios of proteins required for efficient viral particle assembly: too much or too little frameshifting alters this ratio, with detrimental consequences. Based on these studies, it has been proposed that –1 may be a viable target for prevention of viral propagation (reviewed in Dinman et al. 1998).

5.4 Different Models, Different Assay Systems, Different Results

A number of models have been proposed to describe the mechanism by which –1 PRF occurs (Farabaugh 1996; Jacks et al. 1988a; Leger et al. 2007; Namy et al. 2006; Plant et al. 2003; Takyar et al. 2005; Weiss et al. 1989). All the models posit that the stimulatory element causes a pause in translation and that base-pairing is required at the non-wobble positions of at least two tRNA molecules to the mRNA after the frameshift. Differences arise in the timing of the frameshift: after peptide bond formation, before peptide bond formation, and before accommodation of the aminoacyl-tRNA. While there is strong genetic and biochemical evidence supporting a co-accommodation/prepeptidyltransfer model of –1 PRF, a co-translocational model is more intuitive, and there is both genetic and structural evidence supporting this concept as well (reviewed in Dinman and Berry 2006). When –1 PRF is kinetically modeled in the context of the translation elongation cycle, it becomes clear that –1 PRF is simply an endpoint that is potentially achievable through a number of different kinetic pathways. Indeed, the identification of two distinct frameshift products by protein sequencing supports the hypothesis that frameshifting can occur in least two distinct phases of the elongation cycle (Jacks et al. 1988b). A complicating issue is the use of a large variety of assay systems that are used by different laboratories to monitor –1 PRF. For example, prokaryotic, yeast, plant and mammalian ribosomes decipher coronavirus frameshift signals quite differently from one another (Brierley et al. 1997; Plant and Dinman 2006). Thus a suitable system must be used to draw purposeful conclusions from in vitro analyses of –1 PRF.

5.5 The Biology of –1 PRF in SARS-CoV is Different

In contrast to the examples discussed above, the genomic organization of coronaviruses is different. Instead of encoding viral structural proteins, the upstream ORF appears to encode immediate early proteins with enzymatic functions (e.g., proteases) thought to be involved in preparing the infected cell for takeover by the virus. These gene arrangements can be viewed in SARS-CoV genome maps elsewhere in this book. Similar to the viruses discussed above, a frameshift from this ORF directs ribosomes into a downstream ORF that encodes the viral RNA-dependent RNA polymerase (RDRP)) along with other enzymes thought to play roles in replication. Once synthesized, this RDRP is used to produce a subgenomic mRNA (in addition to the antigenomic and subsequently new genomic RNAs), from which the viral structural proteins are synthesized. Thus, the biology of frameshifting is significantly different in coronaviruses. At present, we do not know if frameshifting rates play an important role in virus replication, and if so, we do not know what that role may be. The availability of a fully infectious reverse genetics system (Almazan et al. 2006; Yount et al. 2003) and of a noninfectious replicon (Almazan et al. 2006) are currently serving as the foundations for research into these questions in our laboratory. One hypothesis that we are currently testing is that, similar to a model proposed for barley yellow dwarf virus (Barry and Miller 2002), –1 PRF in SARS-CoV may be involved in regulating the frequency of elongating ribosomes from the 3′ end of the mRNA or (+) strand, which in turn would affect the availability of this end to the viral replicase for (−) strand synthesis. Thus, we speculate that agents that alter –1 PRF efficiency in SARS-CoV may also have antiviral activities by interfering with the switch from protein synthesis to viral genome replication.

5.6 A Unique Feature of the SARS-CoV Frameshift Signal: A Three-Stemmed mRNA Pseudoknot

As discussed above, –1 PRF signals are usually composed of a slippery site and a stimulatory structure separated by a short spacer region. The rules describing slippery sites appear to be conserved among all eukaryotic viruses, and it also appears that the nucleotides surrounding the heptameric slippery site may affect frameshifting efficiencies, albeit to a lesser extent. Experiments altering the spacer region between the slippery site and stimulatory element affect frameshifting frequencies, suggesting that there might be some optimal spacer sequence.

Where the coronaviruses in general, and SARS-CoV in particular, deviate from other viral –1 PRF signals is in the nature of their downstream stimulatory signals. While an mRNA stem-loop structure was first postulated to stimulate –1 PRF in HIV-1 (Jacks et al. 1988b), studies of IBV provided the first evidence for an H-type pseudoknot structure as a stimulatory elements (Brierley et al. 1989). Subsequently H-type pseudoknots were identified as the general rule in a number of frameshift signals of other plant and animal viruses. H-type pseudoknots are so called because they are composed of two coaxially stacked stem-loops where the second stem is formed by base-pairing between sequence in the loop of the first stem-loop, and additional downstream sequence (Fig. 5.2a). Given that SARS-CoV is a coronavirus, that coronaviruses use –1 PRF, and the general consensus that viral –1 PRF signals contain H-type pseudoknots, the first published analysis of the SARS-CoV –1 PRF signal depicted an H-type pseudoknot (Thiel et al. 2003). Simultaneously, our laboratory applied a computational analysis pipeline designed to identify –1 PRF signals from genomic databases to the SARS-CoV –1 PRF signal. This analysis identified the presence of a third stem-loop structure located 3′ of the end of the first stem, and 5′ of the 3′ half of the second stem (Fig. 5.2b). Molecular genetics, biochemical and biophysical studies confirmed the existence of this "three-stemmed pseudoknot," and phylogenetic analyses revealed that this general feature is conserved among most coronaviruses (Plant et al. 2005). At the same time, two other groups of researchers used computational and molecular genetics

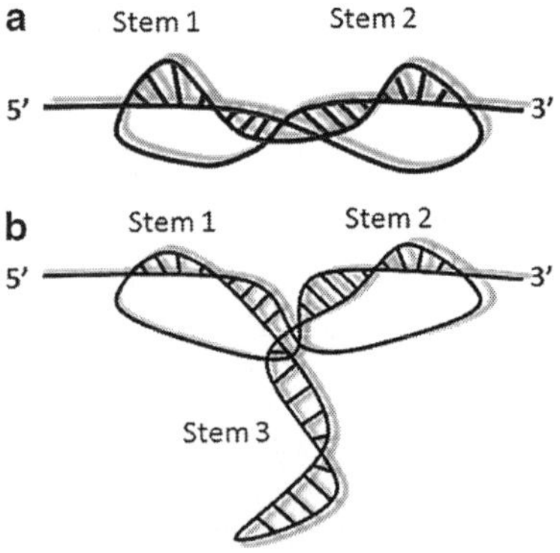

Fig. 5.2 Comparison of typical H-type mRNA pseudoknot (**a**) and three-stemmed mRNA pseudoknot structure of the SARS-CoV (b)

methods to identify the three-stemmed pseudoknot in the SARS-CoV −1 PRF signal (Baranov et al. 2005; Ramos et al. 2004).

Interestingly, efficient frameshifting was observed when the third stem was deleted from the SARS-CoV pseudoknot, or when a similar region was deleted from the IBV stimulatory structure, suggesting that these regions are not required for −1 PRF per se (Brierley et al. 1991; Plant et al. 2005). However, it is clear from mutational analyses that when the third stem is present, it has an effect on −1 PRF (Baranov et al. 2005; Plant et al. 2005). It is also possible that the third stem may function as a binding site for a *trans*-acting factor that may have a role in regulating −1 PRF efficiency, or in some other aspect of the viral life cycle. Current research in our laboratory is focused on determining the biological role of the third stem with regard to the biology of SARS-CoV. Preliminary studies from our laboratory suggest that stabilization of the terminal loop of this stem, perhaps through dimerization, is critical for maintaining high levels of frameshifting. In addition, it has been shown that sequence upstream of the core frameshift signal has been shown to have an inhibitory effect on −1 PRF efficiency in SARS-CoV (Su et al. 2005). This sequence has the potential to form an extensive secondary structure. Although the effect of this region has been suggested to directly affect −1 PRF rates, it is also possible that its role may be indirect, e.g., by causing a fraction of translating ribosomes to dissociate from the mRNA before reaching the −1 PRF signal. Indeed, unpublished data from our laboratory indicate that this is indeed correct. To summarize, although core essential elements of the frameshift signal have been defined, we have just begun to scratch the surface with regard to our understanding of the influence of additional features influencing −1 PRF and the biology of SARS-CoV.

5.7 A Second PRF Signal in SARS-CoV?

A second potential PRF signal has also been identified in SARS-CoV (Wang et al. 2006). This is found in a series of variants of the ORF3a containing stretches of 7 −U residues located 14 nt downstream of initiation codon. Interestingly, runs of both 7 and 8 uridines were able to promote efficient −1 and +1 frameshifting respectively. Substitution of these polyU sequences for the native −1 PRF signal between ORF1a and ORF1b promoted efficient frameshifting even in the absence of a functional downstream pseudoknot structure, suggesting that hepta- and octo-uridine stretches can function as efficient frameshift elements by themselves. While interesting, these observations should be considered in light of two caveats. First, the notion that long stretches of identical nucleotides can promote ribosomal slippage is not new (reviewed in Atkins et al. 1991). Second, the hepta- and octa-U variants of ORF3a may simply be cloning artifacts consequent to reverse transcriptase slippage during the generation of RT-PCR products, and thus may not have any true biological significance. Thus, while provocative, the jury is still out with respect to the presence of a second PRF signal in SARS-CoV.

5.8 Summary and Perspectives

While progress has been made on elucidating the mechanism of −1 PRF and the RNA sequences involved in coronaviruses, the requirement for −1 PRF in the life cycle remains obscure. For other viruses −1 PRF efficiency directly determines the ratios between viral structural and enzymatic proteins. In contrast, −1 PRF in SARS-CoV would appear to affect the relative ratios between immediate early proteins, e.g., proteases and other uncharacterized proteins, with the viral replicase and enzymes that modify RNA. The downstream effect of −1 PRF on the abundance of coat proteins compared to viral RNA in coronaviruses has yet to be determined.

The reason for the presence of a third stem in the SARS-CoV −1 PRF signal also remains a mystery. Unpublished observations in our laboratory suggest that some alterations to the third stem in the SARS coronavirus pseudoknot inhibit virus infectivity without dramatically affecting frameshifting frequency. We have also identified a few proteins encoded by the SARS-CoV subgenomic RNAs that appear to interact with the SARS pseudoknot. This suggests that this region is vital for some part of the virus life cycle other than −1 PRF. Ongoing studies are focused on addressing this question.

An additional research challenge is the production of mutations that have a moderate affect on −1 PRF; these will provide the tools to probe more deeply into the biology of frameshifting in SARS-CoV. As these mutant viruses and replicons become available we will be able to correlate the efficiency of frameshifting with production of genomic and subgenomic RNAs and viral titers. It is expected that some of these mutations will result in defects that will give insight into the function of the extraneous sequence within the frameshift signal, and that this insight will provide an alternative starting point for dissecting the coronavirus replication system.

Acknowledgments The rapid emergence and severity of the SARS-associated coronavirus led to the sharing of unpublished information at conferences and meetings which, in turn, added vigor to the field. Because of the competitive nature of research, this sharing of resources is dependent on the ethical behavior of the researchers. We thank all those who are part of the community and have contributed important information that has the potential to prevent or control virus spread should a similar outbreak occur in the future. This work was supported by NIH RO1 grant AI064307.

References

Almazan F, DeDiego ML, Galan C, Escors D, Alvarez E, Ortego J, Sola I, Zuniga S, Alonso S, Moreno JL, Nogales A, Capiscol C, Enjuanes L (2006) Construction of a severe acute respiratory syndrome coronavirus infectious cDNA clone and a replicon to study coronavirus RNA synthesis. J Virol 80:10900–10906

Atkins JF, Weills RB, Thompson S, Gesteland RFE (1991) Towards a genetic dissection of the basis of triplet decoding, and its natural subversion: programmed reading frame shifts and hops. Annu Rev Genet 25:201–228

Baranov PV, Gesteland RF, Atkins JF (2002) Recoding: translational bifurcations in gene expression. Gene 286:187–201

Baranov PV, Henderson CM, Anderson CB, Gesteland RF, Atkins JF, Howard MT (2005) Programmed ribosomal frameshifting in decoding the SARS-CoV genome. Virology 332:498–510

Barry JK, Miller WA (2002) A -1 ribosomal frameshift element that requires base pairing across four kilobases suggests a mechanism of regulating ribosome and replicase traffic on a viral RNA. Proc Natl Acad Sci USA 99:11133–11138

Boursnell ME, Brown TD, Foulds IJ, Green PF, Tomley FM, Binns MM (1987) Completion of the sequence of the genome of the coronavirus avian infectious bronchitis virus. J Gen Virol 68 (Pt 1):57–77

Brierley I (1995) Ribosomal frameshifting on viral RNAs. J Gen Virol 76:1885–1892

Brierley I, Boursnell ME, Binns MM, Bilimoria B, Blok VC, Brown TD, Inglis SC (1987) An efficient ribosomal frame-shifting signal in the polymerase-encoding region of the coronavirus IBV. EMBO J 6:3779–3785

Brierley IA, Dingard P, Inglis SC (1989) Characterization of an efficient coronavirus ribosomal frameshifting signal: requirement for an RNA pseudoknot. Cell 57:537–547

Brierley IA, Rolley NJ, Jenner AJ, Inglis SC (1991) Mutational analysis of the RNA pseudoknot component of a coronavirus ribosomal frameshifting signal. J Mol Biol 220:889–902

Brierley I, Meredith MR, Bloys AJ, Hagervall TG (1997) Expression of a coronavirus ribosomal frameshift signal in Escherichia coli: influence of tRNA anticodon modification on frameshifting. J Mol Biol 270:360–373

Dinman JD (1995) Ribosomal frameshifting in yeast viruses. Yeast 11:1115–1127

Dinman JD, Berry MJ (2006) Regulation of Termination and Recoding. In: Mathews MB, Sonenberg N, Hershey JWB (eds) Translational control in biology and medicine. Cold Spring Harbor Press, Cold Spring Harbor, NY

Dinman JD, Wickner RB (1992) Ribosomal frameshifting efficiency and Gag/Gag-pol ratio are critical for yeast M_1 double-stranded RNA virus propagation. J Virol 66:3669–3676

Dinman JD, Ruiz-Echevarria MJ, Peltz SW (1998) Translating old drugs into new treatments: identifying compounds that modulate programmed −1 ribosomal frameshifting and function as potential antiviral agents. Trends Biotechnol 16:190–196

Farabaugh PJ (1996) Programmed translational frameshifting. Microbiol Rev 60:103–134

Farabaugh PJ (1997) Programmed alternative reading of the genetic code. R.G. Landes, Austin TX

Holmes KV (2003) SARS coronavirus: a new challenge for prevention and therapy. J Clin Invest 111:1605–1609

Jacks T, Varmus HE (1985) Expression of the Rous Sarcoma Virus pol gene by ribosomal frameshifting. Science 230:1237–1242

Jacks T, Madhani HD, Masiraz FR, Varmus HE (1988a) Signals for ribosomal frameshifting in the Rous Sarcoma Virus gag-pol region. Cell 55:447–458

Jacks T, Power MD, Masiarz FR, Luciw PA, Barr PJ, Varmus HE (1988b) Characterization of ribosomal frameshifting in HIV-1 *gag-pol* expression. Nature 331:280–283

Kawakami K, Paned S, Faioa B, Moore DP, Boeke JD, Farabaugh PJ, Strathern JN, Nakamura Y, Garfinkel DJ (1993) A rare tRNA-Arg(CCU) that regulates Ty*1* element ribosomal frameshifting is essential for Ty*1* retrotransposition in *Saccharomyces cerevisiae*. Genetics 135:309–320

Lai MM (2003) SARS virus: the beginning of the unraveling of a new coronavirus. J Biomed Sci 10:664–675

Leger M, Dulude D, Steinberg SV, Brakier-Gingras L (2007) The three transfer RNAs occupying the A, P and E sites on the ribosome are involved in viral programmed -1 ribosomal frameshift. Nucleic Acids Res 35:5581–5592

Lio P, Goldman N (2004) Phylogenomics and bioinformatics of SARS-CoV. Trends Microbiol 12:106–111

Marra MA, Jones SJ, Astell CR, Holt RA, Brooks-Wilson A, Butterfield YS, Khattra J, Asano JK, Barber SA, Chan SY, Cloutier A, Coughlin SM, Freeman D, Girn N, Griffith OL, Leach SR,

Mayo M, McDonald H, Montgomery SB, Pandoh PK, Petrescu AS, Robertson AG, Schein JE, Siddiqui A, Smailus DE, Stott JM, Yang GS, Plummer F, Andonov A, Artsob H, Bastien N, Bernard K, Booth TF, Bowness D, Drebot M, Fernando L, Flick R, Garbutt M, Gray M, Grolla A, Jones S, Feldmann H, Meyers A, Kabani A, Li Y, Normand S, Stroher U, Tipples GA, Tyler S, Vogrig R, Ward D, Watson B, Brunham RC, Krajden M, Petric M, Skowronski DM, Upton C, Roper RL (2003) The genome sequence of the SARS-Associated Coronavirus. Science 300:1399–1404

Namy O, Moran SJ, Stuart DI, Gilbert RJ, Brierley I (2006) A mechanical explanation of RNA pseudoknot function in programmed ribosomal frameshifting. Nature 441:244–247

Plant EP, Dinman JD (2006) Comparative study of the effects of heptameric slippry site composition on −1 frameshifting among different translational assay systems. RNA 12:666–673

Plant EP, Jacobs KLM, Harger JW, Meskauskas A, Jacobs JL, Baxter JL, Petrov AN, Dinman JD (2003) The 9-angstrom solution: how mRNA pseudoknots promote efficient programmed −1 ribosomal frameshifting. RNA 9:168–174

Plant EP, Perez-Alvarado GC, Jacobs JL, Mukhopadhyay B, Hennig M, Dinman JD (2005) A three-stemmed mRNA pseudoknot in the SARS coronavirus frameshift signal. PLoS Biol 3:1012–1023

Ramos FD, Carrasco M, Doyle T, Brierley I (2004) Programmed -1 ribosomal frameshifting in the SARS coronavirus. Biochem Soc Trans 32:1081–1083

Rota PA, Oberste MS, Monroe SS, Nix WA, Campagnoli R, Icenogle JP, Penaranda S, Bankamp B, Maher K, Chen MH, Tong S, Tamin A, Lowe L, Frace M, DeRisi JL, Chen Q, Wang D, Erdman DD, Peret TC, Burns C, Ksiazek TG, Rollin PE, Sanchez A, Liffick S, Holloway B, Limor J, McCaustland K, Olsen-Rassmussen M, Fouchier R, Gunther S, Osterhaus AD, Drosten C, Pallansch MA, Anderson LJ, Bellini WJ (2003) Characterization of a novel Coronavirus associated with severe acute respiratory syndrome. Science 300: 1394–1399

Snijder EJ, Bredenbeek PJ, Dobbe JC, Thiel V, Ziebuhr J, Poon LL, Guan Y, Rozanov M, Spaan WJ, Gorbalenya AE (2003) Unique and conserved features of genome and proteome of SARS-coronavirus, an early split-off from the coronavirus group 2 lineage. J Mol Biol 331:991–1004

Stadler K, Masignani V, Eickmann M, Becker S, Abrignani S, Klenk HD, Rappuoli R (2003) SARS–beginning to understand a new virus. Nat Rev Microbiol 1:209–218

Su MC, Chang CT, Chu CH, Tsai CH, Chang KY (2005) An atypical RNA pseudoknot stimulator and an upstream attenuation signal for -1 ribosomal frameshifting of SARS coronavirus. Nucleic Acids Res 33:4265–4275

Takyar S, Hickerson RP, Noller HF (2005) mRNA helicase activity of the ribosome. Cell 120: 49–58

Thiel V, Ivanov KA, Putics A, Hertzig T, Schelle B, Bayer S, Weissbrich B, Snijder EJ, Rabenau H, Doerr HW, Gorbalenya AE, Ziebuhr J (2003) Mechanisms and enzymes involved in SARS coronavirus genome expression. J Gen Virol 84:2305–2315

Wang XX, Liao Y, Wong SM, Liu DX (2006) Identification and characterization of a unique ribosomal frameshifting signal in SARS-CoV ORF3a. Adv Exp Med Biol 581:89–92

Weiss RB, Dunn DM, Shuh M, Atkins JF, Gesteland RF (1989) *E. coli* ribosomes re-phase on retroviral frameshift signals at rates ranging from 2 to 50 percent. New Biol 1:159–169

Wickner RB (1996) Prions and RNA viruses of *Saccharomyces cerevisiae*. Annu Rev Genet 30:109–139

Yount B, Curtis KM, Fritz EA, Hensley LE, Jahrling PB, Prentice E, Denison MR, Geisbert TW, Baric RS (2003) Reverse genetics with a full-length infectious cDNA of severe acute respiratory syndrome coronavirus. Proc Natl Acad Sci USA 100:12995–13000

Part III
Viral Proteins

Chapter 6
Expression and Functions of SARS Coronavirus Replicative Proteins

Rachel Ulferts, Isabelle Imbert, Bruno Canard, and John Ziebuhr

Abstract The discovery of a previously unknown coronavirus as the causative agent of the SARS epidemic in 2002/2003 stimulated a large number of studies into the molecular biology of SARS coronavirus (SARS-CoV) and related viruses. This research has provided significant new insight into the functions and activities of the coronavirus replicase–transcriptase complex, a multiprotein complex that directs coordinated processes of both continuous and discontinuous RNA synthesis to replicate and transcribe the large coronavirus genome, a single-stranded, positive-sense RNA of ~30 kb. In this chapter, we review our current understanding of the expression and functions of key replicative enzymes, such as RNA polymerases, helicase, ribonucleases, ribose-2′-*O*-methyltransferase and other replicase gene-encoded proteins involved in genome expression, virus–host interactions and other processes. Collectively, these recent studies reveal fascinating details of an enzymatic machinery that, in the RNA virus world, is unparalleled in terms of the number and nature of virally encoded activities involved in virus replication and host interactions.

6.1 Introduction

Coronaviruses and their closest relatives from the order *Nidovirales* have exceptionally large RNA genomes of about 30 kb and synthesize an extensive set of 5′-leader-containing, subgenome-length RNAs. The synthesis of these RNAs is mediated by the viral replicase–transcriptase complex (RTC), a large multisubunit complex that is comprised of more than a dozen proteins encoded by the viral replicase gene and other proteins. The RTC includes the key replicative proteins of

J. Ziebuhr (✉)
Centre for Infection and Immunity, School of Medicine, Dentistry and Biomedical Sciences, The Queen's University of Belfast, Belfast, United Kingdom
e-mail: j.ziebuhr@qub.ac.uk

S.K. Lal (ed.), *Molecular Biology of the SARS-Coronavirus*,
DOI 10.1007/978-3-642-03683-5_6,

the virus, such as RNA-dependent RNA polymerase (RdRp) and helicase activities, as well as enzymes that are thought to be involved in the processing and modification of viral and/or cellular RNAs, such as primase, endoribonuclease, exoribonuclease and ribose-2′-*O*-methyltransferase activities (for recent reviews, see Masters 2006; Ziebuhr 2005, 2008). The RTC is anchored through three replicase-gene encoded integral membrane proteins to intracellular membranes derived from the endoplasmic reticulum (Knoops et al. 2008; Masters 2006; Ziebuhr and Snijder 2007). Besides proteins encoded by the replicase gene, the RTC contains the nucleocapsid (N) protein (Almazan et al. 2004; Schelle et al. 2005) and several cellular factors, which have not been characterized in great detail (Shi and Lai 2005; van Hemert et al. 2008). It is becoming increasingly clear that the coronavirus replicase gene also encodes proteins that are not essential for viral RNA synthesis but are involved in viral pathogenicity and specific virus–host interactions (see below).

The single-stranded, positive-sense RNA genome of coronaviruses fulfills a double role, serving as an mRNA for the expression of the replicase gene and a template for the synthesis of genome- and subgenome-length minus-strand RNAs. A large part of the genome (about two-thirds) is occupied by the replicase gene, which consists of two (slightly) overlapping open reading frames (ORF) located in the 5′-proximal region of the viral genome RNA. The remaining 3′-terminal third of the genome is dedicated to encoding the four major structural proteins of the virus (S, E, M, and N) and a varying number of so-called accessory proteins (Fig. 6.1). Compared to other coronaviruses, SARS coronavirus (SARS-CoV) encodes an unusually large number of accessory proteins, some of which are involved in counteracting antiviral host responses (Chap. 4). The ORFs located downstream of the SARS-CoV replicase gene are expressed from eight subgenomic (sg) RNAs (Thiel et al. 2003) (Fig. 6.1). A peculiarity of the sgRNAs produced by coronaviruses and some other nidoviruses is that both the plus-strand genome RNA and all plus-strand sgRNAs share a short, so-called leader sequence at their 5′ ends, whose functional relevance is currently unknown (Spaan et al. 1983). It has been speculated that the minus-strand complement of the leader sequence (present at the 3′ ends of all minus-strand RNAs) might facilitate initiation of plus-strand RNA synthesis. Alternatively, the leader might promote the translation of sgRNAs, for example through the presence of sequence elements required for efficient recognition by the viral capping apparatus or other mechanisms enhancing translation initiation. The precise mechanism used to join the "leader" and "body" sequences of sgRNAs (which are located more than 20 kb apart in the genome) is unknown but is generally accepted to involve a discontinuous step during sg minus-strand RNA synthesis and be guided by complementary base-pairing between transcription-regulating sequences (for recent reviews, see Pasternak et al. 2006; Sawicki et al. 2007). The specific requirements for maintaining genome RNAs of an unparalleled size and synthesizing sg minus-strand RNAs in a discontinuous manner are reflected by an exceptional complexity of the protein machinery involved in viral RNA synthesis, which will be the topic of this review. With few exceptions, the proteins and mechanisms involved in genome replication and transcription are

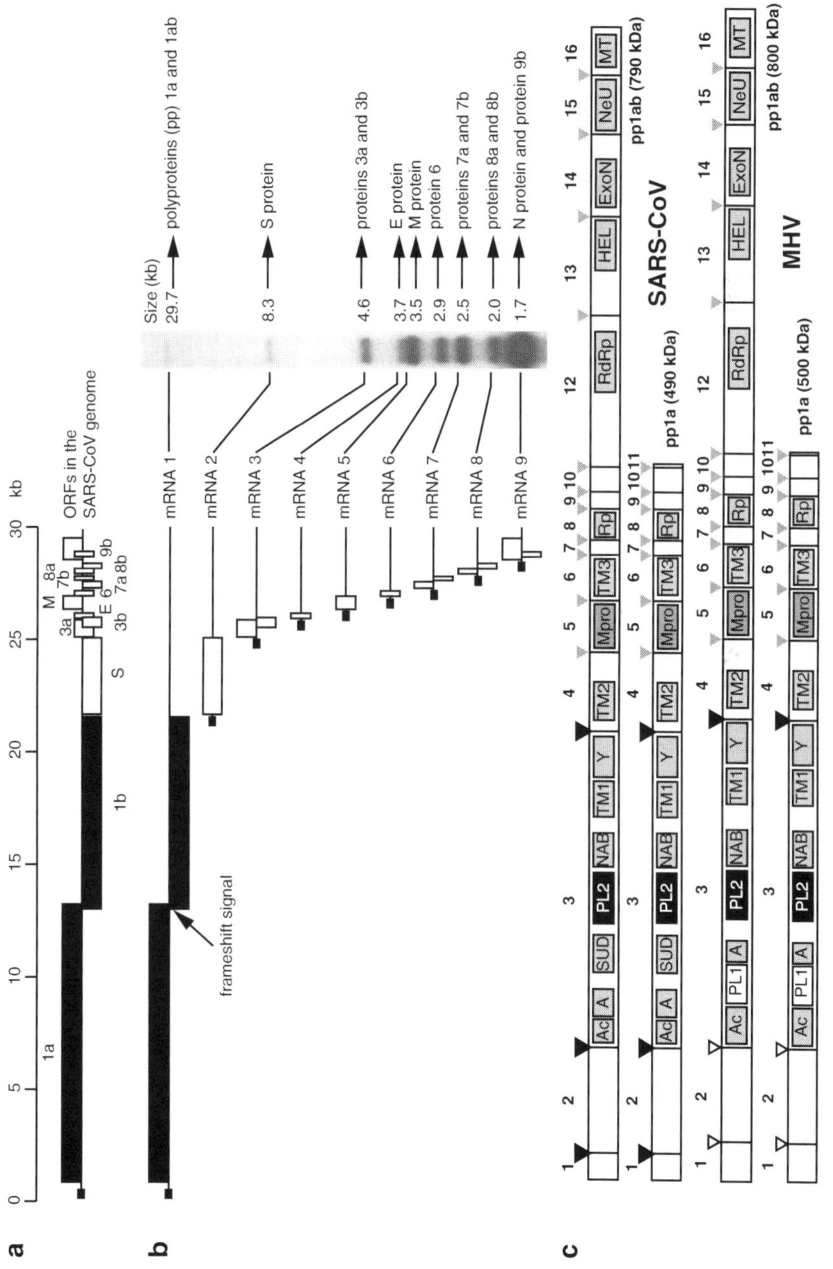
a
b
c
kb
ORFs in the SARS-CoV genome
1a
1b
S
3a
3b
E
M
6
7a
7b
8a
8b
9b
frameshift signal
mRNA 1
mRNA 2
mRNA 3
mRNA 4
mRNA 5
mRNA 6
mRNA 7
mRNA 8
mRNA 9
Size (kb)
29.7
8.3
4.6
3.7
3.5
2.9
2.5
2.0
1.7
polyproteins (pp) 1a and 1ab
S protein
proteins 3a and 3b
E protein
M protein
protein 6
proteins 7a and 7b
proteins 8a and 8b
N protein and protein 9b
Ac
A
SUD
PL1
PL2
NAB
TM1
Y
TM2
Mpro
TM3
Rp
RdRp
HEL
ExoN
NeU
MT
pp1ab (790 kDa)
pp1a (490 kDa)
SARS-CoV
pp1ab (800 kDa)
pp1a (500 kDa)
MHV

conserved among SARS-CoV and other coronaviruses. In this chapter, we will review recent work on the SARS-CoV RTC and refer to related work on other coronaviruses as appropriate.

6.2 Organization and Expression of the Coronavirus Replicase Gene

The SARS-CoV replicase gene contains more than 21,000 nucleotides and is comprised of two ORFs called ORF1a and ORF1b (Fig. 6.1) (Marra et al. 2003; Rota et al. 2003; Thiel et al. 2003). Upon infection of the cell, ORFs 1a and 1b are translated into two large polyproteins, pp1a and pp1ab, of approximately 490 and 790 kDa, respectively. Polyprotein 1ab is a C-terminally extended version of pp1a; pp1ab expression requires a programmed (–1) ribosomal frameshift prior to termination of ORF1a translation, resulting in continuation of translation in the ORF1b reading frame (Thiel et al. 2003). Frameshifting depends on two *cis*-acting RNA elements, a "slippery" heptanucleotide sequence and an RNA pseudoknot structure (see Chap. 6 for details). Based on in vitro translation experiments, it has been estimated that frameshifting occurs in roughly 30% of translation events (Brierley 1995; Dos Ramos et al. 2004; Herold and Siddell 1993; Namy et al. 2006). This results in a considerably higher amount of ORF1a-encoded proteins compared to ORF1b-encoded proteins, which is likely to be of biological relevance as the frameshifting mechanism is conserved in coronaviruses and other nidoviruses.

SARS-CoV pp1a and pp1ab are co- and post-translationally processed by two proteases, a papain-like protease (PL^{pro}) and the main protease (M^{pro}, nsp5), resulting in 16 mature products called nonstructural proteins (nsps) 1–16

Fig. 6.1 SARS coronavirus (SARS-CoV) genome organization and expression. (**a**) Open reading frames in the SARS-CoV genome. The replicase gene is comprised of two large open reading frames (ORFs) which are shown in *black*. (**b**) RNAs produced in SARS-CoV-infected cells. The 5′-terminal ORF(s) expressed from specific RNAs is/are shown as *boxes*. The *small black box* indicates the 5′-leader sequence present on each of the viral RNAs. The replicase gene (ORFs 1a and 1b) are shown in *black*. To the *right*, SARS-CoV genomic and subgenomic RNAs as detected by Northern blotting using a 3′ end-specific probe are shown together with information on sizes (in kilobases) and names of proteins expressed from these RNAs. (**c**) Overview of the domain organization and proteolytic processing of the SARS-CoV replicase polyproteins, pp1a (486 kDa) and pp1ab (790 kDa). For comparison, the mouse hepatitis virus (MHV)-A59 replicase polyproteins are shown. The processing end-products of pp1a are designated nonstructural proteins (nsps) 1 to nsp11 and those of pp1ab are designated nsp1 to nsp10 and nsp12 to nsp16. Cleavage sites that are processed by the viral main protease, M^{pro}, are indicated by *grey arrowheads*; sites that are processed by the papain-like proteases 1 and 2 (PL1 and PL2), are indicated by *white* and *black arrowheads*, respectively. Ac, acidic domain; A, ADRP (ADP-ribose-1″-phosphatase); SUD, SARS-CoV unique domain; NAB, nucleic acid-binding domain; Rp, noncanonical RNA polymerase (putative primase); RdRp, RNA-dependent RNA polymerase; TM1, TM2, TM3, transmembrane domains 1, 2, and 3, respectively; HEL, helicase; ExoN, 3′-to-5′ exoribonuclease; NeU, NendoU (nidoviral uridylate-specific endoribonuclease); MT, ribose-2′-*O*-methyltransferase

(Harcourt et al. 2004; Prentice et al. 2004; Snijder et al. 2003; Thiel et al. 2003). The PL^{pro} domain is part of nsp3 and processes the nsp1|2, nsp2|3 and nsp3|4 cleavage sites. The M^{pro} cleaves the C-terminal part of pp1a/1ab at the remaining 11 sites (Fig. 6.1), releasing the most conserved proteins of the coronavirus RTC. The individual cleavage sites are thought to be processed with varying efficiencies, resulting in the presence of some reasonably stable intermediates. This has been studied in some detail for the proteins released by the PL^{pro} activity (Harcourt et al. 2004). While the nsp1|2 and nsp3|4 sites are rapidly processed, the nsp2|3 site was found to be cleaved at a lower rate, giving rise to a relatively long-lived nsp2–3 intermediate in infected cells. The rapid cleavage of the nsp3|4 site results in fast separation of the domains processed by PL^{pro} and those processed by M^{pro}. Even though an initial study of the nonstructural proteins in SARS-CoV-infected cells failed to detect processing intermediates (Prentice et al. 2004), the cleavage of interdomain junctions by M^{pro} is also thought to occur at varying rates (Fan et al. 2004). This is supported by data obtained for mouse hepatitis virus (MHV), showing that nsp4–10 is a relatively stable processing intermediate (Schiller et al. 1998). In addition to the sequence of the individual cleavage sites, the efficiency of processing appears to be influenced by the secondary structure of the substrate (Fan et al. 2005, 2004). Thus, cleavage of peptides with a high propensity to form β-sheets is enhanced, consistent with the substrate binding mode observed in crystal structures of M^{pro} in complex with inhibitors (Anand et al. 2003; Yang et al. 2003, 2005).

It is generally accepted that processing intermediates fulfil important functions that, in some cases, differ from those of the fully processed proteins. This strategy is commonly employed by plus-strand RNA viruses to expand the functional repertoire of the relatively small number of protein domains present in viral polyproteins (Dougherty and Semler 1993; Palmenberg 1990). The idea of different functions being associated with precursors and final processing products, respectively, is supported by the fact that coronavirus RNA synthesis requires ongoing protein synthesis, with minus-strand synthesis declining more rapidly than positive-strand synthesis after inhibition of translation by cycloheximide (Sawicki and Sawicki 1986).

6.3 Functions and Activities of Replicase Gene-Encoded Nonstructural Proteins

6.3.1 ORF1a-Encoded Nonstructural Proteins 1–11

The proteins processed from the N-terminal regions of the coronavirus replicative polyproteins 1a/1ab are highly divergent among coronaviruses. nsp1 proteins from group 1 and group 2 coronaviruses, respectively, are not evidently related to each other, whereas group 3 coronaviruses lack an nsp1-counterpart altogether, making nsp2 the most N-terminal processing product in this case. The significant sequence diversity is also evident when nsp1 proteins from the same group (1 or 2) are compared with each other (Almeida et al. 2007; Connor and Roper 2007; Snijder et al. 2003). A nuclear magnetic resonance structure was reported for SARS-CoV nsp1 (Almeida et al. 2007). The protein has a complex β-barrel fold flanked by disordered N- and C-terminal domains (see Chap.18 for details) and lacks statistically significant structural similarity to other cellular or viral proteins. Although the protein is not required for replication of SARS-CoV and MHV in cell culture (Brockway and Denison 2005;Wathelet et al. 2007; Züst et al. 2007), there is increasing evidence to suggest that the protein has important functions in vivo. SARS-CoV nsp1 was suggested to counteract cellular innate immune responses by inhibiting IFN signaling pathways, and recombinant SARS-CoV with an attenuating mutation in nsp1 was found to decrease the ability of the virus to replicate in cells with an intact IFN response (Wathelet et al. 2007). Narayanan and co-workers (Narayanan et al. 2008) reported that another SARS-CoV nsp1 mutant, but not the wild-type virus, induced high levels of INF-β, suggesting that nsp1 may also have a role in type I IFN induction, which contradicts findings reported earlier by others (Wathelet et al. 2007; Züst et al. 2007). Characterization of an MHV mutant expressing a C-terminally truncated form of nsp1 showed that the protein is a major pathogenicity factor of this virus (Züst et al. 2007). There is evidence to suggest that MHV nsp1 is involved in counteracting type I IFN signaling and/or the antiviral activities of IFN-induced proteins, whereas a role in suppression of IFN induction seems less likely.

Furthermore, SARS-CoV nsp1 was shown to (1) promote host mRNA degradation, (2) inhibit cellular translation (Kamitani et al. 2006; Narayanan et al. 2008) and (3) affect cell cycling (Wathelet et al. 2007). MHV nsp1 was reported to colocalize (and interact) with other subunits of the viral replicase–transcriptase at the site of viral RNA synthesis but appears to migrate to virion assembly sites at a later stage of infection (Brockway et al. 2004). It remains to be seen if this relocalization is linked to distinct functions of nsp1 at different stages of the replication cycle of MHV and, possibly, other coronaviruses.

For nsp2, no specific function has yet been identified. It has been suggested that nsp2 is a component of SARS-CoV virions (Neuman et al. 2008). The protein is not essential for replication in vitro as deletion of the entire nsp2 coding sequence was tolerated in both SARS-CoV and MHV (Graham et al. 2005). nsp2 deletion mutants grew to slightly reduced titers and RNA synthesis was diminished by about 50%, with all RNA species being equally affected. More recently, it was shown that the replication defects observed in MHV nsp2 deletion mutants could not be compensated by nsp2 expressed from either a subgenomic RNA or from a C-proximal position in the replicase polyprotein, when inserted between nsp13 and nsp14 (Gadlage et al. 2008).

nsp3 is a large, membrane-anchored multidomain protein. Despite significant sequence diversity among coronavirus nsp3 proteins, up to 16 putative functional domains, including ubiquitin-like, metal-binding, nucleic acid-binding, RNA chaperone-like, poly(ADP)-ribose-binding, protease, transmembrane (TM) and other conserved domains have been identified in nsp3 (Neuman et al. 2008; Ziebuhr et al. 2001). nsp3 has been proposed to be a major "hub" for protein–protein and protein–RNA interactions between viral and (possibly) cellular macromolecules (Imbert et al. 2008; Neuman et al. 2008). Many of the domains are common to all coronaviruses while some are group- or species-specific (Neuman et al. 2008; Snijder et al. 2003; Ziebuhr et al. 2001).

Among the conserved domains are papain-like proteases (PL^{pro}s) that cleave the N-terminal part of the polyproteins at up to three sites (Fig. 6.1). While most coronaviruses possess two PL^{pro}s ($PL1^{pro}$ and $PL2^{pro}$), SARS-CoV and IBV encode only one active PL^{pro}, which are orthologues of the $PL2^{pro}$ of other coronaviruses (Snijder et al. 2003; Thiel et al. 2003; Ziebuhr et al. 2001). The SARS-CoV PL^{pro} employs a Cys–His–Asp catalytic triad and exhibits a narrow substrate specificity, with all three processing sites conforming to the consensus sequence LXGG (Harcourt et al. 2004; Thiel et al. 2003). Besides its important role in pp1a/pp1ab processing, the SARS-CoV PL^{pro} has deubiquitinating activity (Barretto et al. 2005; Lindner et al. 2005) and shares structural features with cellular ubiquitin-specific proteases (see Chap. 18 for details) (Ratia et al. 2006; Sulea et al. 2005). The $PL2^{pro}$ homolog of HCoV-NL63, a group 1b coronavirus, also has ubiquitin-specific protease activity (Chen et al. 2007b), suggesting that this activity may be of general biological relevance to the coronavirus life cycle. SARS-CoV PL^{pro} removes ubiquitin and ubiquitin-like modifiers (Ubl) from fusion proteins and debranches polyubiquitin chains, with a marked preference for Lys-48- over Lys-63-conjugated chains (Lindner et al. 2007). The C-terminal residues of ubiquitin and the ubiquitin-like modifier ISG15, LRLRGG, match the consensus sequence of SARS-CoV PL^{pro} cleavage sites in pp1a/pp1ab very well (Harcourt et al. 2004; Thiel et al. 2003). SARS-CoV PL^{pro} is able to distinguish between ISG15 and ubiquitin and preferentially cleaves ISG15-modified proteins. The exact mode of recognition of Ubl modifiers by PL^{pro} is not known but might involve interactions between the Zn^{2+} ribbon domain connecting the α and β subdomains of the PL^{pro} and the β-grasp fold of ubiquitin and additional, currently unknown, interactions. The deubiquitinating and deISGylating activities of PL^{pro} have been speculated to be involved in (1) protecting viral and/or cellular proteins from degradation and (2) counteracting innate immune responses (Lindner et al. 2007; Sulea et al. 2005). Recently, PL^{pro} was shown to inhibit IFN signaling by binding to IRF-3 and interfering with its hyperphosphorylation, dimerization and nuclear translocation (Devaraj et al. 2007). Surprisingly, the inhibition of IFN response was not abrogated by substitution of PL^{pro} active-site residues, suggesting that the protease/deubiquitylase activity of PL^{pro} was not involved. Slightly contrasting data were reported by Zheng and co-workers (Zheng et al. 2008). In this case, the reduction of the type I INF response induced by vesicular stomatitis virus was dependent on the proteolytic activity of the SARS-CoV PL^{pro}.

Another nsp3 domain is the ADP-ribose 1″ phosphatase (ADRP) domain (also called X domain or *macro* domain) (Fig. 6.1). The domain is conserved across members of the genera *Coronavirus*, *Torovirus* and *Bafinivirus* (Draker et al. 2006; Schütze et al. 2006) and several other plus-strand RNA viruses

(Gorbalenya et al. 1991). Viral ADRPs are related to a large family of *macro* domain proteins found in many cellular organisms. The protein family is named after the nonhistone domain of *macro*H2A histones. *Macro* domain proteins share a conserved fold and bind to (and, in some cases, process) a range of substrates related to ADP-ribose (Karras et al. 2005). Several members of the *macro* domain family have been shown to (1) bind poly(ADP)-ribose and/or poly(A) RNA and (2) hydrolyze ADP-ribose-1″-phosphate (Egloff et al. 2006; Karras et al. 2005; Neuvonen and Ahola 2009; Putics et al. 2005, 2006; Saikatendu et al. 2005; Xu et al. 2009). The ADRP domains of HCoV-229E, SARS-CoV and transmissible gastroenteritis virus (TGEV) have been shown to hydrolyze ADP-ribose-1″-phosphate to yield ADP-ribose and inorganic phosphate (Egloff et al. 2006; Karras et al. 2005; Neuvonen and Ahola 2009; Putics et al. 2005, 2006; Saikatendu et al. 2005; Xu et al. 2009). The molecular mechanisms of the dephosphorylation reaction have not been elucidated but are thought to differ from other phosphatases as there is no sequence and/or structural similarity between the respective enzymes (Allen et al. 2003; Egloff et al. 2006). Based on the structure of coronavirus ADRPs in complex with ADP-ribose, two conserved residues (Asn and His) were proposed to be important for activity (Egloff et al. 2006; Saikatendu et al. 2005; Xu et al. 2009). Substitutions of these residues by alanine abolished ADRP activity activity in an in vitro assay, supporting their critical role (Egloff et al. 2006; Putics et al. 2005).

The biological role of the ADRP in the coronavirus life cycle has not been established. Characterization of an HCoV-229E mutant (HCoV-229E-N1305A) expressing an inactive ADRP revealed no apparent defects in virus reproduction and RNA synthesis in a cell culture system (Putics et al. 2005). Similar observations were made for the corresponding MHV-A59 mutant (MHV-N1348A) (Eriksson et al. 2008), confirming that ADRP activity is dispensable for coronavirus replication in vitro. However, when characterized in vivo, the MHV-N1348A mutant was found to be attenuated. At high infection doses, the mutant and wild-type viruses replicated to similar titers in the liver of C57BL/6 mice but, in contrast to wild-type virus, the N1348A mutant failed to cause severe liver pathology. Similar observations were made in C57BL/6 IFNAR$^{-/-}$ mice, arguing against a major role of the ADRP activity in counteracting IFN-α host responses. The underlying mechanisms for the observed low pathogenicity of the MHV-N1348A mutant in the liver have not been identified conclusively but may be linked to a reduced expression of proinflammatory cytokines, in particular IL-6. The data support the biological relevance of the catalytic activity of coronavirus ADRP domains which, on the basis of in vitro observations, has previously been questioned by others (Egloff et al. 2006; Neuvonen and Ahola 2009). The biologically relevant substrate(s) of the ADRP domains of SARS-CoV and other coronaviruses remain(s) to be identified. There is increasing evidence to suggest that *macro* domains have evolved different functionalities using a conserved fold and may have more than one activity which might not necessarily be mutually exclusive in any one domain. A striking example to support this idea is the recent identification of another *macro* domain within the SARS-CoV nsp3, located downstream of the ADRP (Chatterjee et al. 2009) and representing the central domain (SUD-M) of the SARS-CoV unique domain (SUD). On the basis of the NMR structure of SUD-M the ADRP was identified as the closest structural homolog of SUD-M, even though the sequence similarity between the two domains is very low (5% identical residues). It is tempting to speculate that the SUD-M domain may have evolved from ADRP through gene duplication, similar to what was previously proposed for the two paralogous PLpro domains found in many coronaviruses (Ziebuhr et al. 2001). SUD-M was found to bind to single-stranded poly(A)-RNA while others (Tan et al. 2007) had previously reported a poly(dG)/poly(G)-binding activity for SUD. The reasons for these differences are currently unclear but might be due to the different domain boundaries of the proteins characterized in the two studies. The full-length SUD has also been shown to have metal-binding activity (Neuman et al. 2008) which likely resides in the N-terminal part of the protein and involves some of the six conserved cysteine residues (SARS-CoV nsp3 positions 393, 456, 492, 507, 550, and 623) and two conserved histidine residues (positions 539 and 613).

Additional domains with nucleic acid-binding activity are present in SARS-CoV nsp3 and probably other coronaviruses. The bacterially-expressed N-terminal domain of nsp3 consistently copurified with nucleic acids (Serrano et al. 2007). The N-terminal region of nsp3 is particularly rich in acidic residues and is therefore also referred to as acidic (Ac) domain (Ziebuhr et al. 2001). The domain is conserved in all coronaviruses but exhibits a low degree of sequence identity (Serrano et al. 2007). The NMR structure of the Ac domain revealed an N-terminal ubiquitin-like fold (residues 1–112) followed by a disordered tail particularly rich in glutamic acid residues (Serrano et al. 2007). The N-terminal globular domain differs from the ubiquitin-like fold by an elongated $\alpha 2$ helix and the presence of two additional helices, $\alpha 3$ and 3_{10}, which are important for RNA binding. The protein was found to preferentially bind (G)AU(A) sequences.

A further domain linked to RNA binding was identified in SARS-CoV nsp3 just downstream of the PLpro (Neuman et al. 2008). This domain, named nucleic acid-binding domain (NAB), is conserved in group 2 and 3 (but not in group 1) coronaviruses. Bacterially-expressed NAB exhibited ATP-independent double-stranded nucleic acid-unwinding activity, consistent with a possible chaperone function.

Coronavirus replication takes place at virus-induced double membrane vesicles (DMVs) (Gosert et al. 2002; Knoops et al. 2008). The TM domain present in nsp3 (Neuman et al. 2008; Snijder et al. 2003; Ziebuhr et al. 2001), along with TM domains in nsp4 and nsp6, is thought to be involved in tethering the viral replication–transcription complex to intracellular membranes. The topology of the SARS-CoV nsp3 TM domain was recently determined (Oostra et al. 2008). The data suggest that the N- and C-termini are located in the cytoplasm, thus placing the PLpro at the same face of the membrane as all its cleavage sites in pp1a/pp1ab. The domain traverses the membrane twice, while the third, central, predicted transmembrane helix does not appear to function as such. By analogy with the TM domains located in the equine arteritis virus nsp2 and nsp3 proteins (Snijder et al. 2001), whose role in DMV formation has been established earlier, Oostra and co-workers

(Oostra et al. 2007, 2008) suggest that DMV formation in SARS-CoV infected cells is mediated by nsp3 and nsp4. They further hypothesize that the central non-TM hydrophobic domain might play an important role in this process by dipping into the membrane and inducing curvature of the membranes.

nsp4 is a tetra-spanning membrane protein (Oostra et al. 2008). Both termini are located in the cytoplasm (Oostra et al. 2007). TM helices 1 and 2 are connected by a long lumenal loop that is N-glycosylated (Oostra et al. 2007). The presumed critical role of nsp4 in DMV formation is supported by studies using a temperature-sensitive mutant of MHV-A59 in which substitution of the nsp4 Asn-258 residue with Thr led to impaired DMV formation, while polyprotein processing was not affected (Clementz et al. 2008). The critical role of nsp4, including its transmembrane-spanning regions 1–3, in coronavirus replication has been established in a recent reverse genetics study using MHV-A59 (Sparks et al. 2007). The study also showed that the C-terminal nsp4 residues Lys-398 to Thr-492 are dispensable for MHV replication.

nsp5 is the viral main protease (M^{pro}). It cleaves at 11 sites in the central and C-terminal pp1a/pp1ab regions (Thiel et al. 2003; Ziebuhr et al. 1995; Ziebuhr et al. 2000). Because of its prominent role in pp1a/pp1ab processing and the large body of structural and biochemical information available for this enzyme, the M^{pro} is considered an attractive target for the development of antivirals. The enzyme is distantly related to the 3C proteases of picornaviruses (hence its traditional name "3C-like protease," $3CL^{pro}$) but diverged significantly from these viral homologs (Gorbalenya et al. 1989b). For example, in place of the typical Cys–His–Asp/Glu catalytic triad of picornaviral 3C proteases, the coronavirus M^{pro} employs a Cys–His catalytic dyad (Anand et al. 2003, 2002; Tan et al. 2005; Xue et al. 2008; Yang et al. 2003, 2005; Zhao et al. 2008). A water molecule was found to occupy the position of the third member of the catalytic triad in coronavirus M^{pro} structures and it has been suggested that this water molecule might stabilize the protonated His during catalysis. In common with many cellular and viral chymotrypsin-like proteases, coronavirus M^{pro} has a two-β-barrel fold which, in the coronavirus enzymes, is linked to a unique α-helical domain at the C-terminus (Anand et al. 2002). M^{pro} forms homodimers (Anand et al. 2003, 2002; Yang et al. 2003). Dimerization mainly occurs through interactions between the C-terminal domains of the two protomers in the dimer as well as a stretch of N-terminal residues (N-finger) (see Chap. 9 for details). Dimerization is generally believed to be a prerequisite for *trans*-processing activity. The dimeric structure is another feature that sets the coronavirus M^{pro} apart from 3C proteases which function as monomers. Over the past few years a large number of studies have provided significant insight into the structural details and dynamics of M^{pro} dimerization and their functional consequences. For a review of this work, the reader is referred to Chap. 9.

nsp6 is another TM protein. Both the N- and C-terminus are located cytoplasmically (Oostra et al. 2008), indicating an even number of TM helices and thus placing the M^{pro} on the same face of the membrane as all its substrates. The protein has seven putative TM helices. To satisfy the observed N_{endo}–C_{endo} localization of the protein, Oostra and co-workers (2008) proposed that only six of the predicted TM helices function as such in the context of the full-length protein and helix 6 or possibly helix 7 might not traverse the membrane. Further, it has been suggested that the nonmembrane-spanning helix may act as interaction platform for other replicase components or aid in the formation of DMVs. The function of nsp6 remains to be characterized.

nsp8 was recently reported to be a second "noncanonical" RNA-dependent RNA polymerase (Imbert et al. 2006) (see Chap. 18). The protein synthesizes short oligonucleotides of up to 6 nts and requires an RNA template and metal ions for activity. RNA synthesis was sequence specific, with a preference for the internal 5′-(G/U)CC-3′ trinucleotide sequence as site of initiation, but exhibited a relatively low fidelity and processivity. nsp8 is well conserved among coronaviruses. The noncanonical RdRp is therefore suggested to be an essential enzymatic activity involved in RNA synthesis in all coronaviruses and, possibly, other related nidoviruses. Imbert and co-workers (Imbert et al. 2006) hypothesized that nsp8 could act as a primase that produces RNA primers which are then extended by the main RdRp, nsp12, in a mechanism reminiscent of DNA replication.

nsp8 was shown to interact with nsp7 and together these proteins form a hexadecameric ring structure consisting of eight copies of each protein as shown by X-ray crystallography (Zhai et al. 2005). The complex encircles a channel with a diameter of ~30 Å that is lined with positively-charged residues (for details, see Chap. 18). Addition of nsp7 to the primase activity assay did not increase the primase activity. However, nsp8 exhibits a low thermostability and nsp7 might therefore act as a stabilizing mortar (Imbert et al. 2006; Zhai et al. 2005).

The structure of SARS-CoV nsp9 was solved by two groups (Egloff et al. 2004; Sutton et al. 2004). A distant structural relationship with domain II of the M^{pro} was noted, suggesting that both proteins may be evolutionary related and thus represent another example of domain duplication in the coronavirus replicase, similar to what was discussed above for coronavirus PL^{pro} and *macro* domains. Furthermore, nsp9 is structurally related to proteins containing an oligosaccharide/oligonucleotide-binding (OB) fold, although the connectivity of the individual secondary structural elements differs in nsp9. A large proportion of OB-fold proteins bind nucleic acids and SARS-CoV nsp9 was confirmed to also bind ssRNA. Binding of ssRNA by nsp9 was unspecific but specificity might possibly be attained by interaction with other viral or cellular proteins. Sutton and co-workers (2004) showed that SARS-CoV nsp9 interacts with nsp8 in an analytical ultracentrifugation analysis. The experiment further suggested that this interaction might help stabilize an otherwise disordered domain of nsp8. nsp9 forms dimers through the interaction of parallel α-helices that contain a GXXXG protein–protein interaction motif (Miknis et al. 2009). Substitutions of either of the Gly residues with Glu disrupted dimer formation while RNA binding was only marginally affected. Viable SARS-CoV mutants carrying either mutation could not be recovered, suggesting that nsp9 dimerization is critical for virus replication.

nsp10 is another small replicase protein with RNA-binding activity. Two studies analyzing the structure of nsp10 described a single-domain protein that contains two Zn^{2+}-fingers (Joseph et al. 2006; Su et al. 2006). nsp10 was shown to bind single- and double-stranded RNA and DNA with low affinity and no apparent specificity. nsp10 interacts with nsp9 as shown by cross-linking (Joseph et al. 2006) and in a GST-pulldown assay (Imbert et al. 2008). As nsp9, in turn, interacts with nsp8 and nsp8 forms a complex with nsp7 and all these proteins are involved in homotypic interactions (see above and Chap. 18), it is tempting to believe that these proteins form a multiprotein complex that, in the course of infection, undergoes structural rearrangements (possibly due to M^{pro}-mediated cleavages), thereby activating, modulating or inactivating specific RTC function(s) as required at the various steps of viral RNA synthesis.

Consistent with the presumed key role of M^{pro} processing in the formation of a functional RTC, viable MHV mutants could not be recovered if cleavage at the nsp7|8, nsp8|9 or nsp10|11(12) sites was abolished (Deming et al. 2007). Disruption of proteolytic processing at the nsp9|10 site gave rise to an attenuated, but viable phenotype. The MHV nsp9|10 cleavage site mutant restored near wild-type growth kinetics after serial passaging which was not caused by restoration of processing at the nsp9|10 site but by a number of compensatory mutations at distant positions in the viral genome. This suggests that nsp9–10 can function as a fusion protein, though efficient replication requires adaptation of the virus. The functional role of the specific changes identified in some of the recovered viruses remain to be characterized.

nsp11 forms the C-terminal part of pp1a. SARS-CoV nsp11 is an oligopeptide of 13 residues. It is produced if no programmed frameshift occurs at the "slippery sequence" (see Chap. 6), leading to translation termination at the ORF1a stop codon. nsp11 shares its N-terminal amino acids (upstream of the frameshift site) with nsp12. Most of the nsp11 coding sequence overlaps with the RNA sequence involved in frameshifting and the nsp12 coding sequence, posing severe constraints on the nsp11 sequence and arguing against a functional role of nsp11. The protein has also not been detected in infected cells.

6.3.2 *ORF1b-Encoded Nonstructural Proteins 12–16*

The RNA-dependent RNA polymerase (RdRp, nsp12) is the most conserved protein of the coronavirus replicase–transcriptase. Expression of nsp12 (and all other ORF1b-encoded proteins) requires ribosomal frameshifting, implying that nsp12–16 are produced at significantly lower levels compared to ORF1a-encoded functions. SARS-CoV nsp12 is a protein of 932 residues. The actual catalytic domain containing the conserved RdRp motifs (Gorbalenya et al. 1989b; Koonin 1991) occupies the C-terminal region (C-terminal domain, CTD) of nsp12 while the N-terminal domain (NTD) spanning the first 375 amino acids has no known counterpart in other RdRps (Xu et al. 2003). Xu and co-workers (2003) proposed a three-dimensional (3D) homology model for the SARS-CoV nsp12 CTD. This model showed the characteristic cupped right hand palm–finger–thumb structure encircling a nucleic acid-binding tunnel. The RdRp activity of nsp12 was recently confirmed by showing that bacterially-expressed nsp12 is able to extend an oligo (U) primer hybridized to a poly(A)-template (Cheng et al. 2005).

SARS-CoV nsp13 is a multidomain protein of 601 amino acid residues. The N-terminal region contains a zinc-binding domain (ZBD) while a helicase domain featuring the typical conserved morifs of superfamily 1 helicases is present in the C-terminal half (Gorbalenya et al. 1989a, 1989b). The ZBD contains 12 conserved cysteine and histidine residues that are predicted to form a binuclear Zn^{2+}-binding cluster (Seybert et al. 2005; van Dinten et al. 2000). It is conserved in all nidoviruses and is critical for helicase activity in vitro (Seybert et al. 2005) and RNA synthesis of HCoV-229E in cell culture (Hertzig and Ziebuhr, unpublished). Coronavirus helicases (including SARS-CoV nsp13) were shown to unwind RNA and DNA duplexes in a 5′-to-3′ direction with respect to the single-stranded RNA they initially bind to (Ivanov and Ziebuhr 2004; Ivanov et al. 2004a; Seybert et al. 2000; Tanner et al. 2003). Translocation of nsp13 along RNA (and concomitant duplex unwinding) is fueled by NTP or dNTP hydrolysis. Consistent with many other helicases, the nsp13-associated NTPase/dNTPase activity is stimulated by nucleic acids (Heusipp et al. 1997). Additionally, nsp13 exhibits RNA 5′-triphosphatase activity which was proposed to catalyze the first step of the 5′-capping reaction of viral RNAs (Ivanov and Ziebuhr 2004; Ivanov et al. 2004a).

The N-terminal part of nsp14 contains a 3′-to-5′ exoribonuclease (ExoN) domain. ExoN is related to the DEDD superfamily of exonucleases (Zuo and Deutscher 2001) but carries an additional putative Zn^{2+}-finger structure that is inserted between the conserved motifs II and III (Snijder et al. 2003). The ExoN activity of SARS-CoV nsp14 was demonstrated in vitro and shown to be specific for single-stranded and double-stranded RNA (Minskaia et al. 2006). The protein was shown to require metal ions for activity and isothermic titration calorimetry data suggest that nsp14 binds two Mg^{2+} ions per molecule (Chen et al. 2007a). This data, together with the profound reduction of activity upon substitution of putative metal ion-coordinating residues (Minskaia et al. 2006), suggests that catalysis occurs through a two-metal-ion mechanism similar to that used by many cellular enzymes mediating phosphoryltransfer reactions (Beese and Steitz 1991).

It has been proposed that coronaviruses and other nidoviruses with genome sizes of about 30 kb have evolved specific mechanisms to replicate their large RNA genomes. The argument is that, in the absence of such mechanisms, the intrinsically error-prone RdRps of RNA viruses would cross a (postulated) threshold of nucleotide misincorporations above which the survival of a given virus population becomes impossible. The question of how coronaviruses are able to maintain their exceptionally large RNA genomes has not been resolved but a number of recent observations suggest that nsp14 was critically involved in the evolution of these large genomes. First, coronaviruses were shown to encode a 3′-to-5′ ExoN that is related to cellular enzymes involved in proof-reading mechanisms during cellular DNA replication. Second, the ExoN activity is not conserved in nidoviruses with much smaller genomes (i.e., *Arteriviridae*), even though the pp1a/pp1ab domain organization is otherwise very well conserved among small and large nidoviruses (Gorbalenya et al. 2006; Snijder et al. 2003).

Third, ExoN activity was found to be essential for HCoV-229E replication in cell culture (Minskaia et al. 2006). Transfection of genome-length RNAs containing substitutions of ExoN active-site residues failed to produce viable virus. Northern blot analysis of viral RNA in transfected cells revealed a severe reduction in genome replication, an altered molar ratio of sgRNAs and aberrant migration of two sgRNAs in some of the mutants (Minskaia et al. 2006). Consistent with these data, deletion of the ExoN domain from a SARS-CoV replicon was reported to abolish RNA synthesis and substitution of a putative ExoN active-site residue reduced genome replication and transcription about 10-fold in this system (Almazan et al. 2006). Fourth, the characterization of MHV-A59 ExoN active-site mutants revealed defects in viral RNA synthesis and rapid accumulation of mutations across the genome (Eckerle et al. 2007). The authors calculated that, during passage (and under strong selection pressure), MHV-A59 ExoN mutants accumulated approximately 15-fold more substitutions than the wild-type virus. The data obtained in these studies are consistent with the proposed role of ExoN in increasing the fidelity of the coronavirus RdRp, although there is still no direct proof for this specific function.

Coronaviruses encode a second conserved ribonuclease, NendoU (Nidoviral endoribonuclease, specific for U), which is located within nsp15 (Snijder et al. 2003). The domain is conserved not only in coronaviruses but also in all other members of the order *Nidovirales*. NendoU homologs could not be identified in any other RNA virus, which makes the domain a genetic marker of nidoviruses (Ivanov et al. 2004b). SARS-CoV nsp15 was shown to be an endoribonuclease that cleaves preferentially 3' of uridylates and generates 2'–3' cyclic phosphate ends (Bhardwaj et al. 2004; Ivanov et al. 2004b). The activity is significantly enhanced by Mn^{2+} while addition of Mg^{2+} or Ca^{2+} only had minor effects on activity. The structure of SARS-CoV nsp15 revealed a novel fold (see Chap. 18). In spite of unrelated structures and lack of sequence similarity, nsp15 and RNaseA were proposed to use the same catalytic mechanism (Ricagno et al. 2006). Residues forming the catalytic triad of bovine RNase A (His-12, Lys-41, His-119) could be superimposed with His/Lys residues known to be critical for activity of SARS-CoV nsp15 (Ivanov et al. 2004b) and both RNaseA and SARS-CoV NendoU generate 2'–3' cyclic phosphate ends, suggesting that endoribonucleolytic cleavage by NendoU proceeds through the same catalytic mechanism (Ricagno et al. 2006). The proposed RNaseA-like catalytic mechanism does not involve metal ions and therefore the observed stimulatory effects of Mn^{2+} on NendoU activities of several (but not all) coronaviruses (Bhardwaj et al. 2004; Cao et al. 2008; Ivanov et al. 2004b; Xu et al. 2006) cannot readily be reconciled with a role in catalysis. Based on intrinsic tryptophan fluorescence data, Bhardwaj and co-workers (2004) suggested that Mn^{2+} ions induce specific conformational changes in the protein that might promote its nuclease activity. However, the available crystal structure information does not provide evidence for the presence of Mn^{2+} ion-binding sites in any of the coronavirus NendoUs studied (Ricagno et al. 2006; Xu et al. 2006). More recently, it was suggested that Mn^{2+} ions increase the RNA-binding activity of NendoU (Bhardwaj et al. 2006).

Coronavirus NendoUs form hexamers consisting of a dimer of trimers in crystals and in solution (see Chap. 18). Amino acid substitutions that interfere with hexamerization were reported to reduce the nucleolytic activity and RNA affinity of SARS-CoV NendoU, suggesting that hexamerization is critically involved in activity (Guarino et al. 2005). However, the presence of six independent active sites in the NendoU crystal structure, together with relatively minor differences in the K_m and k_{cat} values determined for monomeric and hexameric MHV NendoU (Xu et al. 2006) and the fact that maltose-binding protein–NendoU fusion proteins that are unable to form hexamers possess endoribonucleolytic activity (Ivanov et al. 2004b) suggest that hexamerization may not be essential for activity. It therefore remains possible that NendoU may be active prior to its proteolytic release from pp1ab, for example in the context of NendoU-containing processing intermediates or the full-length polyprotein.

The role of NendoU in the coronavirus life cycle is not well understood. Reverse genetics data obtained for HCoV-229E (Ivanov et al. 2004b) and MHV (Kang et al. 2007) suggest that NendoU activity is not essential for viral replication. Substitutions of NendoU active-site His and Lys residues in MHV were shown to cause subtle defects in sgRNA accumulation and reduce virus titers by about 10-fold. Substitutions of a conserved Asp residue abolished viral RNA synthesis, both in HCoV-229E and MHV (Ivanov et al. 2004b; Kang et al. 2007). Both structural information and biochemical data suggest that this particular residue may have an important structural role, suggesting that the observed defects in viral replication may be due to misfolding of nsp15 or other pp1a/pp1ab subunits rather than caused by the lack of NendoU activity. More studies are needed to understand the function of the conserved NendoU activity in nidoviral replication.

NendoU belongs to a family of proteins that is prototyped by a cellular endoribonuclease, called XendoU, from *Xenopus laevis* (Laneve et al. 2003). XendoU is a poly(U)-specific endoribonuclease that, together with ExoN and methyltransferase (MT) activities, is involved in the processing of intron-encoded small nucleolar (sno) RNAs and site-specific RNA methylation pathways in *X. laevis* oocytes (Laneve et al. 2003). It has been suggested (but not yet explored experimentally) that a similar set of activities associated with coronavirus nsp14 (ExoN), nsp15 (NendoU) and nsp16 (MT), which are coexpressed in the C-proximal pp1ab region, may be involved in related RNA-processing and methylation pathways (Snijder et al. 2003).

The most C-terminal processing product of pp1ab, nsp16, has been proposed to be related to the RrmJ/FtsJ family of S-adenosyl-methionine-dependent ribose-2′-O-methyltransferases (Feder et al. 2003; Snijder et al. 2003). The predicted MT activity was recently confirmed for nsp16 of feline coronavirus (FCoV) (Decroly et al. 2008). FCoV nsp16 was shown to methylate 7MeGpppAC$_n$ at the ribose-2′-O moiety of the adenosine, converting a cap-0 to a cap-1 structure. The domain is critically involved in coronavirus replication as deletion or ablation of expression of nsp16 abolished RNA synthesis in SARS-CoV (Almazan et al. 2006) and HCoV-229E (Hertzig, Schelle and Ziebuhr, unpublished data) while substitution of one of the residues forming the catalytic tetrad reduced sgRNA synthesis about

10-fold in a SARS-CoV replicon system (Almazan et al. 2006). The 5′-cap structure is known to be critically important for the stability of cellular mRNAs and translation initiation (reviewed in Shuman 2001). The cellular capping apparatus is located in the nucleus, implying that viruses that replicate in the cytoplasm need to either provide all the enzymes required to produce RNA cap structures or employ alternative mechanisms, such as cap snatching (Plotch et al. 1981). In eukaryotic cells, the production of $^{7Me}GpppG_{2'OMe}N_{(2'OMe)}$ cap-1 (and cap-2) structures is achieved through four consecutive enzymatic reactions: (1) removal of the RNA 5′ γ-phosphate by an RNA 5′-triphosphatase, (2) transfer of GMP to the remaining 5′ diphosphate end by a guanylyltransferase, (3) methylation of the guanine at the N7 position by a guanine-N7-methyltransferase, resulting in a cap-0 structure, (4) methylation of ribose-2′-O-moieties of the first (and second) nucleotide of the mRNA by a ribose-2′-O-methyltransferase, resulting in cap-1 and cap-2 structures, respectively (Langberg and Moss 1981; Shuman 2001). Coronaviral RNAs are modified at the 5′-end, probably with a cap structure (Lai et al. 1982), and it seems reasonable to suggest that the nsp16-associated ribose-2′-O-methyltransferase activity catalyzes the conversion of cap-0 into cap-1 structures whereas the nsp13-associated RNA 5′-triphosphatase activity might catalyze the first step of 5′-cap formation (Ivanov and Ziebuhr 2004; Ivanov et al. 2004a). Homologs of cellular guanylyltransferase and guanine-N7-methyltransferases have not been identified in coronaviruses and it remains to be seen what viral or cellular proteins/activities mediate the two remaining reactions, namely GMP transfer and guanine-N7 methylation, to synthesize RNA cap structures on coronavirus RNAs. While some + RNA viruses, such as alphaviruses, encode all four required activities (reviewed in Salonen et al. 2005), other viruses use one and the same domain to perform two different reactions. For example, the West Nile virus MT methylates at both the guanine N7 position and the ribose-2′-O moiety using an interesting substrate-repositioning mechanism and a single S-adenosylmethionine-binding site (Dong et al. 2008). While FCoV nsp16 had no guanine-N7-methyltransferase activity and did not bind unmethylated $GpppAC_n$ (Decroly et al. 2008) the authors point out that recognition of unmethylated, guanylylated RNA might depend on regulatory RNA elements located further downstream in the RNA, similar to what was described for flaviviruses (Dong et al. 2007; Ray et al. 2006).

6.4 Future Directions

The emergence of SARS-CoV led to a renewed interest in coronavirology and recent years saw a significant increase in research involving this family of viruses. Many of the previously predicted coronavirus replicase gene-encoded enzyme activities were characterized by biochemical and structural approaches using viral proteins expressed in heterologous systems (Ziebuhr 2008). In a few cases, structural studies informed subsequent biochemical studies, resulting in the identification of RNA-binding domains and other functions (Egloff et al. 2004; Joseph et al. 2006; Zhai et al. 2005).

Most of the biochemical and structural studies reported over the past years involved isolated nsps or subdomains of these proteins rather than multidomain complexes. Although these studies provided invaluable new insight into structure–function relationships of many of these proteins, there is hardly any information regarding the quarternary structure(s) and subunit compositions of replicase and/or transcriptase complexes catalyzing specific reactions during viral RNA synthesis. For example, very little is known about the special factors and macromolecular interactions involved in discontinuous minus-strand RNA synthesis, a unique feature of coronaviruses and several other nidoviruses. Similarly, the replication and maintenance of the exceptionally large coronavirus genome is likely to involve the concerted action of replicase gene-encoded nsps, possibly including processing precursors and intermediates as pointed out above. The physical and functional interactions between the various components of the replicase–transcriptase probably undergo significant changes in the course of the viral replication cycle, further complicating the characterization of these complexes. Despite these major technical challenges, recent studies increasingly try to elucidate the functions and structures of complexes involving two or more proteins (Zhai et al. 2005). Also, the availability of reverse genetics systems for several coronaviruses has greatly facilitated the characterization of specific protein functions and critical interactions between domains encoded by very different regions of the replicase gene (Donaldson et al. 2007).

The recently established method for purification of entire functional (membrane-bound) RTCs (van Hemert et al. 2008) presents exciting new possibilities to study coronavirus RNA synthesis. Advanced imaging techniques including cryo-electron microscopy have provided fascinating new insight into key structures involved in virus replication. For example, cryo-electron microscopy of ribosomes stalled at a specific stage of frameshifting provided a glimpse of how the structure of the mRNA template affects ribosome function to mediate this frameshift (Namy et al. 2006), and electron tomography of SARS-CoV-infected cells provided a 3D view of the unique continuous reticulovesicular network, the formation of which is induced by the virus (Knoops et al. 2008). However, much remains to be studied to understand how the virus coaxes the cell to produce these structures. The precise role of these membrane structures in virus replication and, possibly, immune evasion need to be characterized in more detail.

Further progress has been made in our understanding of the interactions of the virus with the host cell, especially with regard to innate immune responses to coronavirus infections and the role of specific viral proteins in counteracting these antiviral host responses. The understanding of these pathways in combination with biochemical, structural and genetic approaches will continue to provide exciting insights into virus replication and virus–host interactions and form the basis for the development of antiviral drugs and new and better coronavirus vaccines.

References

Allen MD, Buckle AM, Cordell SC, Lowe J, Bycroft M (2003) The crystal structure of AF1521 a protein from Archaeoglobus fulgidus with homology to the non-histone domain of macroH2A. J Mol Biol 330:503–511

Almazan F, Galan C, Enjuanes L (2004) The nucleoprotein is required for efficient coronavirus genome replication. J Virol 78:12683–12688

Almazan F, Dediego ML, Galan C, Escors D, Alvarez E, Ortego J, Sola I, Zuniga S, Alonso S, Moreno JL, Nogales A, Capiscol C, Enjuanes L (2006) Construction of a severe acute respiratory syndrome coronavirus infectious cDNA clone and a replicon to study coronavirus RNA synthesis. J Virol 80:10900–10906

Almeida MS, Johnson MA, Herrmann T, Geralt M, Wuthrich K (2007) Novel beta-barrel fold in the NMR Structure of the replicase nonstructural protein 1 from the SARS coronavirus. J Virol 81:3151–3161

Anand K, Palm GJ, Mesters JR, Siddell SG, Ziebuhr J, Hilgenfeld R (2002) Structure of coronavirus main proteinase reveals combination of a chymotrypsin fold with an extra alpha-helical domain. EMBO J 21:3213–3224

Anand K, Ziebuhr J, Wadhwani P, Mesters JR, Hilgenfeld R (2003) Coronavirus main proteinase ($3CL^{pro}$) structure: basis for design of anti-SARS drugs. Science 300:1763–1767

Barretto N, Jukneliene D, Ratia K, Chen Z, Mesecar AD, Baker SC (2005) The papain-like protease of severe acute respiratory syndrome coronavirus has deubiquitinating activity. J Virol 79:15189–15198

Beese LS, Steitz TA (1991) Structural basis for the 3′-5′ exonuclease activity of Escherichia coli DNA polymerase I: a two metal ion mechanism. EMBO J 10:25–33

Bhardwaj K, Guarino L, Kao CC (2004) The severe acute respiratory syndrome coronavirus Nsp15 protein is an endoribonuclease that prefers manganese as a cofactor. J Virol 78:12218–12224

Bhardwaj K, Sun J, Holzenburg A, Guarino LA, Kao CC (2006) RNA recognition and cleavage by the SARS coronavirus endoribonuclease. J Mol Biol 361:243–256

Brierley I (1995) Ribosomal frameshifting viral RNAs. J Gen Virol 76:1885–1892

Brockway SM, Denison MR (2005) Mutagenesis of the murine hepatitis virus nsp1-coding region identifies residues important for protein processing, viral RNA synthesis, and viral replication. Virology 340:209–223

Brockway SM, Lu XT, Peters TR, Dermody TS, Denison MR (2004) Intracellular localization and protein interactions of the gene 1 protein p28 during mouse hepatitis virus replication. J Virol 78:11551–11562

Cao J, Wu CC, Lin TL (2008) Turkey coronavirus non-structure protein NSP15–an endoribonuclease. Intervirology 51:342–351

Chatterjee A, Johnson MA, Serrano P, Pedrini B, Joseph JS, Neuman BW, Saikatendu K, Buchmeier MJ, Kuhn P, Wuthrich K (2009) Nuclear magnetic resonance structure shows that the severe acute respiratory syndrome coronavirus-unique domain contains a macrodomain fold. J Virol 83:1823–1836

Chen P, Jiang M, Hu T, Liu Q, Chen XS, Guo D (2007a) Biochemical characterization of exoribonuclease encoded by SARS coronavirus. J Biochem Mol Biol 40:649–655

Chen Z, Wang Y, Ratia K, Mesecar AD, Wilkinson KD, Baker SC (2007b) Proteolytic processing and deubiquitinating activity of papain-like proteases of human coronavirus NL63. J Virol 81:6007–6018

Cheng A, Zhang W, Xie Y, Jiang W, Arnold E, Sarafianos SG, Ding J (2005) Expression, purification, and characterization of SARS coronavirus RNA polymerase. Virology 335:165–176

Clementz MA, Kanjanahaluethai A, O'Brien TE, Baker SC (2008) Mutation in murine coronavirus replication protein nsp4 alters assembly of double membrane vesicles. Virology 375:118–129

Connor RF, Roper RL (2007) Unique SARS-CoV protein nsp1: bioinformatics, biochemistry and potential effects on virulence. Trends Microbiol 15:51–53

Decroly E, Imbert I, Coutard B, Bouvet M, Selisko B, Alvarez K, Gorbalenya AE, Snijder EJ, Canard B (2008) Coronavirus nonstructural protein 16 is a cap-0 binding enzyme possessing (nucleoside-2'O)-methyltransferase activity. J Virol 82:8071–8084

Deming DJ, Graham RL, Denison MR, Baric RS (2007) Processing of open reading frame 1a replicase proteins nsp7 to nsp10 in murine hepatitis virus strain A59 replication. J Virol 81:10280–10291

Devaraj SG, Wang N, Chen Z, Chen Z, Tseng M, Barretto N, Lin R, Peters CJ, Tseng CT, Baker SC, Li K (2007) Regulation of IRF-3-dependent innate immunity by the papain-like protease domain of the severe acute respiratory syndrome coronavirus. J Biol Chem 282:32208–32221

Donaldson EF, Graham RL, Sims AC, Denison MR, Baric RS (2007) Analysis of murine hepatitis virus strain A59 temperature-sensitive mutant TS-LA6 suggests that nsp10 plays a critical role in polyprotein processing. J Virol 81:7086–7098

Dong H, Ray D, Ren S, Zhang B, Puig-Basagoiti F, Takagi Y, Ho CK, Li H, Shi PY (2007) Distinct RNA elements confer specificity to flavivirus RNA cap methylation events. J Virol 81:4412–4421

Dong H, Ren S, Zhang B, Zhou Y, Puig-Basagoiti F, Li H, Shi PY (2008) West Nile virus methyltransferase catalyzes two methylations of the viral RNA cap through a substrate-repositioning mechanism. J Virol 82:4295–4307

Dos Ramos F, Carrasco M, Doyle T, Brierley I (2004) Programmed -1 ribosomal frameshifting in the SARS coronavirus. Biochem Soc Trans 32:1081–1083

Dougherty WG, Semler BL (1993) Expression of virus-encoded proteinases: functional and structural similarities with cellular enzymes. Microbiol Rev 57:781–822

Draker R, Roper RL, Petric M, Tellier R (2006) The complete sequence of the bovine torovirus genome. Virus Res 115:56–68

Eckerle LD, Lu X, Sperry SM, Choi L, Denison MR (2007) High fidelity of murine hepatitis virus replication is decreased in nsp14 exoribonuclease mutants. J Virol 81:12135–12144

Egloff MP, Ferron F, Campanacci V, Longhi S, Rancurel C, Dutartre H, Snijder EJ, Gorbalenya AE, Cambillau C, Canard B (2004) The severe acute respiratory syndrome-coronavirus replicative protein nsp9 is a single-stranded RNA-binding subunit unique in the RNA virus world. Proc Natl Acad Sci USA 101:3792–3796

Egloff MP, Malet H, Putics A, Heinonen M, Dutartre H, Frangeul A, Gruez A, Campanacci V, Cambillau C, Ziebuhr J, Ahola T, Canard B (2006) Structural and functional basis for ADP-ribose and poly(ADP-ribose) binding by viral macro domains. J Virol 80:8493–8502

Eriksson KK, Cervantes-Barragan L, Ludewig B, Thiel V (2008) Mouse hepatitis virus liver pathology is dependent on ADP-ribose-1″-phosphatase, a viral function conserved in the alpha-like supergroup. J Virol 82:12325–12334

Fan K, Wei P, Feng Q, Chen S, Huang C, Ma L, Lai B, Pei J, Liu Y, Chen J, Lai L (2004) Biosynthesis, purification, and substrate specificity of severe acute respiratory syndrome coronavirus 3C-like proteinase. J Biol Chem 279:1637–1642

Fan K, Ma L, Han X, Liang H, Wei P, Liu Y, Lai L (2005) The substrate specificity of SARS coronavirus 3C-like proteinase. Biochem Biophys Res Commun 329:934–940

Feder M, Pas J, Wyrwicz LS, Bujnicki JM (2003) Molecular phylogenetics of the RrmJ/fibrillarin superfamily of ribose 2'-O-methyltransferases. Gene 302:129–138

Gadlage MJ, Graham RL, Denison MR (2008) Murine coronaviruses encoding nsp2 at different genomic loci have altered replication, protein expression, and localization. J Virol 82: 11964–11969

Gorbalenya AE, Koonin EV, Donchenko AP, Blinov VM (1989a) Two related superfamilies of putative helicases involved in replication, recombination, repair and expression of DNA and RNA genomes. Nucleic Acids Res 17:4713–4730

Gorbalenya AE, Koonin EV, Donchenko AP, Blinov VM (1989b) Coronavirus genome: prediction of putative functional domains in the non-structural polyprotein by comparative amino acid sequence analysis. Nucleic Acids Res 17:4847–4861

Gorbalenya AE, Koonin EV, Lai MM (1991) Putative papain-related thiol proteases of positive-strand RNA viruses. Identification of rubi- and aphthovirus proteases and delineation of a novel conserved domain associated with proteases of rubi-, alpha- and coronaviruses. FEBS Lett 288:201–205

Gorbalenya AE, Enjuanes L, Ziebuhr J, Snijder EJ (2006) Nidovirales: evolving the largest RNA virus genome. Virus Res 117:17–37

Gosert R, Kanjanahaluethai A, Egger D, Bienz K, Baker SC (2002) RNA replication of mouse hepatitis virus takes place at double-membrane vesicles. J Virol 76:3697–3708

Graham RL, Sims AC, Brockway SM, Baric RS, Denison MR (2005) The nsp2 replicase proteins of murine hepatitis virus and severe acute respiratory syndrome coronavirus are dispensable for viral replication. J Virol 79:13399–13411

Guarino LA, Bhardwaj K, Dong W, Sun J, Holzenburg A, Kao C (2005) Mutational analysis of the SARS virus Nsp15 endoribonuclease: identification of residues affecting hexamer formation. J Mol Biol 353:1106–1117

Harcourt BH, Jukneliene D, Kanjanahaluethai A, Bechill J, Severson KM, Smith CM, Rota PA, Baker SC (2004) Identification of severe acute respiratory syndrome coronavirus replicase products and characterization of papain-like protease activity. J Virol 78:13600–13612

Herold J, Siddell SG (1993) An 'elaborated' pseudoknot is required for high frequency frameshifting during translation of HCV 229E polymerase mRNA. Nucleic Acids Res 21:5838–5842

Heusipp G, Harms U, Siddell SG, Ziebuhr J (1997) Identification of an ATPase activity associated with a 71-kilodalton polypeptide encoded in gene 1 of the human coronavirus 229E. J Virol 71:5631–5634

Imbert I, Guillemot JC, Bourhis JM, Bussetta C, Coutard B, Egloff MP, Ferron F, Gorbalenya AE, Canard B (2006) A second, non-canonical RNA-dependent RNA polymerase in SARS coronavirus. EMBO J 25:4933–4942

Imbert I, Snijder EJ, Dimitrova M, Guillemot JC, Lecine P, Canard B (2008) The SARS-Coronavirus PLnc domain of nsp3 as a replication/transcription scaffolding protein. Virus Res 133:136–148

Ivanov KA, Ziebuhr J (2004) Human coronavirus 229E nonstructural protein 13: characterization of duplex-unwinding, nucleoside triphosphatase, and RNA 5′-triphosphatase activities. J Virol 78:7833–7838

Ivanov KA, Thiel V, Dobbe JC, van der Meer Y, Snijder EJ, Ziebuhr J (2004a) Multiple enzymatic activities associated with severe acute respiratory syndrome coronavirus helicase. J Virol 78:5619–5632

Ivanov KA, Hertzig T, Rozanov M, Bayer S, Thiel V, Gorbalenya AE, Ziebuhr J (2004b) Major genetic marker of nidoviruses encodes a replicative endoribonuclease. Proc Natl Acad Sci USA 101:12694–12699

Joseph JS, Saikatendu KS, Subramanian V, Neuman BW, Brooun A, Griffith M, Moy K, Yadav MK, Velasquez J, Buchmeier MJ, Stevens RC, Kuhn P (2006) Crystal structure of nonstructural protein 10 from the severe acute respiratory syndrome coronavirus reveals a novel fold with two zinc-binding motifs. J Virol 80:7894–7901

Kamitani W, Narayanan K, Huang C, Lokugamage K, Ikegami T, Ito N, Kubo H, Makino S (2006) Severe acute respiratory syndrome coronavirus nsp1 protein suppresses host gene expression by promoting host mRNA degradation. Proc Natl Acad Sci USA 103:12885–12890

Kang H, Bhardwaj K, Li Y, Palaninathan S, Sacchettini J, Guarino L, Leibowitz JL, Kao CC (2007) Biochemical and genetic analyses of murine hepatitis virus Nsp15 endoribonuclease. J Virol 81:13587–13597

Karras GI, Kustatscher G, Buhecha HR, Allen MD, Pugieux C, Sait F, Bycroft M, Ladurner AG (2005) The macro domain is an ADP-ribose binding module. EMBO J 24:1911–1920

Knoops K, Kikkert M, Worm SH, Zevenhoven-Dobbe JC, van der Meer Y, Koster AJ, Mommaas AM, Snijder EJ (2008) SARS-coronavirus replication is supported by a reticulovesicular network of modified endoplasmic reticulum. PLoS Biol 6:e226

Koonin EV (1991) The phylogeny of RNA-dependent RNA polymerases of positive-strand RNA viruses. J Gen Virol 72:2197–2206

Lai MM, Patton CD, Stohlman SA (1982) Further characterization of mRNA's of mouse hepatitis virus: presence of common 5'-end nucleotides. J Virol 41:557–565

Laneve P, Altieri F, Fiori ME, Scaloni A, Bozzoni I, Caffarelli E (2003) Purification, cloning, and characterization of XendoU, a novel endoribonuclease involved in processing of intron-encoded small nucleolar RNAs in Xenopus laevis. J Biol Chem 278:13026–13032

Langberg SR, Moss B (1981) Post-transcriptional modifications of mRNA. Purification and characterization of cap I and cap II RNA (nucleoside-2'-)-methyltransferases from HeLa cells. J Biol Chem 256:10054–10060

Lindner HA, Fotouhi-Ardakani N, Lytvyn V, Lachance P, Sulea T, Menard R (2005) The papain-like protease from the severe acute respiratory syndrome coronavirus is a deubiquitinating enzyme. J Virol 79:15199–15208

Lindner HA, Lytvyn V, Qi H, Lachance P, Ziomek E, Menard R (2007) Selectivity in ISG15 and ubiquitin recognition by the SARS coronavirus papain-like protease. Arch Biochem Biophys 466:8–14

Marra MA, Jones SJ, Astell CR, Holt RA, Brooks-Wilson A, Butterfield YS, Khattra J, Asano JK, Barber SA, Chan SY, Cloutier A, Coughlin SM, Freeman D, Girn N, Griffith OL, Leach SR, Mayo M, McDonald H, Montgomery SB, Pandoh PK, Petrescu AS, Robertson AG, Schein JE, Siddiqui A, Smailus DE, Stott JM, Yang GS, Plummer F, Andonov A, Artsob H, Bastien N, Bernard K, Booth TF, Bowness D, Czub M, Drebot M, Fernando L, Flick R, Garbutt M, Gray M, Grolla A, Jones S, Feldmann H, Meyers A, Kabani A, Li Y, Normand S, Stroher U, Tipples GA, Tyler S, Vogrig R, Ward D, Watson B, Brunham RC, Krajden M, Petric M, Skowronski DM, Upton C, Roper RL (2003) The Genome sequence of the SARS-associated coronavirus. Science 300:1399–1404

Masters PS (2006) The molecular biology of coronaviruses. Adv Virus Res 66:193–292

Miknis ZJ, Donaldson EF, Umland TC, Rimmer RA, Baric RS, Schultz LW (2009) SARS-CoV nsp9 dimerization is essential for efficient viral growth. J Virol Jan 19 (Epub ahead of print)

Minskaia E, Hertzig T, Gorbalenya AE, Campanacci V, Cambillau C, Canard B, Ziebuhr J (2006) Discovery of an RNA virus 3'->5' exoribonuclease that is critically involved in coronavirus RNA synthesis. Proc Natl Acad Sci USA 103:5108–5113

Namy O, Moran SJ, Stuart DI, Gilbert RJ, Brierley I (2006) A mechanical explanation of RNA pseudoknot function in programmed ribosomal frameshifting. Nature 441:244–247

Narayanan K, Huang C, Lokugamage K, Kamitani W, Ikegami T, Tseng CT, Makino S (2008) Severe acute respiratory syndrome coronavirus nsp1 suppresses host gene expression, including that of type I interferon, in infected cells. J Virol 82:4471–4479

Neuman BW, Joseph JS, Saikatendu KS, Serrano P, Chatterjee A, Johnson MA, Liao L, Klaus JP, Yates JR 3 rd, Wuthrich K, Stevens RC, Buchmeier MJ, Kuhn P (2008) Proteomics analysis unravels the functional repertoire of coronavirus nonstructural protein 3. J Virol 82:5279–5294

Neuvonen M, Ahola T (2009) Differential activities of cellular and viral macro domain proteins in binding of ADP-ribose metabolites. J Mol Biol 385:212–225

Oostra M, te Lintelo EG, Deijs M, Verheije MH, Rottier PJ, de Haan CA (2007) Localization and membrane topology of coronavirus nonstructural protein 4: involvement of the early secretory pathway in replication. J Virol 81:12323–12336

Oostra M, Hagemeijer MC, van Gent M, Bekker CP, te Lintelo EG, Rottier PJ, De Haan CA (2008) Topology and membrane anchoring of the coronavirus replication complex: not all hydrophobic domains of nsp3 and nsp6 are membrane spanning. J Virol 82:12392–12405

Palmenberg AC (1990) Proteolytic processing of picornaviral polyprotein. Annu Rev Microbiol 44:603–623

Pasternak AO, Spaan WJM, Snijder EJ (2006) Nidovirus transcription: how to make sense...? J Gen Virol 87:1403–1421

Plotch SJ, Bouloy M, Ulmanen I, Krug RM (1981) A unique cap(m7G pppXm)-dependent influenza virion endonuclease cleaves capped RNAs to generate the primers that initiate viral RNA transcription. Cell 23:847–858

Prentice E, McAuliffe J, Lu X, Subbarao K, Denison MR (2004) Identification and characterization of severe acute respiratory syndrome coronavirus replicase proteins. J Virol 78:9977–9986

Putics A, Filipowicz W, Hall J, Gorbalenya AE, Ziebuhr J (2005) ADP-ribose-1″-monophosphatase: a conserved coronavirus enzyme that is dispensable for viral replication in tissue culture. J Virol 79:12721–12731

Putics A, Gorbalenya AE, Ziebuhr J (2006) Identification of protease and ADP-ribose 1″-monophosphatase activities associated with transmissible gastroenteritis virus non-structural protein 3. J Gen Virol 87:651–656

Ratia K, Saikatendu KS, Santarsiero BD, Barretto N, Baker SC, Stevens RC, Mesecar AD (2006) Severe acute respiratory syndrome coronavirus papain-like protease: structure of a viral deubiquitinating enzyme. Proc Natl Acad Sci USA 103:5717–5722

Ray D, Shah A, Tilgner M, Guo Y, Zhao Y, Dong H, Deas TS, Zhou Y, Li H, Shi PY (2006) West Nile virus 5′-cap structure is formed by sequential guanine N-7 and ribose 2'-O methylations by nonstructural protein 5. J Virol 80:8362–8370

Ricagno S, Egloff MP, Ulferts R, Coutard B, Nurizzo D, Campanacci V, Cambillau C, Ziebuhr J, Canard B (2006) Crystal structure and mechanistic determinants of SARS coronavirus nonstructural protein 15 define an endoribonuclease family. Proc Natl Acad Sci USA 103:11892–11897

Rota PA, Oberste MS, Monroe SS, Nix WA, Campagnoli R, Icenogle JP, Penaranda S, Bankamp B, Maher K, Chen MH, Tong S, Tamin A, Lowe L, Frace M, DeRisi JL, Chen Q, Wang D, Erdman DD, Peret TC, Burns C, Ksiazek TG, Rollin PE, Sanchez A, Liffick S, Holloway B, Limor J, McCaustland K, Olsen-Rasmussen M, Fouchier R, Gunther S, Osterhaus AD, Drosten C, Pallansch MA, Anderson LJ, Bellini WJ (2003) Characterization of a novel coronavirus associated with severe acute respiratory syndrome. Science 300:1394–1399

Saikatendu KS, Joseph JS, Subramanian V, Clayton T, Griffith M, Moy K, Velasquez J, Neuman BW, Buchmeier MJ, Stevens RC, Kuhn P (2005) Structural basis of severe acute respiratory syndrome coronavirus ADP-ribose-1″-phosphate dephosphorylation by a conserved domain of nsP3. Structure 13:1665–1675

Salonen A, Ahola T, Kaariainen L (2005) Viral RNA replication in association with cellular membranes. Curr Top Microbiol Immunol 285:139–173

Sawicki SG, Sawicki DL (1986) Coronavirus minus-strand RNA synthesis and effect of cycloheximide on coronavirus RNA synthesis. J Virol 57:328–334

Sawicki SG, Sawicki DL, Siddell SG (2007) A contemporary view of coronavirus transcription. J Virol 81:20–29

Schelle B, Karl N, Ludewig B, Siddell SG, Thiel V (2005) Selective replication of coronavirus genomes that express nucleocapsid protein. J Virol 79:6620–6630

Schiller JJ, Kanjanahaluethai A, Baker SC (1998) Processing of the coronavirus MHV-JHM polymerase polyprotein: identification of precursors and proteolytic products spanning 400 kilodaltons of ORF1a. Virology 242:288–302

Schütze H, Ulferts R, Schelle B, Bayer S, Granzow H, Hoffmann B, Mettenleiter TC, Ziebuhr J (2006) Characterization of White bream virus reveals a novel genetic cluster of nidoviruses. J Virol 80:11598–11609

Serrano P, Johnson MA, Almeida MS, Horst R, Herrmann T, Joseph JS, Neuman BW, Subramanian V, Saikatendu KS, Buchmeier MJ, Stevens RC, Kuhn P, Wuthrich K (2007) Nuclear magnetic resonance structure of the N-terminal domain of nonstructural protein 3 from the severe acute respiratory syndrome coronavirus. J Virol 81:12049–12060

Seybert A, Hegyi A, Siddell SG, Ziebuhr J (2000) The human coronavirus 229E superfamily 1 helicase has RNA and DNA duplex-unwinding activities with 5′-to-3′ polarity. RNA 6:1056–1068

Seybert A, Posthuma CC, van Dinten LC, Snijder EJ, Gorbalenya AE, Ziebuhr J (2005) A complex zinc finger controls the enzymatic activities of nidovirus helicases. J Virol 79:696–704

Shi ST, Lai MM (2005) Viral and cellular proteins involved in coronavirus replication. Curr Top Microbiol Immunol 287:95–131

Shuman S (2001) Structure, mechanism, and evolution of the mRNA capping apparatus. Prog Nucleic Acid Res Mol Biol 66:1–40

Snijder EJ, van Tol H, Roos N, Pedersen KW (2001) Non-structural proteins 2 and 3 interact to modify host cell membranes during the formation of the arterivirus replication complex. J Gen Virol 82:985–994

Snijder EJ, Bredenbeek PJ, Dobbe JC, Thiel V, Ziebuhr J, Poon LL, Guan Y, Rozanov M, Spaan WJ, Gorbalenya AE (2003) Unique and conserved features of genome and proteome of SARS-coronavirus, an early split-off from the coronavirus group 2 lineage. J Mol Biol 331:991–1004

Spaan W, Delius H, Skinner M, Armstrong J, Rottier P, Smeekens S, van der Zeijst BA, Siddell SG (1983) Coronavirus mRNA synthesis involves fusion of non-contiguous sequences. EMBO J 2:1839–1844

Sparks JS, Lu X, Denison MR (2007) Genetic analysis of Murine hepatitis virus nsp4 in virus replication. J Virol 81:12554–12563

Su D, Lou Z, Sun F, Zhai Y, Yang H, Zhang R, Joachimiak A, Zhang XC, Bartlam M, Rao Z (2006) Dodecamer structure of severe acute respiratory syndrome coronavirus nonstructural protein nsp10. J Virol 80:7902–7908

Sulea T, Lindner HA, Purisima EO, Menard R (2005) Deubiquitination, a new function of the severe acute respiratory syndrome coronavirus papain-like protease? J Virol 79:4550–4551

Sutton G, Fry E, Carter L, Sainsbury S, Walter T, Nettleship J, Berrow N, Owens R, Gilbert R, Davidson A, Siddell S, Poon LL, Diprose J, Alderton D, Walsh M, Grimes JM, Stuart DI (2004) The nsp9 replicase protein of SARS-coronavirus, structure and functional insights. Structure 12:341–353

Tan J, Verschueren KH, Anand K, Shen J, Yang M, Xu Y, Rao Z, Bigalke J, Heisen B, Mesters JR, Chen K, Shen X, Jiang H, Hilgenfeld R (2005) pH-dependent conformational flexibility of the SARS-CoV main proteinase (M(pro)) dimer: molecular dynamics simulations and multiple X-ray structure analyses. J Mol Biol 354:25–40

Tan J, Kusov Y, Mutschall D, Tech S, Nagarajan K, Hilgenfeld R, Schmidt CL (2007) The "SARS-unique domain" (SUD) of SARS coronavirus is an oligo(G)-binding protein. Biochem Biophys Res Commun 364:877–882

Tanner JA, Watt RM, Chai YB, Lu LY, Lin MC, Peiris JS, Poon LL, Kung HF, Huang JD (2003) The severe acute respiratory syndrome (SARS) coronavirus NTPase/helicase belongs to a distinct class of 5′ to 3′ viral helicases. J Biol Chem 278:39578–39582

Thiel V, Ivanov KA, Putics A, Hertzig T, Schelle B, Bayer S, Weissbrich B, Snijder EJ, Rabenau H, Doerr HW, Gorbalenya AE, Ziebuhr J (2003) Mechanisms and enzymes involved in SARS coronavirus genome expression. J Gen Virol 84:2305–2315

van Dinten LC, van Tol H, Gorbalenya AE, Snijder EJ (2000) The predicted metal-binding region of the arterivirus helicase protein is involved in subgenomic mRNA synthesis, genome replication, and virion biogenesis. J Virol 74:5213–5223

van Hemert MJ, van den Worm SH, Knoops K, Mommaas AM, Gorbalenya AE, Snijder EJ (2008) SARS-coronavirus replication/transcription complexes are membrane-protected and need a host factor for activity in vitro. PLoS Pathog 4:e1000054

Wathelet MG, Orr M, Frieman MB, Baric RS (2007) Severe acute respiratory syndrome coronavirus evades antiviral signaling: role of nsp1 and rational design of an attenuated strain. J Virol 81:11620–11633

Xu X, Liu Y, Weiss S, Arnold E, Sarafianos SG, Ding J (2003) Molecular model of SARS coronavirus polymerase: implications for biochemical functions and drug design. Nucleic Acids Res 31:7117–7130

Xu X, Zhai Y, Sun F, Lou Z, Su D, Xu Y, Zhang R, Joachimiak A, Zhang XC, Bartlam M, Rao Z (2006) New antiviral target revealed by the hexameric structure of mouse hepatitis virus nonstructural protein nsp15. J Virol 80:7909–7917

Xu Y, Cong L, Chen C, Wei L, Zhao Q, Xu X, Ma Y, Bartlam M, Rao Z (2009) Crystal structures of two coronavirus ADP-ribose-1″-monophosphatases and their complexes with ADP-Ribose: a systematic structural analysis of the viral ADRP domain. J Virol 83:1083–1092

Xue X, Yu H, Yang H, Xue F, Wu Z, Shen W, Li J, Zhou Z, Ding Y, Zhao Q, Zhang XC, Liao M, Bartlam M, Rao Z (2008) Structures of two coronavirus main proteases: implications for substrate binding and antiviral drug design. J Virol 82:2515–2527

Yang H, Yang M, Ding Y, Liu Y, Lou Z, Zhou Z, Sun L, Mo L, Ye S, Pang H, Gao GF, Anand K, Bartlam M, Hilgenfeld R, Rao Z (2003) The crystal structures of severe acute respiratory syndrome virus main protease and its complex with an inhibitor. Proc Natl Acad Sci USA 100:13190–13195

Yang H, Xie W, Xue X, Yang K, Ma J, Liang W, Zhao Q, Zhou Z, Pei D, Ziebuhr J, Hilgenfeld R, Yuen KY, Wong L, Gao G, Chen S, Chen Z, Ma D, Bartlam M, Rao Z (2005) Design of wide-spectrum inhibitors targeting coronavirus main proteases. PLoS Biol 3:e324

Zhai Y, Sun F, Li X, Pang H, Xu X, Bartlam M, Rao Z (2005) Insights into SARS-CoV transcription and replication from the structure of the nsp7-nsp8 hexadecamer. Nat Struct Mol Biol 12:980–986

Zhao Q, Li S, Xue F, Zou Y, Chen C, Bartlam M, Rao Z (2008) Structure of the main protease from a global infectious human coronavirus, HCoV-HKU1. J Virol 82:8647–8655

Zheng D, Chen G, Guo B, Cheng G, Tang H (2008) PLP2, a potent deubiquitinase from murine hepatitis virus, strongly inhibits cellular type I interferon production. Cell Res 18:1105–1113

Ziebuhr J (2005) The coronavirus replicase. Curr Top Microbiol Immunol 287:57–94

Ziebuhr J (2008) Coronavirus replicative proteins. In: Perlman S, Gallagher T, Snijder EJ (eds) Nidoviruses. ASM, Washington, DC, pp 65–81

Ziebuhr J, Snijder EJ (2007) The coronavirus replicase gene: special enzymes for special viruses. In: Thiel V (ed) Coronaviruses – molecular and cellular biology. Caister Academic, Norfolk, UK, pp 33–63

Ziebuhr J, Herold J, Siddell SG (1995) Characterization of a human coronavirus (strain 229E) 3C-like proteinase activity. J Virol 69:4331–4338

Ziebuhr J, Snijder EJ, Gorbalenya AE (2000) Virus-encoded proteinases and proteolytic processing in the Nidovirales. J Gen Virol 81:853–879

Ziebuhr J, Thiel V, Gorbalenya AE (2001) The autocatalytic release of a putative RNA virus transcription factor from its polyprotein precursor involves two paralogous papain-like proteases that cleave the same peptide bond. J Biol Chem 276:33220–33232

Zuo Y, Deutscher MP (2001) Exoribonuclease superfamilies: structural analysis and phylogenetic distribution. Nucleic Acids Res 29:1017–1026

Züst R, Cervantes-Barragan L, Kuri T, Blakqori G, Weber F, Ludewig B, Thiel V (2007) Coronavirus non-structural protein 1 is a major pathogenicity factor: implications for the rational design of coronavirus vaccines. PLoS Pathog 3:e109

Chapter 7
SARS Coronavirus Replicative Enzymes: Structures and Mechanisms

Isabelle Imbert, Rachel Ulferts, John Ziebuhr, and Bruno Canard

Abstract The SARS coronavirus (SARS-CoV) replicase gene encodes 16 non-structural proteins (nsps) with multiple enzymatic activities. Several of these enzymes are common components of replication complexes of other plus-strand RNA viruses, such as picornavirus 3C-like protease, papain-like protease, RNA-dependent RNA polymerase, RNA helicase, and ribose $2'$-*O*-methyltransferase activities, while others such as exoribonuclease, endoribonuclease, and adenosine diphosphate-ribose $1''$-phosphatase activities, are rarely or not conserved in viruses outside the order *Nidovirales*. The latter enzymes are believed to be involved in unique metabolic pathways used by coronaviruses to (1) replicate and transcribe their extremely large RNA genomes, and (2) interfere with cellular functions and antiviral host responses.

Since the global outbreak of SARS in 2003, major efforts have been made to elucidate the structures of the protein components of the SARS-CoV replication/transcription complex. Thus, in less than 5 years, the structures of as many as 16 SARS-CoV proteins or functional domains have been determined. Remarkably, eight of these 16 structures had novel folds, illustrating the uniqueness of the coronavirus replicative machinery. Furthermore, several new protein functions and potential drug targets have been identified in these studies. Current structural studies mainly focus on the few remaining proteins for which no structural information is available and larger protein complexes comprised of different nsps. The studies aim at obtaining detailed information on the functions and macromolecular assembly of the coronavirus replication/transcription machinery which, over a long period of time, may be used to develop selective antiviral drugs. This chapter reviews structural information on the SARS-CoV *macro* domain (ADRP) as well

B. Canard (✉)
Case 925 Ecole d'Ingénieurs de Luminy, Architecture et Fonction des Macromolécules Biologiques, Unité Mixte de Recherche 6098, Centre National de la Recherche Scientifique and Universités d'Aix-Marseille I et II, 163 Avenue de Luminy, 13288, Marseille Cedex 9, France
e-mail: Bruno.Canard@afmb.univ-mrs.fr

S.K. Lal (ed.), *Molecular Biology of the SARS-Coronavirus*,
DOI 10.1007/978-3-642-03683-5_7,

as nsps 7, 8, 9, and 15 and summarizes our current knowledge of active-site residues and intermolecular interactions of these proteins.

7.1 The ADP-Ribose-1″-Phosphatase Domain

The SARS coronavirus (SARS-CoV) ADP-ribose-1″-phosphatase (ADRP) is part of nonstructural protein nsp3, a large multidomain protein of 1,922 amino acids (Snijder et al. 2003; Thiel et al. 2003). The protein is thought to contain at least seven domains: (1) N-terminal Glu/Asp-rich domain (acidic domain, AD); (2) ADRP domain (also called *macro* domain or X domain); (3) SUD (for "SARS-CoV unique domain," not conserved in other coronaviruses); (4) papain-like protease (PL^{pro}), which cleaves the viral polyproteins at three N-proximal sites and also has deubiquitinating activity (Barretto et al. 2005; Harcourt et al. 2004; Lindner et al. 2005; Ratia et al. 2006; Thiel et al. 2003); (5) an uncharacterized domain possibly extending the papain-like protease domain (termed PLnc for papain-like noncanonical); (6) a transmembrane domain (Kanjanahaluethai et al. 2007) in the N-terminal region of the Y domain (Ziebuhr et al. 2001); and (7) the remainder of the Y domain which includes a number of conserved metal-binding residues.

Macro domains are conserved in a wide variety of bacteria, archaea and eukaryotes. The name *macro* domain refers to the early finding that members of this group of proteins are related to the nonhistone domain of the histone macroH2A (Pehrson and Fried 1992). *Macro* domains are also conserved in a number of positive-strand RNA viruses, including specific genera of the *Nidovirales*, members of the *Togaviridae*, Rubella virus and Hepatitis E virus (HEV) (Gorbalenya et al. 1991; Snijder et al. 2003). The biological role of *macro* domains in the life cycle of positive-strand RNA viruses has not been resolved. ADRP is a side-product of cellular pre-tRNA splicing, a pathway seemingly unrelated to viral RNA replication. ADRP activity was first identified for a *macro* domain homolog from yeast using a proteomics approach (Martzen et al. 1999) and subsequently demonstrated for other related proteins including homologs from three coronaviruses, HCoV-229E, TGEV, and SARS-CoV, as well as alphaviruses and HEV (Egloff et al. 2006; Putics et al. 2005, 2006). Inactivation of *macro* domain-associated ADRP activity in HCoV-229E did not affect viral genome replication and subgenomic RNA synthesis in cell culture, suggesting that the activity, which is conserved across members of the *Coronaviridae*, may have important functions *in vivo*, for example in evading antiviral host responses (Putics et al. 2005).

The structure of the SARS-CoV *macro* domain has been determined at 1.8 Å resolution, both in the apo form and in complex with ADP-ribose (Egloff et al. 2006; Saikatendu et al. 2005) (Fig. 7.1). To date, the catalytic residues of the coronavirus *macro* domain have not been well defined. Only one absolutely conserved Asn residue (N41, Fig. 7.1a) could be implicated in ADRP activity whereas other residues in the vicinity of the catalytic center are not conserved and therefore less likely to be critically involved in catalysis. Similar to archaebacterial *macro*

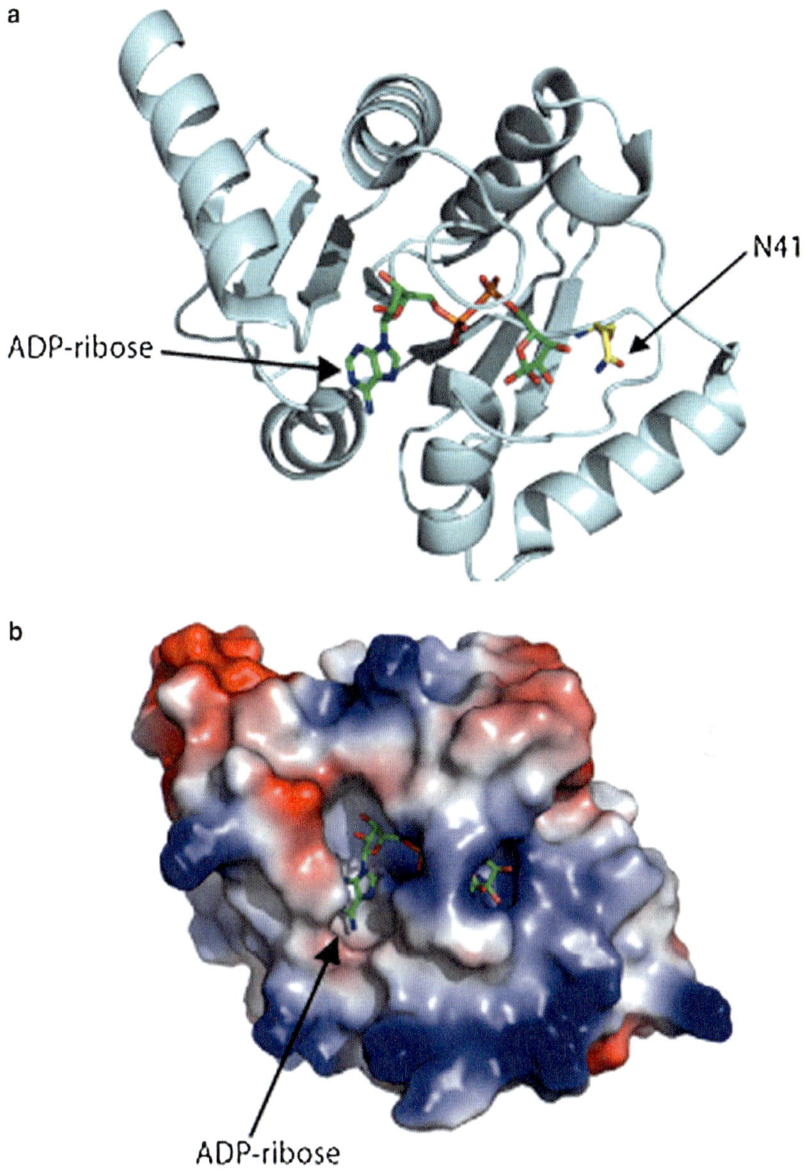

Fig. 7.1 Crystal structure of the SARS-CoV nsp3 *macro* domain (from Egloff et al. 2006; PDB 2FAV). (**a**) Ribbon representation of the SARS-CoV *macro* domain in complex with ADP-ribose. Shown is the only conserved residue whose mutation abrogates activity in all homologs studied so far (N41). (**b**) Surface potential analysis (*blue*, positive charge; *red*, negative charge) of the whole domain showing the presumed active site accommodating an ADP-ribose molecule. Images were generated using PYMOL

domains, SARS-CoV and several other RNA virus *macro* domains were shown to have poly(ADP-ribose)-binding activities (Egloff et al. 2006). A possible binding mode for poly(ADP-ribose) which does not appear to involve a charged groove (Fig. 7.1b) has been modeled by Egloff et al. (2006). The significance of the observed poly(ADP-ribose)-binding activity and the biologically relevant substrates of RNA virus *macro* domains are currently not clear and remain to be studied in more detail.

7.2 The nsp7–nsp8–nsp9–nsp10 Cistron

The nsp7 to nsp10 proteins are encoded by the 3′-terminal ORF1a region. The proteins are highly conserved amongst members of the genus *Coronavirus* and, albeit to a lesser extent, members of other genera within the *Coronaviridae* family. Interestingly, the proteins encoded in the equivalent 3′-terminal ORF1a region of arterivirus genomes do not appear to be closely related to the coronavirus nsp7 to nsp10 proteins (Pasternak et al. 2006), which may indicate differences in the structures, functions and subunit compositions of replication/transcription complexes between the various families within the order *Nidovirales*. Using a reverse genetics approach, nsp 7, 8, 9, and 10 were shown to be essential for replication of murine hepatitis virus (MHV) in cell culture (Deming et al. 2007). Deletion of any of the four proteins abolished RNA synthesis and, consequently, production of infectious virus progeny (Deming et al. 2007). The precise functions of the proteins in viral replication remain to be characterized.

7.2.1 The nsp7 Protein

Nsp7 is a small protein of about 10 kDa that is well conserved in coronaviruses but has no detectable homolog outside the *Coronaviridae*. The SARS-CoV nsp7 structure was solved by NMR (Peti et al. 2005). The protein features a single domain with a novel fold comprised of five helical secondary structures (Fig. 7.2a) whose mutual positions are stabilized by a number of interhelical side-chain interactions. The residues involved in these interactions are predominantly hydrophobic; they form two interdigitated layers that hold the helices together and thus stabilize the fold. The surface charge distribution is asymmetrical (Fig. 7.2b) and both surfaces may be involved in protein–protein interactions.

In infected cells, coronavirus nsp7 and/or nsp7-containing precursors localize to cytoplasmic membrane structures which are thought to be the sites of viral replication (Bost et al. 2000; Ng et al. 2001). SARS-CoV nsp7 was shown to dimerize and interact with nsp5, nsp8 (see below), nsp9 and nsp13 (von Brunn et al. 2007; Zhai et al. 2005). Furthermore, the MHV nsp7 was shown to interact specifically with nsp1 and nsp10 (Brockway et al. 2004).

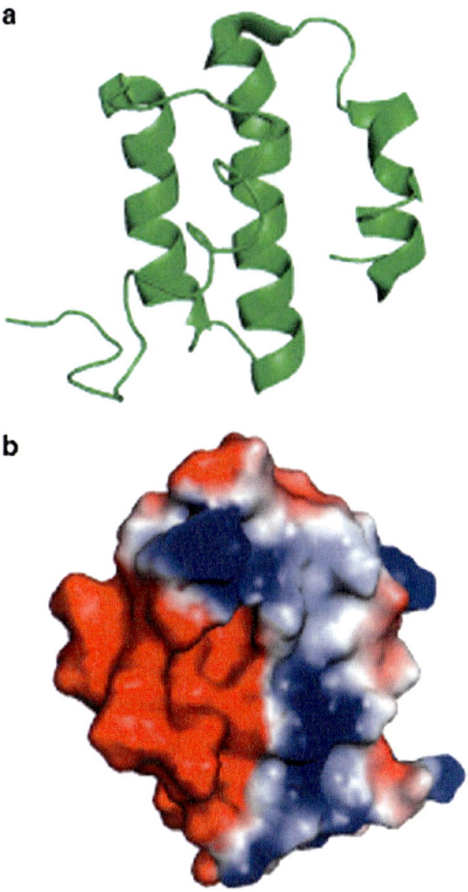

Fig. 7.2 Views of the solution structure of SARS-CoV nsp7 (from Peti et al. 2005; PDB 1YSY). (**a**) Ribbon presentation of the SARS-CoV nsp7 NMR structure. (**b**) The surface shows the electrostatic potential (*blue*, positive charge; *red*, negative charge). Images were generated using PYMOL

7.2.2 The nsp8 Protein

The protein has a molecular mass of about 22 kDa and is conserved among the various genera of the family *Coronaviridae*. It has interesting functional and structural properties. Initial crystallization trials of bacterially expressed SARS-CoV nsp8 remained unsuccessful until 2005, when the crystal structure of SARS-CoV nsp8 in complex with SARS-CoV nsp7 was solved (Zhai et al. 2005). The crystal structure revealed a hexadecameric complex composed of eight molecules of nsp8 and eight molecules of nsp7 (Fig. 7.3a). This so-called nsp7–nsp8 super-complex has an intricate hollow cylindrical structure with an inner diameter of about 30 Å. Most of the nsp8 residues that face the interior of the channel are highly conserved among coronaviruses. The inner dimensions and electrostatic properties enable the cylindrical nsp7–nsp8 structure to encircle and interact with nucleic acid. Interactions with dsRNA were demonstrated for nsp8 (but not nsp7) by

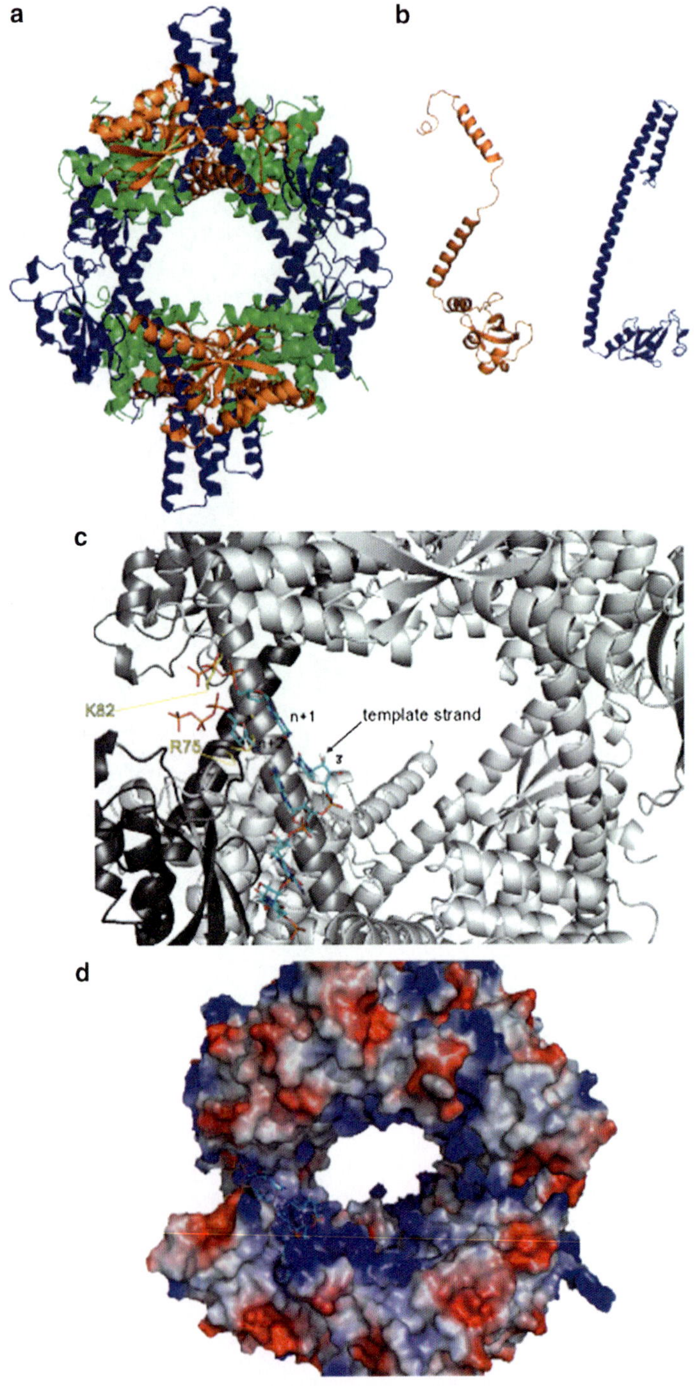
a
b
c
K82
R75
n+1
n+2
template strand
3'
d

electrophoretic mobility shift assays and critical residues involved in RNA binding were identified by mutagenesis.

The α-helical fold seen in the previously reported NMR solution structure of nsp7 (Peti et al. 2005) was also seen in the crystal structure, where nsp7 was part of a complex with Nsp8. nsp8 was found to possess a novel fold with two slightly different conformations. One of the conformations was described as a "golf-club"-like structure composed of an N-terminal α-helical "shaft" domain and a C-terminal, mixed α/β "head" domain (Fig. 7.3b). The second conformation was described as resembling a golf club with a bent shaft, whereas the rest of the structure (particularly, the head domain) is very similar to the first structure (Fig. 7.3b). The eight nsp8 molecules constitute the framework of a large protein complex made up of "bricks" (nsp8) and "mortar" (nsp7), with nsp7 filling some of the remaining space between individual nsp8 molecules, thereby stabilizing the structure of the complex. Despite a number of intricate interactions between nsp7 and nsp8, the absence of nsp7 is not generally thought to markedly change the overall shape of the structure. On the basis of the architecture and electrostatic properties of the complex it was suggested that nsp7–nsp8 might function as a processivity factor, similar to bacterial (β_2-clamp) or eukaryotic DNA polymerase processivity factors, such as PCNA (proliferating cell nuclear antigen).

This hypothesis was supported and significantly extended by data demonstrating that SARS-CoV nsp8 represents a second, noncanonical RNA-dependent RNA polymerase (RdRp) (Imbert et al. 2006). A recombinant form of the SARS-CoV nsp8 proved to be capable of synthesizing short oligonucleotides (<6 residues). It had relatively low fidelity and strongly preferred RNA with a 5′-(G/U)CC-3′ trinucleotide consensus sequence as a template. Three charged/polar residues, Lys-58, Lys-82 and Ser-85, all of them located in the N-terminal domain, were found to be essential for activity and may be part of a network of residues that catalyzes the phosphoryl transfer reaction. A structure and activity-based model of the nsp8-associated RdRp activity is presented in Fig. 7.3c,d. The structure of the C-terminal "head" domain of nsp8 appears to be distantly related to the catalytic palm subdomain of RNA virus RdRps. Moreover, it was shown that the "canonical" coronavirus RdRp domain residing in nsp12 contains a conserved sequence, called motif G, that is usually found in primer-dependent RNA polymerases (Gorbalenya et al. 2002). Taken together, the available information suggests that nsp8 may function as a primase to catalyze the synthesis of RNA primers to be used by the primer-dependent coronavirus nsp12 RdRp.

Fig. 7.3 Different views of the SARS-CoV nsp7–nsp8 supercomplex (from Zhai et al. 2005; PDB 2AHM). (**a**) Structure of the nsp7–nsp8 hexadecamer supercomplex. Nsp7 and the two conformations of nsp8 are colored *green*, *blue*, and *orange*, respectively. (**b**) The two alternative nsp8 conformations. (**c**) Model of two nsp8 monomers in complex with a ssRNA template (5′-UAGC-3′) and two nucleotides (GTP and CTP). RNA template, GTP, and CTP are shown by a stick model. Two amino acid residues, Lys-58 and Arg-75, whose substitution abolished activity are represented in *yellow*. The two first NTPs incorporated (GTP in +1 and CTP in +2) are indicated. (**d**) The surface is colored according to electrostatic potential (*blue*, positive charge; *red*, negative charge). Images were generated using PYMOL

The presumed central role of nsp8 in coronavirus replication gains additional support by data suggesting that nsp8 interacts with a large number of SARS-CoV replicative proteins, including nsp2, nsp5 (3CL^{pro}), nsp6, nsp7, nsp8, nsp9, nsp12 (RdRp), nsp13 (helicase), and nsp14 (exoribonuclease) (Imbert et al. 2008; von Brunn et al. 2007). Finally, an elegant MHV reverse genetics study provided evidence to suggest interactions between nsp8 (and also nsp9) with the 3′untranslated region, which is consistent with the proposed role of these proteins in coronavirus RNA synthesis, possibly the initiation of minus-strand RNA synthesis (Zust et al. 2008).

7.2.3 The nsp9 Protein

Crystal structures of nsp9, a protein of about 13 kDa, were solved simultaneously in two laboratories in 2004, one to a resolution of 2.7 Å (Egloff et al. 2004) and the other to 2.8 Å (Sutton et al. 2004). The studies consistently established that nsp9 is a single-strand RNA/DNA binding protein. The structure of SARS-CoV nsp9 has a central core comprised of seven β-strands. The core is flanked by a C-terminal α-helix and an N-terminal extension (Fig. 7.4a). X-ray crystallography, dynamic light scattering, analytical ultracentrifugation and GST pull-down experiments indicate that nsp9 forms dimers in solution (Campanacci et al. 2003; Imbert et al. 2008; Sutton et al. 2004). RNA binding by nsp9 has been suggested to involve the loops of the β-barrel domain while the C-terminal β-hairpin and helix, which are well conserved across coronaviruses, are probably involved in dimerization and interactions with other proteins. Database searches did not reveal structural homologs for nsp9 (Egloff et al. 2004). However, the short six-stranded β-barrel of nsp9 includes an open five-stranded barrel that is reminiscent of the five-stranded β-barrel structure of oligosaccharide/oligonucleotide-binding (OB)-fold proteins. About two-thirds of the proteins belonging to this protein superfamily are nucleic acid-binding proteins (Arcus 2002). As in OB-fold proteins, SARS-CoV nsp9 appears to bind nucleic acids by using a network of positively charged amino acids for binding the phosphate backbone of the substrate, whereas several exposed aromatic residues probably make additional stacking interactions with nucleobases (Fig. 7.4b). The nucleic acid-binding properties of nsp9 were subsequently confirmed by surface plasmon resonance, fluorescence quenching studies (Egloff et al. 2004) and electrophoretic mobility shift assays (Sutton et al. 2004). To date, there is no evidence to suggest that the nucleic acid-binding activity of nsp9 is sequence-specific.

SARS-CoV nsp9 was shown to interact with nsp6, nsp7 and nsp8 (von Brunn et al. 2007). Nsp9/nsp8 interactions were confirmed by a number of methods, including analytical ultracentrifugation, yeast two-hybrid data, GST pull-down and co-immunoprecipitation experiments, suggesting quite stable interactions between the two proteins. Moreover, the structural disorder of the nsp8 N-terminal region in solution has been reported to decrease when nsp9 was added to a solution containing nsp8 (Sutton et al. 2004). On the basis of the nsp7–nsp8 hexadecameric

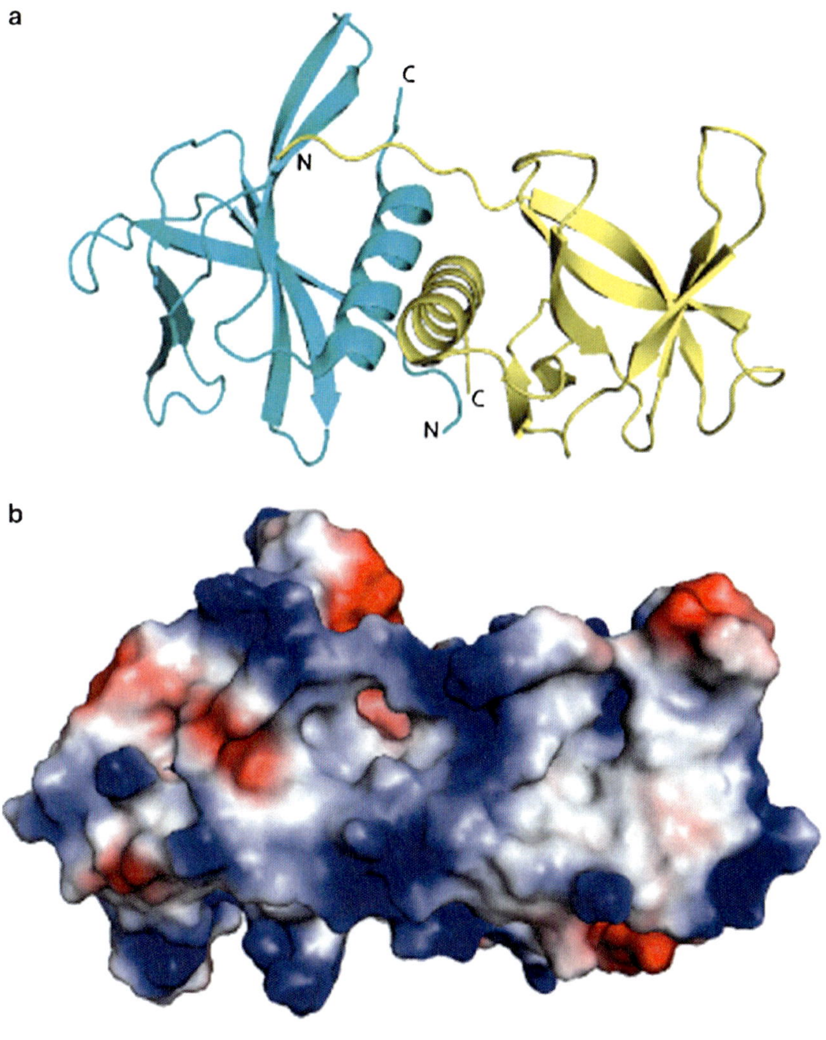

Fig. 7.4 Ribbon representation of SARS-CoV nsp9 (from Egloff et al. 2004; PDB 1QZ8). (**a**) One molecule of the dimer is *yellow* and the other is *blue*. (**b**) Electrostatic surface potential of nsp9 is colored according to electrostatic potential (*blue*, positive charge; *red*, negative charge). Images were generated using PYMOL

structure (see Sect. 7.2.2), it was suggested that the most probable nsp9 binding site is formed by the N-terminal 50 residues of nsp8. This part of the structure is close to the entrance of the central pore and appears to be quite flexible, as indicated by the lack of electron density for these residues. Both the nsp9 structure and tryptophan fluorescence quenching data suggest that ssRNA is wrapped around the nsp9 dimer (Egloff et al. 2004). It has been speculated that nsp9 dimers bind to single-stranded nascent and template strands as they emerge from the channel of the nsp7–nsp8 complex at a time when stable secondary structures have not yet formed, thereby protecting ssRNAs from ribonucleolytic cleavage.

7.2.4 The nsp10 Protein

Nsp10 is well conserved among coronaviruses. In the polyprotein 1a, the protein is located upstream of nsp11, a short 13-residue peptide encoded by the 3′-terminal nucleotides of ORF1a. Nsp10 is thought to have an essential role in viral replication. It was shown that a MHV temperature-sensitive mutant carrying a nonsynonymous mutation in the nsp10 coding sequence has a defect in minus-strand RNA synthesis at the nonpermissive temperature (Sawicki et al. 2005; Siddell et al. 2001). Crystallization of nsp10 revealed monomers and homodimers (Joseph

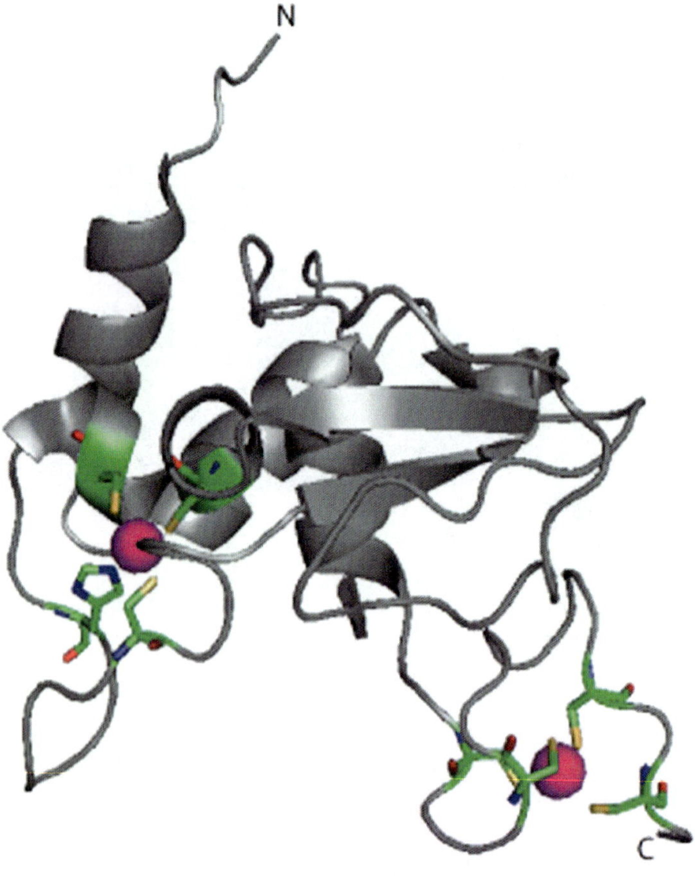

Fig. 7.5 Ribbon diagram of SARS-CoV nsp10 (from Joseph et al. 2006; PDB 2FYG). Residues coordinating the two Zn^{2+} ions are shown as sticks and pink balls, respectively. Atoms are colored as follows: *green*, carbon; *blue*, nitrogen; *red*, oxygen; *orange*, sulfur

et al. 2006) whereas a complex dodecameric structure was observed when nsp10 was expressed and crystallized as a fusion with nsp11 (Su et al. 2006). The nsp10 monomer consists of a pair of antiparallel N-terminal helices stacked against an irregular β-sheet, a coil-rich C-terminus, and two zinc fingers (Fig. 7.5). Nsp10 represents a novel fold and is the first structural representative of this family of zinc finger proteins. The zinc finger motifs are strictly conserved in coronaviruses, supporting their essential role in viral replication.

It is currently unclear whether or not the nsp10 dodecameric structure seen in one of the structural studies is of biological relevance. The relevance has been questioned for several reasons. First, site-directed mutagenesis of the nsp10–nsp11 cleavage site in the MHV genome generated nonviable viruses, indicating that $3CL^{pro}$-mediated cleavage at this site is essential for viral replication (Deming et al. 2007). Second, the monomer structure has an intact second zinc finger which appears to stabilize the structure of the C-terminal tail of nsp10. By contrast, in the dodecamer structure, the second zinc finger lacks the C-proximal cysteine residue, resulting in local disorder at the nsp10 carboxyl terminus. Finally, substitutions of residues predicted to be crucial for the dodecamer formation did not cause a lethal phenotype in MHV (Donaldson et al. 2007).

Gel shift assays indicate that nsp10 binds single- and double-stranded RNA and DNA with low affinity and without obvious sequence specificity (Joseph et al. 2006). SARS-CoV nsp10 was shown to interact with nsp9 (Su et al. 2006). It also interacts strongly with nsp14 and nsp16 and, to a lesser extent, with nsp8 (Imbert et al. 2008). The precise role of nsp10 within the coronavirus replication/transcription complex remains to be identified.

7.3 The Endoribonuclease, nsp15

The uridylate-specific endoribonuclease (nsp15 or NendoU) is conserved in all members of the *Nidovirales* but no other virus, which makes the protein a major genetic marker of this group of viruses (Gorbalenya et al. 2006; Ivanov et al. 2004a). Distantly-related homologs of NendoU were also identified in some prokaryotes and eukaryotes, where they form a small protein family prototyped by XendoU, a *Xenopus laevis* endoribonuclease involved in the nucleolytic processing of intron-encoded box C/D U16 small nucleolar RNAs (Laneve et al. 2003; Snijder et al. 2003). Mainly on the basis of this sequence relationship, it has been speculated that NendoU may (also) have cellular substrates and it should be interesting to investigate whether NendoU is involved in producing small regulatory RNAs (similar to the small nucleolar RNAs produced by XendoU). Nsp15 activity is significantly stimulated by manganese ions and the enzyme generates 2′–3′ cyclic phosphate ends (Ivanov et al. 2004a). It is generally accepted that nsp15 functions as a homohexamer, even though the enzyme has some activity as a monomer (Guarino et al. 2005; Ivanov et al. 2004a; Joseph et al. 2007; Ricagno et al. 2006). NendoU cleaves downstream of uridylates, both in single and double-stranded RNA. The sequence

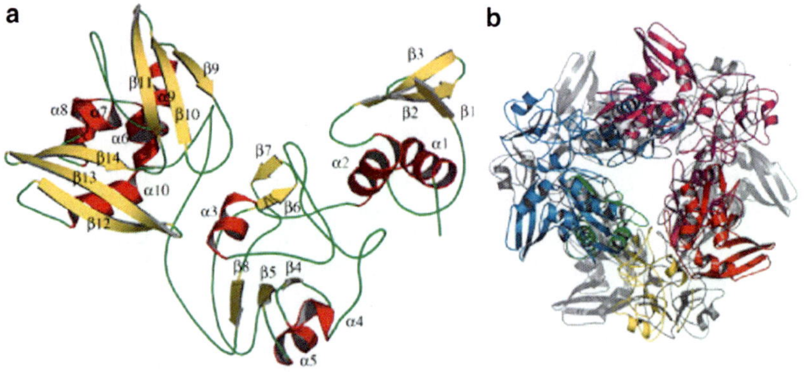

Fig. 7.6 Structure of SARS-CoV nsp15 endoribonuclease (from Ricagno et al. 2006; PDB 2H85). (**a**) Cartoon representation of the nsp15 monomer. The structure consists of three domains: N-terminal domain (α1–α2), central domain (α3–β8), and C-terminal domain (α6–β14). Secondary structures are colored as follows: *red*, α-helices; *yellow*, β-sheets; and *green*, loops. (**b**) A view of the nsp15 hexamer. In the trimer shown in the foreground, the subdomains of one of the monomers are colored as follows: *green*, N-terminal domain; *yellow*, central domain; and *red*, C-terminal domain. The other two molecules are shown in *blue* and *magenta*

context of the uridylate has been shown to affect cleavage efficiency and differential cleavage efficiencies have been reported for base-paired and nonbase-paired uridylates, respectively (Bhardwaj et al. 2006; Ivanov et al. 2004a).

Nsp15 crystal structures have been reported for SARS-CoV (Fig. 7.6) (Joseph et al. 2007; Ricagno et al. 2006; Bhardwaj et al. 2008) and MHV (Xu et al. 2006). Nsp15 exhibits a unique fold and assembles into a toric hexameric structure (Fig. 7.6b) with a central pore and six potentially active catalytic sites at the periphery. Unlike the situation in the nsp7–nsp8 complex, the diameter of the central pore seen in the nsp15 hexamer is too small to accommodate RNA. Furthermore, this part of the structure does not appear to interact with RNA (Bhardwaj et al. 2006; Ricagno et al. 2006). Each protomer contains nine α-helices and 21 β-strands (Fig. 7.6a). Alanine substitutions of highly conserved residues demonstrated that the C-terminal domain contains the active site. There are striking similarities between the active site residues of nsp15 and RNase A, suggesting that the enzymes use common catalytic mechanisms, although they are not related in their tertiary structures (Ricagno et al. 2006). The general acid–base mechanism used by coronavirus NendoUs probably involves His-234, His-249, and Lys-289 (SARS-CoV nsp15 numbering). Additional conserved residues likely involved in substrate specificity have been identified by X-ray crytallography and site-directed mutagenesis (Bhardwaj et al. 2008; Ricagno et al. 2006).

The overall architecture of coronavirus NendoUs suggests that the protein might be tethered to other partners connecting this hexameric assembly to the viral replication complex and, potentially, cellular structures. Consistent with this idea, NendoU has been implicated in coronavirus and arterivirus RNA synthesis and/or the production of virus progeny (Ivanov et al. 2004a; Kang et al. 2007; Posthuma et al. 2006) and interactions between SARS-CoV nsp15 and nsp3, nsp8, nsp9, and nsp12 have been reported (Imbert et al. 2008). Although endonucleases are ubiquitous enzymes that are involved in many aspects of RNA metabolism, they are extremely rare in the RNA virus world, with only very few exceptions (e.g., in orthomyxo-, pesti-, and retroviruses). Although the precise role of the enzyme in the viral life cycle and its biologically relevant substrates remain to be identified, it seems reasonable to think that NendoU acts as part of a larger protein complex and that nucleolytic activity and substrate specificity are tightly regulated by additional factors to minimize uncontrolled nucleolytic cleavage of viral and cellular RNAs.

7.4 Conclusion

The SARS outbreak in 2003 sparked significant new interest in the molecular biology and biochemistry of coronavirus replication. A multitude of structural and functional studies into SARS-CoV replicative enzymes have been published over the past few years, giving a major boost to coronavirus and, more generally, nidovirus research. Interestingly, in several instances, crystal structures of coronavirus proteins have uncovered new biochemical and enzymatic activities, thus opening up new vistas for future studies and stimulating exciting biochemical and genetic studies into the biological functions of specific coronavirus nonstructural proteins. Also, the SARS-CoV structural work has been extremely rewarding for crystallographers interested in novel folds and original enzymes. There are still major challenges ahead, mainly with respect to ORF1b-encoded proteins. For example, there is still no crystal structure available for the special type of primer-dependent RNA polymerases encoded by coronaviruses. Also studies of the zinc-binding domain-containing superfamily 1 helicase (Ivanov et al. 2004b) and 3′–5′ exonuclease (Minskaia et al. 2006) domains residing in nsp13 and nsp14, respectively, and the functionally interesting nsp16 2′-*O*-methyltransferase (Decroly et al. 2008) promise to reveal exciting new structural and functional insight. Undoubtedly, these upcoming structures will greatly stimulate coronavirus (nidovirus) research and perhaps even reveal novel and truly unique RNA processing pathways involved in the replication of these viruses.

Acknowledgment The work of I.I. and B.C. was supported by the Structural Proteomics in Europe (SPINE) project of the European Union 5th Framework Research Program (Grant QLRT-2001-00988) and subsequently by the Euro-Asian SARS-DTV Network (SP22-CT-2004-511064) from the European Commission Specific Research and Technological Development Program "Integrating and Strengthening the European Research Area" and by Interdisciplinary CNRS 2007 Program "Infectious Emerging Diseases." The work of R.U. and J.Z. has been supported by the Sino-German Centre for Research Promotion, the German Research Council and the Queen's University of Belfast.

References

Arcus V (2002) OB-fold domains: a snapshot of the evolution of sequence, structure and function. Curr Opin Struct Biol 12:794–801

Barretto N, Jukneliene D, Ratia K, Chen Z, Mesecar AD, Baker SC (2005) The papain-like protease of severe acute respiratory syndrome coronavirus has deubiquitinating activity. J Virol 79:15189–15198

Bhardwaj K, Sun J, Holzenburg A, Guarino LA, Kao CC (2006) RNA recognition and cleavage by the SARS coronavirus endoribonuclease. J Mol Biol 361:243–256

Bhardwaj K, Palaninathan S, Alcantara JM, Yi LL, Guarino L, Sacchettini JC, Kao CC (2008) Structural and functional analyses of the severe acute respiratory syndrome coronavirus endoribonuclease Nsp15. J Biol Chem 283:3655–3664

Bost AG, Carnahan RH, Lu XT, Denison MR (2000) Four proteins processed from the replicase gene polyprotein of mouse hepatitis virus colocalize in the cell periphery and adjacent to sites of virion assembly. J Virol 74:3379–3387

Brockway SM, Lu XT, Peters TR, Dermody TS, Denison MR (2004) Intracellular localization and protein interactions of the gene 1 protein p28 during mouse hepatitis virus replication. J Virol 78:11551–11562

Campanacci V, Egloff MP, Longhi S, Ferron F, Rancurel C, Salomoni A, Durousseau C, Tocque F, Bremond N, Dobbe JC, Snijder EJ, Canard B, Cambillau C (2003) Structural genomics of the SARS coronavirus: cloning, expression, crystallization and preliminary crystallographic study of the Nsp9 protein. Acta Crystallogr D Biol Crystallogr 59:1628–1631

Decroly E, Imbert I, Coutard B, Bouvet M, Selisko B, Alvarez K, Gorbalenya AE, Snijder EJ, Canard B (2008) Coronavirus nonstructural protein 16 is a cap-0 binding enzyme possessing (nucleoside-2'O)-methyltransferase activity. J Virol 82(16):8071–8084

Deming DJ, Graham RL, Denison MR, Baric RS (2007) Processing of open reading frame 1a replicase proteins nsp7 to nsp10 in murine hepatitis virus strain A59 replication. J Virol 81:10280–10291

Donaldson EF, Sims AC, Graham RL, Denison MR, Baric RS (2007) Murine Hepatitis Virus replicase protein nsp10 is a critical regulator of viral RNA synthesis. J Virol 81(12):6356–6368

Egloff MP, Ferron F, Campanacci V, Longhi S, Rancurel C, Dutartre H, Snijder EJ, Gorbalenya AE, Cambillau C, Canard B (2004) The severe acute respiratory syndrome-coronavirus replicative protein nsp9 is a single-stranded RNA-binding subunit unique in the RNA virus world. Proc Natl Acad Sci USA 101:3792–3796

Egloff MP, Malet H, Putics A, Heinonen M, Dutartre H, Frangeul A, Gruez A, Campanacci V, Cambillau C, Ziebuhr J, Ahola T, Canard B (2006) Structural and functional basis for ADP-ribose and poly(ADP-ribose) binding by viral macro domains. J Virol 80:8493–8502

Gorbalenya AE, Koonin EV, Lai MM (1991) Putative papain-related thiol proteases of positive-strand RNA viruses. Identification of rubi- and aphthovirus proteases and delineation of a novel conserved domain associated with proteases of rubi-, alpha- and coronaviruses. FEBS Lett 288:201–205

Gorbalenya AE, Pringle FM, Zeddam JL, Luke BT, Cameron CE, Kalmakoff J, Hanzlik TN, Gordon KH, Ward VK (2002) The palm subdomain-based active site is internally permuted in viral RNA-dependent RNA polymerases of an ancient lineage. J Mol Biol 324:47–62

Gorbalenya AE, Enjuanes L, Ziebuhr J, Snijder EJ (2006) Nidovirales: evolving the largest RNA virus genome. Virus Res 117:17–37

Guarino LA, Bhardwaj K, Dong W, Sun J, Holzenburg A, Kao C (2005) Mutational analysis of the SARS virus Nsp15 endoribonuclease: identification of residues affecting hexamer formation. J Mol Biol 353:1106–1117

Harcourt BH, Jukneliene D, Kanjanahaluethai A, Bechill J, Severson KM, Smith CM, Rota PA, Baker SC (2004) Identification of severe acute respiratory syndrome coronavirus replicase products and characterization of papain-like protease activity. J Virol 78:13600–13612

Imbert I, Guillemot JC, Bourhis JM, Bussetta C, Coutard B, Egloff MP, Ferron F, Gorbalenya AE, Canard B (2006) A second, non-canonical RNA-dependent RNA polymerase in SARS coronavirus. Embo J 25:4933–4942

Imbert I, Snijder EJ, Dimitrova M, Guillemot JC, Lecine P, Canard B (2008) The SARS-Coronavirus PLnc domain of nsp3 as a replication/transcription scaffolding protein. Virus Res 133:136–148

Ivanov KA, Hertzig T, Rozanov M, Bayer S, Thiel V, Gorbalenya AE, Ziebuhr J (2004a) Major genetic marker of nidoviruses encodes a replicative endoribonuclease. Proc Natl Acad Sci USA 101:12694–12699

Ivanov KA, Thiel V, Dobbe JC, van der Meer Y, Snijder EJ, Ziebuhr J (2004b) Multiple enzymatic activities associated with severe acute respiratory syndrome coronavirus helicase. J Virol 78:5619–5632

Joseph JS, Saikatendu KS, Subramanian V, Neuman BW, Brooun A, Griffith M, Moy K, Yadav MK, Velasquez J, Buchmeier MJ, Stevens RC, Kuhn P (2006) Crystal structure of nonstructural protein 10 from the severe acute respiratory syndrome coronavirus reveals a novel fold with two zinc-binding motifs. J Virol 80:7894–7901

Joseph JS, Saikatendu KS, Subramanian V, Neuman BW, Buchmeier MJ, Stevens RC, Kuhn P (2007) Crystal structure of a monomeric form of severe acute respiratory syndrome coronavirus endonuclease nsp15 suggests a role for hexamerization as an allosteric switch. J Virol 81:6700–6708

Kang H, Bhardwaj K, Li Y, Palaninathan S, Sacchettini J, Guarino L, Leibowitz JL, Kao CC (2007) Biochemical and genetic analyses of murine hepatitis virus Nsp15 endoribonuclease. J Virol 81:13587–13597

Kanjanahaluethai A, Chen Z, Jukneliene D, Baker SC (2007) Membrane topology of murine coronavirus replicase nonstructural protein 3. Virology 361:391–401

Laneve P, Altieri F, Fiori ME, Scaloni A, Bozzoni I, Caffarelli E (2003) Purification, cloning, and characterization of XendoU, a novel endoribonuclease involved in processing of intron-encoded small nucleolar RNAs in Xenopus laevis. J Biol Chem 278:13026–13032

Lindner HA, Fotouhi-Ardakani N, Lytvyn V, Lachance P, Sulea T, Menard R (2005) The papain-like protease from the severe acute respiratory syndrome coronavirus is a deubiquitinating enzyme. J Virol 79:15199–15208

Martzen MR, McCraith SM, Spinelli SL, Torres FM, Fields S, Grayhack EJ, Phizicky EM (1999) A biochemical genomics approach for identifying genes by the activity of their products. Science 286:1153–1155

Minskaia E, Hertzig T, Gorbalenya AE, Campanacci V, Cambillau C, Canard B, Ziebuhr J (2006) Discovery of an RNA virus 3′-> 5′ exoribonuclease that is critically involved in coronavirus RNA synthesis. Proc Natl Acad Sci USA 103:5108–5113

Ng LF, Xu HY, Liu DX (2001) Further identification and characterization of products processed from the coronavirus avian infectious bronchitis virus (IBV) 1a polyprotein by the 3C-like proteinase. Adv Exp Med Biol 494:291–298

Pasternak AO, Spaan WJ, Snijder EJ (2006) Nidovirus transcription: how to make sense...? J Gen Virol 87:1403–1421

Pehrson JR, Fried VA (1992) MacroH2A, a core histone containing a large nonhistone region. Science 257:1398–1400

Peti W, Johnson MA, Herrmann T, Neuman BW, Buchmeier MJ, Nelson M, Joseph J, Page R, Stevens RC, Kuhn P, Wuthrich K (2005) Structural genomics of the severe acute respiratory syndrome coronavirus: nuclear magnetic resonance structure of the protein nsP7. J Virol 79:12905–12913

Posthuma CC, Nedialkova DD, Zevenhoven-Dobbe JC, Blokhuis JH, Gorbalenya AE, Snijder EJ (2006) Site-directed mutagenesis of the Nidovirus replicative endoribonuclease NendoU exerts pleiotropic effects on the arterivirus life cycle. J Virol 80:1653–1661

Putics A, Filipowicz W, Hall J, Gorbalenya AE, Ziebuhr J (2005) ADP-ribose-1″-monophosphatase: a conserved coronavirus enzyme that is dispensable for viral replication in tissue culture. J Virol 79:12721–12731

Putics A, Gorbalenya AE, Ziebuhr J (2006) Identification of protease and ADP-ribose 1″-monophosphatase activities associated with transmissible gastroenteritis virus non-structural protein 3. J Gen Virol 87:651–656

Ratia K, Saikatendu KS, Santarsiero BD, Barretto N, Baker SC, Stevens RC, Mesecar AD (2006) Severe acute respiratory syndrome coronavirus papain-like protease: structure of a viral deubiquitinating enzyme. Proc Natl Acad Sci USA 103:5717–5722

Ricagno S, Egloff MP, Ulferts R, Coutard B, Nurizzo D, Campanacci V, Cambillau C, Ziebuhr J, Canard B (2006) Crystal structure and mechanistic determinants of SARS coronavirus nonstructural protein 15 define an endoribonuclease family. Proc Natl Acad Sci USA 103:11892–11897

Saikatendu KS, Joseph JS, Subramanian V, Clayton T, Griffith M, Moy K, Velasquez J, Neuman BW, Buchmeier MJ, Stevens RC, Kuhn P (2005) Structural basis of severe acute respiratory syndrome coronavirus ADP-ribose-1″-phosphate dephosphorylation by a conserved domain of nsP3. Structure 13:1665–1675

Sawicki SG, Sawicki DL, Younker D, Meyer Y, Thiel V, Stokes H, Siddell SG (2005) Functional and genetic analysis of coronavirus replicase-transcriptase proteins. PLoS Pathog 1:e39

Siddell S, Sawicki D, Meyer Y, Thiel V, Sawicki S (2001) Identification of the mutations responsible for the phenotype of three MHV RNA-negative ts mutants. Adv Exp Med Biol 494:453–458

Snijder EJ, Bredenbeek PJ, Dobbe JC, Thiel V, Ziebuhr J, Poon LL, Guan Y, Rozanov M, Spaan WJ, Gorbalenya AE (2003) Unique and conserved features of genome and proteome of SARS-coronavirus, an early split-off from the coronavirus group 2 lineage. J Mol Biol 331:991–1004

Su D, Lou Z, Sun F, Zhai Y, Yang H, Zhang R, Joachimiak A, Zhang XC, Bartlam M, Rao Z (2006) Dodecamer structure of severe acute respiratory syndrome coronavirus nonstructural protein nsp10. J Virol 80:7902–7908

Sutton G, Fry E, Carter L, Sainsbury S, Walter T, Nettleship J, Berrow N, Owens R, Gilbert R, Davidson A, Siddell S, Poon LL, Diprose J, Alderton D, Walsh M, Grimes JM, Stuart DI (2004) The nsp9 replicase protein of SARS-coronavirus, structure and functional insights. Structure 12:341–353

Thiel V, Ivanov KA, Putics A, Hertzig T, Schelle B, Bayer S, Weissbrich B, Snijder EJ, Rabenau H, Doerr HW, Gorbalenya AE, Ziebuhr J (2003) Mechanisms and enzymes involved in SARS coronavirus genome expression. J Gen Virol 84:2305–2315

von Brunn A, Teepe C, Simpson JC, Pepperkok R, Friedel CC, Zimmer R, Roberts R, Baric R, Haas J (2007) Analysis of Intraviral Protein–Protein Interactions of the SARS Coronavirus ORFeome. PLoS ONE 2:e459

Xu X, Zhai Y, Sun F, Lou Z, Su D, Xu Y, Zhang R, Joachimiak A, Zhang XC, Bartlam M, Rao Z (2006) New antiviral target revealed by the hexameric structure of mouse hepatitis virus nonstructural protein nsp15. J Virol 80:7909–7917

Zhai Y, Sun F, Li X, Pang H, Xu X, Bartlam M, Rao Z (2005) Insights into SARS-CoV transcription and replication from the structure of the nsp7-nsp8 hexadecamer. Nat Struct Mol Biol 12:980–986

Ziebuhr J, Thiel V, Gorbalenya AE (2001) The autocatalytic release of a putative RNA virus transcription factor from its polyprotein precursor involves two paralogous papain-like proteases that cleave the same peptide bond. J Biol Chem 276:33220–33232

Zust R, Miller TB, Goebel SJ, Thiel V, Masters PS (2008) Genetic interactions between an essential 3′ cis-acting RNA pseudoknot, replicase gene products, and the extreme 3' end of the mouse coronavirus genome. J Virol 82:1214–1228

Chapter 8
Quaternary Structure of the SARS Coronavirus Main Protease

Gu-Gang Chang

Abstract The maturation of the SARS coronavirus (CoV) involves the autocleavage of polyproteins 1a and 1ab by a main protease and papain-like protease. The functional unit of the main protease is a dimer in which each subunit has a Cys145–His41 catalytic dyad, with His41 acting as a general base. There is also a close correlation between dimer formation and the enzyme catalytic activity. A flip-flop mechanism is proposed for the main protease, in which the two subunits are used alternately in acylation and deacylation. Both the main protease and the papain-like protease are ideal targets for rational drug design strategies against SARS-CoV.

Abbreviations

AUC	Analytical ultracentrifuge
CoV	Coronavirus
Mpro	Main protease
nsp	Nonstructural proteins
PLpro	Papain-like protease
pp	Polyprotein
SARS	Severe acute respiratory syndrome
WT	Wild type

G.-G. Chang
National Yang-Ming University, Taipei, Taiwan
e-mail: ggchang@ym.edu.tw

S.K. Lal (ed.), *Molecular Biology of the SARS-Coronavirus*,
DOI 10.1007/978-3-642-03683-5_8, © Springer-Verlag Berlin Heidelberg 2010

8.1 Introduction

In 2003, an atypical and highly contagious pneumonia, severe acute respiratory syndrome (SARS), caused a global health crisis. After a worldwide intensive investigation, a specific novel form of human coronavirus (CoV), denoted SARS-CoV, was identified as the pathogenic agent behind this epidemic (Peiris et al. 2004). The maturation and production of the infectious progeny of SARS-CoV involves proteolytic processing of the virus polyproteins by a main protease (Mpro) (also called 3CLpro because of its 3′-proximal chymotrypsin-like catalytic domain) and a papain-like protease (PLpro).

This chapter summarizes the present knowledge of the structure and function of the SARS-CoV Mpro protein. Particular attention is paid to the quaternary structure of this protease, as the catalytically active form of this enzyme is a dimer. The structure and function of PLpro is also touched upon briefly. The current state of play in anti-SARS drug development strategies that target Mpro is discussed in the chapter following (Chap. 9).

8.2 Molecular Biology of the SARS-CoV Polyproteins

The genomic organization of SARS-CoV is similar to that of other coronaviruses, but phylogenetic analysis and sequence comparisons of the viral proteins indicate that SARS-CoV is in fact different from any of the previously characterized viruses of this type (Tanner et al. 2003; Eickmann et al. 2003). Coronaviruses are enveloped positive-sense, single-strand RNA viruses. The genome length of SARS-CoV is around 30,000 nucleotides and its replicase gene encodes two overlapping polyproteins, polyprotein 1a (pp1a) (486 kDa) and polyprotein 1ab (1a + 1b) (790 kDa) (Fig. 8.1a). These polyproteins are extensively cleaved by the internally encoded SARS-CoV proteases, Mpro and PLpro.

The 33.8-kDa Mpro plays a major role in the proteolytic processing of the virion polyproteins and cleaves pp1a at seven sites and pp1b at four sites (Fig. 8.1a). The 35-kDa protease PLpro cleaves pp1a at three sites. These autoprocessing reactions result in the maturation of 16 nonstructural proteins (nsp), including those that are common to the replication of plus-strand RNA viruses: the RNA-dependent RNA polymerase at nsp12 and helicase at nsp13. The Mpro and PLpro enzymes are themselves located at nsp5 and nsp3, respectively.

These proteases are involved in the viral life cycle, the maturation of the precapsid, and in the production of infectious virions. Viral protease inhibitors would thus be predicted to have great clinical potential in the treatment of the associated infectious diseases (Krausslich and Wimmer 1988; Tong 2002). The structure–function relationship of these proteases has therefore received much recent attention in the search for an effective anti-SARS-CoV agent (Anand et al. 2003; Yang et al. 2005; Wei et al. 2006; Lai et al. 2006; Liang 2006; Bartlam et al. 2008).

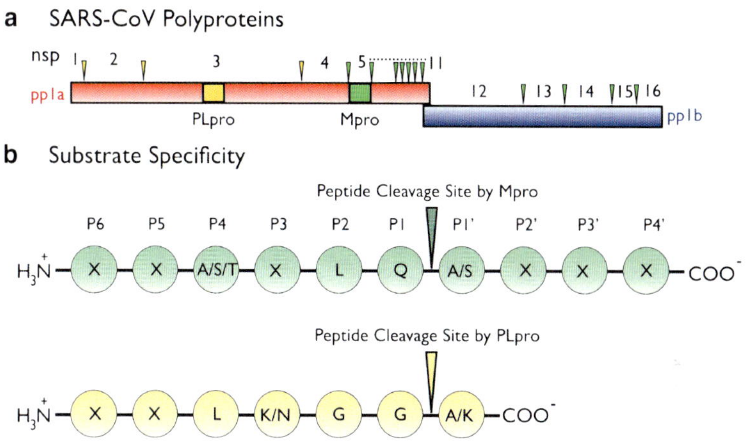

Fig. 8.1 Polyproteins of SARS-CoV. (**a**) Autoprocessing of polyprotein 1a and 1ab (1a + 1b) by Mpro occurs at 11 sites (*green triangles*) and of PLpro at three sites (*yellow triangles*) resulting in the maturation of 16 non-structural proteins (nsp). The locations of the Mpro and PLpro enzymes are at nsp 5 and 3, respectively. (**b**) Substrate specificity of Mpro and PLpro. Redundant residue positions are labeled with an X

8.3 Structure of the SARS-CoV Main Protease

8.3.1 Three-Dimensional Structure of the SARS-CoV Main Protease

Mpro was the first of the SARS-CoV proteins to have its three-dimensional structure solved by crystallography (Fig. 8.2) (Yang et al. 2003; Bartlam et al. 2005, 2007). This protease is a homodimer in which the two subunits are arranged perpendicularly to each other (Fig. 8.3a). Each protomer of SARS-CoV Mpro comprises 13 strands and 11 helices distributed among three distinct structural domains. The first two domains (residues 8–101 for domain I and 102–184 for domain II) have an antiparallel β-barrel structure, which has a folding scaffold similar to other viral chymotrypsin-like proteases (Anand et al. 2002; Hegyi et al. 2002; Ziebuhr et al. 2003). Each subunit also has its own substrate binding site with a His41–Cys145 catalytic dyad located at the interface between domains I and II. However, unlike chymotrypsin, the active site of SARS-CoV Mpro contains a catalytic cysteinyl residue instead of a seryl residue. Furthermore, SARS-CoV Mpro contains an extra domain (III) consisting of five α-helices (residues 201–306), which is a specific feature of coronavirus main proteases. This helical domain III is linked to domain II by a long loop (residues 185–200) (Fig. 8.2a).

The catalytic N-terminal domain (I + II) and C-terminal domain III of SARS-CoV Mpro can fold independently. The N-terminal domain (I + II) without domain

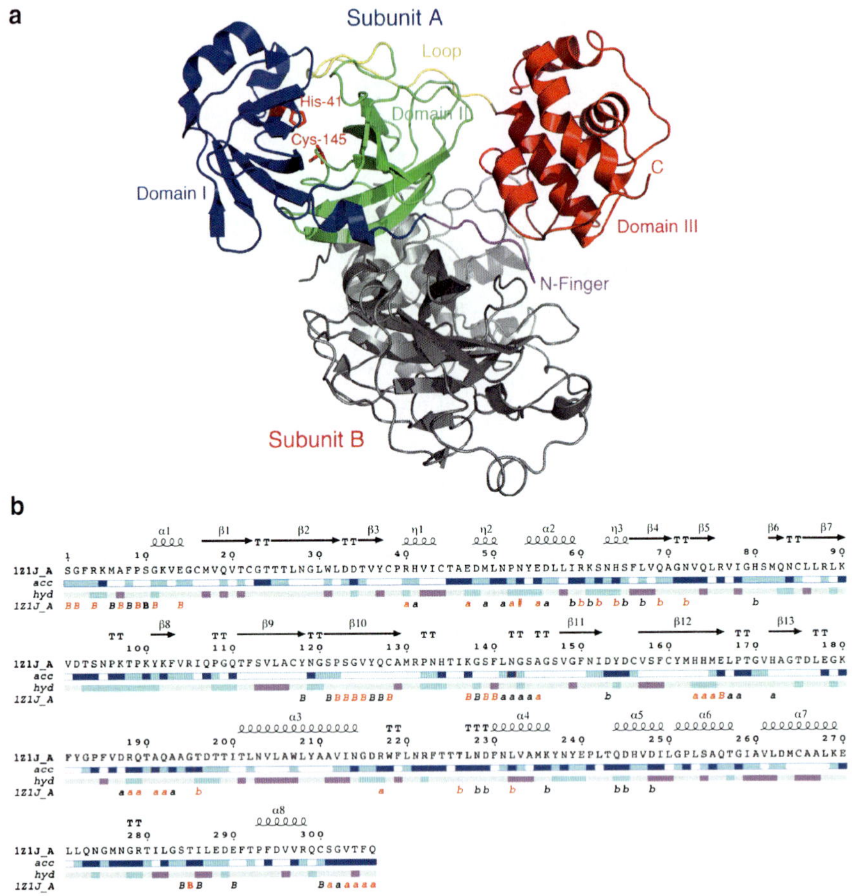

Fig. 8.2 Structure of SARS-CoV Mpro. (**a**) Ribbon presentation of the SARS-CoV Mpro dimer (pdb code: 1Z1J). Domains I (*blue*) and II (*green*) constitute a chymotrypsin-like folding scaffold with a catalytic dyad comprising His41 and Cys145 (alanine in 1Z1J) shown using a bond-and-ball model (*red*). This catalytic domain is linked to an extra domain III (*red*) by a long loop (*yellow*). The N-finger (*magenta*) of subunit *A* protrudes into the active site region of subunit *B*, which is shown in *gray*. (**b**) Structural analysis of Mpro. The primary amino acid sequence is displayed along with secondary structural elements, crystallographic contact, and hydropathy. *Panel* (**a**) was generated using MacPymol (DeLano 2002) and *panel* (**b**) with ENDscript (Gouet et al. 2003)

III folds into a structure that is indistinguishable from the intact chymotrypsin-like fold but is enzymatically inactive (Chang et al. 2007). The extra domain III of SARS-CoV Mpro increases the structural stability of the catalytic domain. This may be achieved by increasing the folding rate of domains I and II, which would thus increase the overall stability of the protein. Furthermore, domain III is related to the quaternary structure of Mpro, which has important functional implications for this enzyme.

8.3.2 Quaternary Structure of the SARS-CoV Main Protease

A coronavirus Mpro (from transmissible gastroenteritis virus) was the first viral protease shown to be dimeric, both in its crystal form and in solution (Anand et al. 2002). Significantly, SARS-CoV Mpro also exists as a dimer in solution (Chou et al. 2004) and both its N- and C-terminal residues are involved in dimer formation. The N-terminal finger (N-finger, containing residues 1–7) of subunit *A* extends from domain I toward domain III and forms intensive interactions with subunit *B*. The side chain of Arg4 at the N-finger fits into a pocket of subunit *B* and forms a salt bridge with Glu290 that constitutes one of the major interactions between the two subunits (Chou et al. 2004). In addition, the subunit interfacial region of the enzyme contains many hydrophobic interactions and hydrogen bonds (Fig. 8.3b). The interactions between the N-terminus and the other monomer play an important role in maintaining the active site integrity of the dimer (Lin et al. 2008; Chen et al. 2008b; Zhong et al. 2008). Importantly, the dimeric form of SARS-CoV Mpro is the biologically functional form of this enzyme (Anand et al. 2002; Shi et al. 2008) and dissociation of the subunits yields enzymatically inactive monomers (Shi et al. 2004; Fan et al. 2004; Chang et al. 2007).

The functional role of the N-terminus and C-terminus of Mpro has been evaluated by truncation and mutation studies. Both N-terminal and C-terminal regions are involved in the activity of this enzyme as well as in its dimerization. N-terminal truncation of the whole N-finger results in almost complete loss of enzymatic activity (Hsu et al. 2005b). Critical N-terminal amino acid residues to Arg4 and C-terminal to Gln299 have been identified as those involved in dimerization, thus generating the correct conformation of the active site (Hsu et al. 2005b; Lin et al. 2008). The C-terminal helical domain interacts with the active site of another protomer in the dimer and switches the enzyme molecule from the inactive form to the active form (Shi et al. 2004). Hsu et al. (2005a) have proposed an autocleavage mechanism, which explains the dimeric nature of the mature enzyme.

8.4 Enzyme Activity-Assay for the SARS-CoV Main Protease

The hydrolytic activity of Mpro can be assayed by its ability to cleave a peptide substrate. A procedure to separate the substrate and product peptides by high performance liquid chromatography has now been developed (Fan et al. 2004). The reaction is monitored by the formation of products peaks from the substrate peak and this method is thus very useful in the identification of cleaved peptide products. However, this procedure is very labor-intensive and thus not suitable for high throughput screening protocols.

Various fluorescence-based methods have also been developed for Mpro in which the enzyme activity can be assayed using an internally quenched fluorogenic

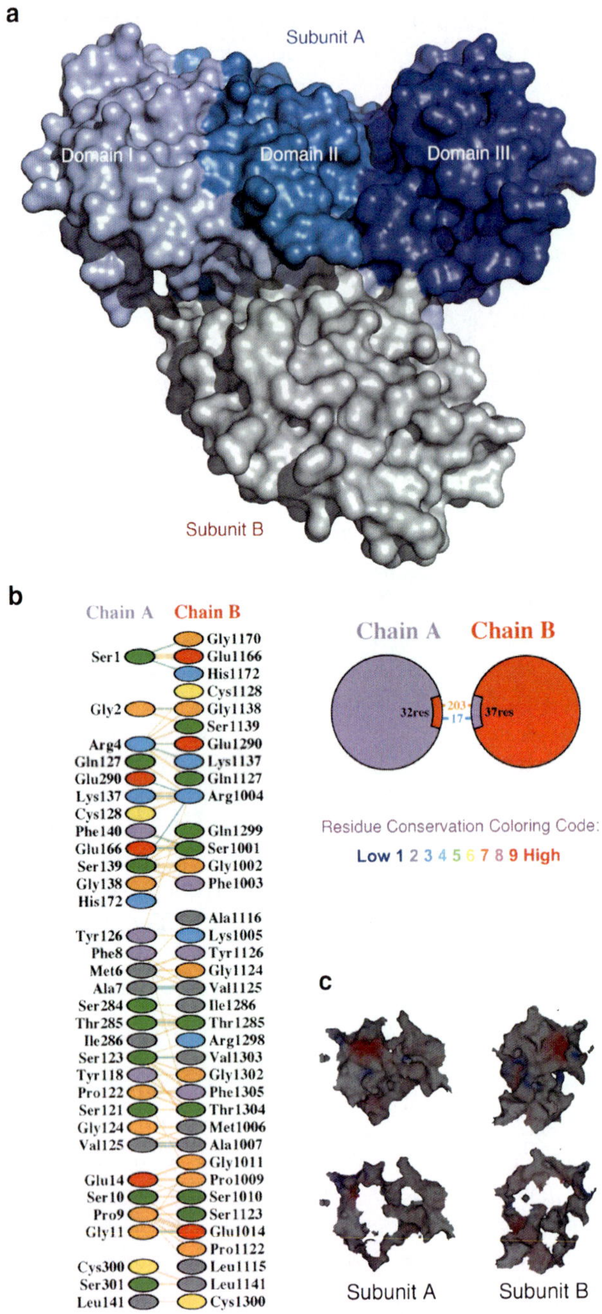

Fig. 8.3 Quaternary structure of SARS-CoV Mpro. (**a**) Surface model of Mpro. (**b**) Subunit contacts at the Mpro dimer interface. Hydrogen bonds (*cyan*), non-bonded interactions (*orange*), and the interacting amino acid residues are shown by residue conservation. The width of the

substrate peptide (e.g., *ortho*-aminobenzoic acid-peptide-2,4-dinitrophenyl amide) (Chou et al. 2004). Enhanced fluorescence due to cleavage of the peptide can be monitored at 420 nm with excitation at 362 nm using a luminescence spectrometer. For precise quantitation, a calibration curve under identical conditions should be constructed with equal amounts of hydrolytic products.

Alternatively, colorimetric methods (e.g., *p*-nitroanilide-peptide based assays) have been adopted for use in an Mpro assay (Huang et al. 2004). The chromophore *p*-nitroanilide has a known absorbance that is conductive to quantitation. In addition, colorimetric assay methods do not require inner filter effect corrections, which are essential for fluorimetric assays, and photometric devices are less expensive.

Fluorimetric analysis is generally at least 10-fold more sensitive than a colorimetric-based assay. However, its intrinsic insensitivity is in fact a distinct advantage of using colorimetric analysis for Mpro. In the case of Mpro activity, to confirm a direct correlation between the quaternary structure of the protease and its enzyme activity, a method that can simultaneously monitor protein dissociation and enzyme inactivation is highly desirable. Analytical ultracentrifugation with a band-forming centerpiece is ideal for this task (Harding and Rowe 1996). The substrate is diluted in buffer and loaded into the ultracentrifuge cell. Deuterium oxide, sucrose, or glycerol can be used to increase the density of this substrate solution. The band-forming centerpiece has small drilled-out holes to accommodate the enzyme solution, which is separated from the substrate before the application of the centrifugal force. Upon commencing centrifugation, the enzyme solution, which is less dense than the substrate solution, will form a thin layer on top of the separation column. The enzymatic reaction will start at the bound interface. During ultracentrifugation, the separation of monomer and dimer bands can be monitored by UV absorption, whereas the enzyme activity levels can be assessed by the absorption at 405 nm if *p*-nitroanilide-peptide is used. The advantage of then using a colorimetric detector is that the k_{cat} value of Mpro is low enough to allow large amounts of protein to be used. If fluorogenic substrates are used, the sensitivity of enzymatic reaction will be such that the protein levels may be below the threshold of detection.

The Mpro protease is an ideal model for the analysis of the correlation between quaternary structure and enzyme activity (Barrila et al. 2006; Shi and Song 2006). This is due partly to the fact that it is a relatively simple dimeric system and that it has only a moderate catalytic efficiency. To date, no activity has been detected for the Mpro monomer.

striped line is proportional to the number of atomic contacts. (**c**) Surface potential of the contacting regions. *Upper panels* show the interfacial region within a 10 Å distance of subunits *A* and *B*. In the *lower panels*, the hollow regions represent the direct contact regions in subunits *A* and *B*. Positively charged areas are shown in *blue* and negatively charged areas in *red*. *Panel* (**a**) was generated using MacPymol (DeLano 2002), *panel* (**b**) with PdbSum (Laskowski et al. 2005), and *panel* (**c**) with Spock (Christopher 1998)

8.5 Catalytic Mechanism of the SARS-CoV Main Protease

SARS-CoV Mpro cleaves pp1a at 11 sites containing the canonical Leu–Gln–↓–(Ala/Ser) sequence (Fig. 8.1b). The first step in this process involves binding of the substrate at the enzyme active center, which forms a Michaelis complex. An electrostatic trigger mediated by Cys145 at the susceptible peptide bond initiates the chemical reaction. Acylation of the sulfhydryl group of this cysteine results in a covalent link between the C-terminal moiety of the substrate and the SH group and the release of the N-terminal moiety (Fig. 8.4a). Finally, deacylation and release of the C-terminal moiety completes the reaction (Solowiej et al. 2008).

The –SH group of Cys145 is ion-paired with a nearby histidine residue (His41). This forms the catalytic dyad (Cys145–His41), which differs from most serine proteases that have a catalytic Ser–His–Asp triad in their active sites. In Mpro, a stable water molecule occupies the Asp position of the typical serine protease triad and this molecule might play a role in stabilizing the imidazolium ring during catalysis (Bartlam et al. 2005). Mutations at the catalytic dyad residues (H41A and C145A) almost completely abolish enzymatic activity and these mutant enzymes exist exclusively as dimers (Huang et al. 2004; Chang et al. 2007). However, mutation of Cys145 to Ser results in a partially active enzyme. These results are consistent with the notion that in the chemical mechanism underlying Mpro activity, His41 acts as a general base during the deacylation step and that the catalytic dyad involving Cys145 and His41 is left uncharged (Huang et al. 2004; Solowiej et al. 2008).

The rate-limiting step for Mpro hydrolysis is the covalent deacylation step. There is a close correlation here between the kinetic parameters and subunit dissociation constant. Mpro subunit dissociation affects catalysis but not substrate binding (Lin et al. 2008). Molecular dynamic simulations have also demonstrated an asymmetric dimer and inactivation of the enzyme after dissociation (Tan et al. 2005; Chen et al. 2006; Zheng et al. 2007). Crystal structures of the monomeric Mpro provide direct structural evidence for the catalytic incompetence of the dissociated monomer (Chen et al. 2008a; Shi et al. 2008).

An association–activation–catalysis–dissociation mechanism has been proposed for Mpro enzyme activity control (Chen et al. 2005). The catalytically competent conformation in one protomer is induced only upon dimer formation. Under physiological conditions, Mpro exists as an asymmetric dimer that might have a half-site acylation–deacylation catalytic cycle; i.e., when one subunit is in the active acylated form, the other is in the deacylated form. The dimer is the essential functional unit of this protease that regulates catalytic turnover.

The proposed flip-flop mechanism for Mpro is shown in Fig. 8.4b, which may account for the kinetics and structural information available for this enzyme. The two subunits are used alternately in acylation and deacylation reactions whereby binding at subunit *A* induces the deacylation at subunit *B* and vice versa. Mpro is thus proposed to be regulated by negative cooperativity.

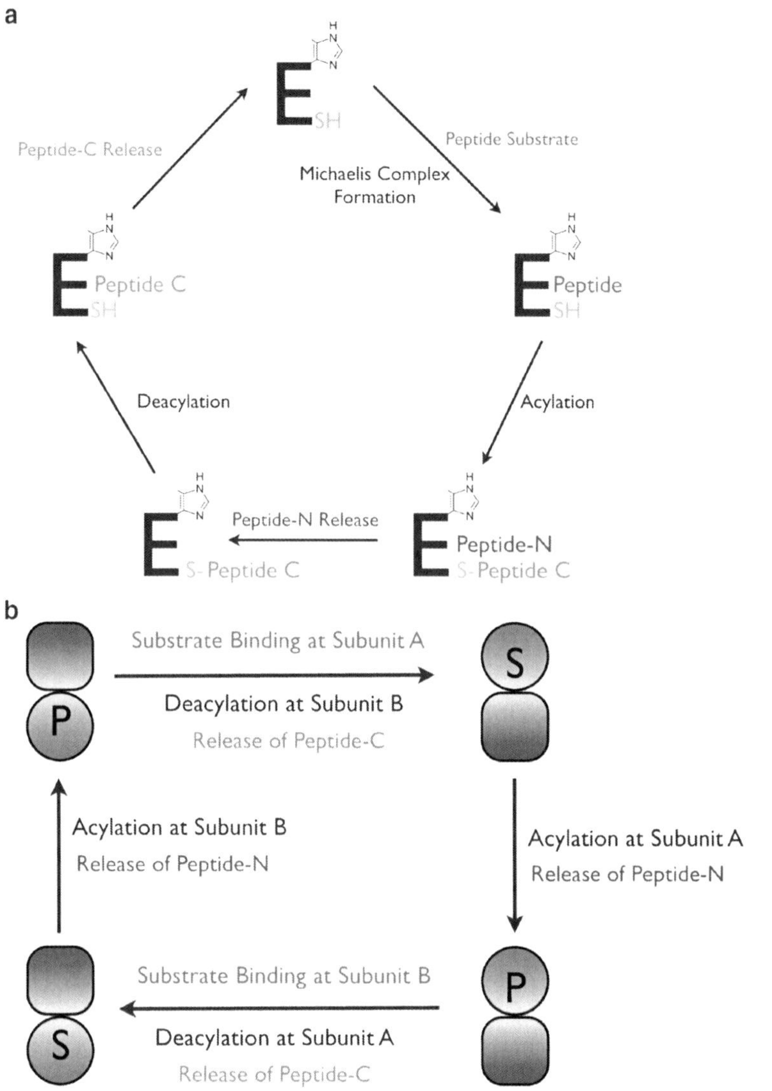

Fig. 8.4 Reaction mechanism of SARS-CoV Mpro. (**a**) Catalytic mechanism of Mpro. The binding of a peptide substrate to the active site forms a Michaelis complex. The peptide substrate is then cleaved at the Gln–Ala (or Gln–Ser) peptide bond. The N-terminal half peptide is released as the first product whereas the C-terminal half acylates the active site Cys145 residue. The acylated intermediate is then deacylated, releasing the C-terminal peptide, and this completes the catalytic cycle. His41 acts as a catalytic general base in the deacylation step. (**b**) A proposed flip-flop mechanism for the possible role of the Mpro quaternary structure in the regulation of its activity. The active subunit is indicated by the *circle*, and the inactive subunit is shown as a *square*. Only one of the two subunits is catalytically active at any one time and the two subunits thus exist in an alternate active–inactive cycle. Substrate binding at one subunit induces the deacylation of the other

8.6 Structure and Function of the SARS-CoV Papain-Like Protease

In addition to Mpro, SARS-CoV expresses a papain-like protease (PLpro) that cleaves polyprotein 1a at three sites harboring the canonical Leu–(Lys/Asn)–Gly–Gly–↓–(Ala/Lys) sequence (Fig. 8.1b). The tertiary structure of PLpro reveals a distant relationship to the papain family of cysteine proteases (Ratia et al. 2006). The catalytic triad of this enzyme (Cys112–His273–Asp287) also has a broad range of pH optima that is characteristic of the thiolate–imidazolium ion pair that exists also in other papain-like cysteine proteases (Storer and Ménard 1994; Han et al. 2005).

The functional unit of PLpro is a monomer comprising four structural domains (Fig. 8.5). A zinc atom is bound at the finger domain and the active site is located at the interface of the palm and thumb domains. A special feature of PLpro is its ubiquitin-like domain, and indeed SARS-CoV PLpro has been shown to possess deubiquitination activity (Barretto et al. 2005; Lindner et al. 2005, 2007). This dual-functional role makes PLpro another viable target for the development of anti-SARS drugs.

Thiocarbonyl-containing analogs (6-mercaptopurine and 6-thioguanine) have been demonstrated to be PLpro active site-directed compounds (Chou et al. 2008) that bind with high affinity, block the essential sulfhydryl group after binding, and thereby prevent subunit acylation and block enzyme activity. These thiopurine compounds are currently used clinically to treat children with acute lymphoblastic or myeloblastic leukemia (Pui and Evans 1998; Elion 1989) and the adverse toxicities of these drugs are well documented. These thiopurine analogs are important potential lead compounds for the development of anti-SARS-CoV agents in the near future.

8.7 Conclusions

The maturation of SARS-CoV involves two viral proteases, Mpro and PLpro. Mpro has a Cys145–His41 catalytic dyad at its active center with His41 acting as a general base. In addition, the functional unit of Mpro is a dimer and there is a close correlation between dimer formation and catalytic activity. A flip-flop mechanism is proposed for Mpro in which its two subunits are alternately used in acylation and deacylation steps. The subunit interfacial region of the main protease is an ideal target for rational drug design in the future treatment of SARS-CoV. Inhibitors of PLpro are also potential avenues for developing anti-SARS therapies.

Acknowledgment This work was supported by the National Science Council, ROC.

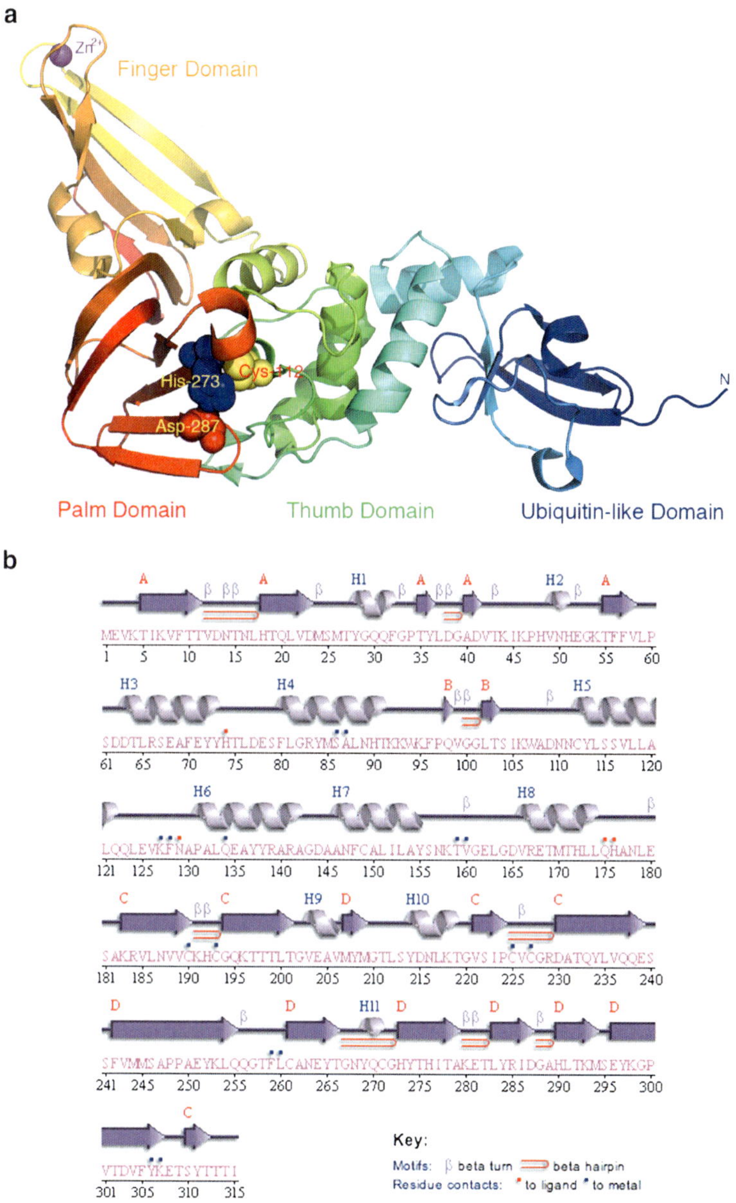

Fig. 8.5 Structure of SARS-CoV PLpro. (**a**) Ribbon diagram of SARS-CoV PLpro (pdb code: 2FE8) is shown in *rainbow colors* from the N-terminus (*blue*) to C-terminus (*red*). The catalytic triad (Cys112–His273–Asp287) and the Zn atom are highlighted using a sphere model. (**b**) Amino acid sequence and other structural annotations. The key for these structural features of PLpro is shown below the *panel*. *Panel* (**a**) was generated using MacPymol (DeLano 2002) and *panel* (**b**) with PdbSum (Laskowski et al. 2005)

References

Anand K, Palm GJ, Mesters JR, Siddell SG, Ziebuhr J, Hilgenfeld R (2002) Structure of coronavirus main proteinase reveals combination of a chymotrypsin fold with an extra alpha-helical domain. EMBO J 21:3213–3224

Anand K, Ziebuhr J, Wadhwani P, Mesters JR, Hilgenfeld R (2003) Coronavirus main proteinase (3CLpro) structure: basis for design of anti-SARS drugs. Science 300:1763–1767

Barretto N, Jukneliene D, Ratia K, Chen Z, Mesecar AD, Baker SC (2005) The papain-like protease of severe acute respiratory syndrome coronavirus has deubiquitinating activity. J Virol 79:15189–15198

Barrila J, Bacha U, Freire E (2006) Long-range cooperative interactions modulate dimerization in SARS 3CLpro. Biochemistry 45:14908–14916

Bartlam M, Yang H, Rao Z (2005) Structural insights into SARS coronavirus proteins. Curr Opin Struct Biol 15:664–672

Bartlam M, Xu Y, Rao Z (2007) Structural proteomics of the SARS coronavirus: a model response to emerging infectious diseases. J Struct Funct Genomics 8:85–97

Bartlam M, Xue X, Rao Z (2008) The search for a structural basis for therapeutic intervention against the SARS coronavirus. Acta Crystallogr A 64:204–213

Chang HP, Chou CY, Chang GG (2007) Reversible unfolding of the severe acute respiratory syndrome coronavirus main protease in guanidinium chloride. Biophys J 92:1374–1383

Chen S, Chen L, Tan J, Chen J, Du L, Sun T, Shen J, Chen K, Jiang H, Shen X (2005) Severe acute respiratory syndrome coronavirus 3C-like proteinase N terminus is indispensable for proteolytic activity but not for enzyme dimerization. Biochemical and thermodynamic investigation in conjunction with molecular dynamics simulations. J Biol Chem 280:164–173

Chen H, Wei P, Huang C, Tan L, Liu Y, Lai L (2006) Only one protomer is active in the dimer of SARS 3C-like proteinase. J Biol Chem 281:13894–13898

Chen S, Hu T, Zhang J, Chen J, Chen K, Ding J, Jiang H, Shen X (2008a) Mutation of Gly-11 on the dimer interface results in the complete crystallographic dimer dissociation of severe acute respiratory syndrome coronavirus 3C-like protease: crystal structure with molecular dynamics simulations. J Biol Chem 283:554–564

Chen S, Zhang J, Hu T, Chen K, Jiang H, Shen X (2008b) Residues on the dimer interface of SARS coronavirus 3C-like protease: dimer stability characterization and enzyme catalytic activity analysis. J Biochem 143:525–536

Chou CY, Chang HC, Hsu WC, Lin TZ, Lin CH, Chang GG (2004) Quaternary structure of the severe acute respiratory syndrome (SARS) coronavirus main protease. Biochemistry 43:14958–14970

Chou CY, Chien CH, Han YS, Prebanda MT, Hsieh HP, Turk B, Chang GG, Chen X (2008) Thiopurine analogues inhibit papain-like protease of severe acute respiratory syndrome coronavirus. Biochem Pharmacol 75:1601–1609

Christopher JA (1998) The SPOCK program manual. http://quorum.tamu.edu/. Cited 14 May 2008

DeLano WL (2002) The PyMOL molecular graphics system, DeLano Scientific, Palo Alto, CA, USA, http://www.pymol.org/. Cited 14 May 2008

Eickmann M, Becker S, Klenk HD, Doerr HW, Stadler K, Censini S, Guidotti S, Masignani V, Scarselli M, Mora M, Donati C, Han JH, Song HC, Abrignani S, Covacci A, Rappuoli R (2003) Phylogeny of the SARS coronavirus. Science 302:1504–1505

Elion GB (1989) The purine path to chemotherapy. Science 244:41–47

Fan K, Wei P, Feng Q, Chen S, Huang C, Ma L, Lai B, Pei J, Liu Y, Chen J, Lai L (2004) Biosynthesis, purification, and substrate specificity of severe acute respiratory syndrome coronavirus 3C-like proteinase. J Biol Chem 279:1637–1642

Gouet P, Robert X, Courcelle E (2003) ESPript/ENDscript: extracting and rendering sequence and 3D information from atomic structures of proteins. Nucleic Acids Res 31:3320–3323; http://espript.ibcp.fr//ESPript/ENDscript/. Cited 14 May 2008

Han YS, Chang GG, Juo CG, Lee HJ, Yeh SH, Hsu JT, Chen X (2005) Papain-like protease 2 (PLP2) from severe acute respiratory syndrome coronavirus (SARS-CoV): expression, purification, characterization, and inhibition. Biochemistry 44:10349–10359

Harding SE, Rowe AJ (1996) Active enzyme centrifugation. In: Engel PC (ed) Enzymology LabFax. Bios Scientific, Oxford, UK, pp 66–75

Hegyi A, Friebe A, Gorbalenya AE, Ziebuhr J (2002) Mutational analysis of the active centre of coronavirus 3C-like proteases. J Gen Virol 83:581–593

Hsu MF, Kuo CJ, Chang KT, Chang HC, Chou CC, Ko TP, Shr HL, Chang GG, Wang AH, Liang PH (2005a) Mechanism of the maturation process of SARS-CoV 3CL protease. J Biol Chem 280:31257–31266

Hsu WC, Chang HC, Chou CY, Tsai PJ, Lin PY, Chang GG (2005b) Critical assessment of important regions in the subunit association and catalytic action of the severe acute respiratory syndrome coronavirus main protease. J Biol Chem 280:22741–22748

Huang C, Wei P, Fan K, Liu Y, Lai L (2004) 3C-like proteinase from SARS coronavirus catalyzes substrate hydrolysis by a general base mechanism. Biochemistry 43:4568–4574

Kraussiich HG, Wimmer E (1988) Viral proteinases. Annu Rev Biochem 57:701–754

Lai L, Han X, Chen H, Wei P, Huang C, Liu S, Fan K, Zhou L, Liu Z, Pei J, Liu Y (2006) Quaternary structure, substrate selectivity and inhibitor design for SARS 3C-like proteinase. Curr Pharm Des 12:4555–4564

Laskowski RA, Chistyakov VV, Thornton JM (2005) PDBsum more: new summaries and analyses of the known 3D structures of proteins and nucleic acids. Nucleic Acids Res 33:D266–D268; http://www.ebi.ac.uk/thornton-srv/databases/pdbsum/. Cited 14 May 2008

Liang PH (2006) Characterization and inhibition of SARS-coronavirus main protease. Curr Top Med Chem 6:361–376

Lin PY, Chou CY, Chang HC, Hsu WC, Chang GG (2008) Correlation between dissociation and catalysis of SARS-CoV main protease. Arch Biochem Biophys 472:34–42

Lindner HA, Fotouhi-Ardakani N, Lytvyn V, Lachance P, Sulea T, Ménard R (2005) The papain-like protease from the severe acute respiratory syndrome coronavirus is a deubiquitinating enzyme. J Virol 79:15199–15208

Lindner HA, Lytvyn V, Qi H, Lachance P, Ziomek E, Ménard R (2007) Selectivity in ISG15 and ubiquitin recognition by the SARS coronavirus papain-like protease. Arch Biochem Biophys 466:8–14

Peiris JS, Guan Y, Yuen KY (2004) Severe acute respiratory syndrome. Nat Med 10:S88–S97

Pui CH, Evans WE (1998) Acute lymphoblastic leukemia. N Engl J Med 339:605–615

Ratia K, Saikatendu KS, Santarsiero BD, Barretto N, Baker SC, Stevens RC, Mesecar AD (2006) Severe acute respiratory syndrome coronavirus papain-like protease: structure of a viral deubiquitinating enzyme. Proc Natl Acad Sci USA 103:5717–5722

Shi J, Song J (2006) The catalysis of the SARS 3C-like protease is under extensive regulation by its extra domain. FEBS J 273:1035–1045

Shi J, Wei Z, Song J (2004) Dissection study on the severe acute respiratory syndrome 3C-like protease reveals the critical role of the extra domain in dimerization of the enzyme: defining the extra domain as a new target for design of highly specific protease inhibitors. J Biol Chem 279:24765–24773

Shi J, Sivaraman J, Song J (2008) Mechanism for controlling dimer-monomer switch and coupling dimerization to catalysis of the severe acute respiratory syndrome coronavirus 3C-like protease. J Virol 82:4620–4629

Solowiej J, Thomson JA, Ryan K, Luo C, He M, Lou J, Murray BW (2008) Steady-state and pre-steady-state kinetic evaluation of severe acute respiratory syndrome coronavirus (SARS-CoV) 3CLpro cysteine protease: development of an ion-pair model for catalysis. Biochemistry 47:2617–2630

Storer AC, Ménard R (1994) Catalytic mechanism in papain family of cysteine peptidases. Methods Enzymol 244:486–500

Tan J, Verschueren KH, Anand K, Shen J, Yang M, Xu Y, Rao Z, Bigalke J, Heisen B, Mesters JR, Chen K, Shen X, Jiang H, Hilgenfeld R (2005) pH-dependent conformational flexibility of the SARS-CoV main proteinase (Mpro) dimer: molecular dynamics simulations and multiple X-ray structure analyses. J Mol Biol 354:25–40

Tanner JA, Watt RM, Chai YB, Lu LY, Lin MC, Peiris JS, Poon LL, Kung HF, Huang JD (2003) The severe acute respiratory syndrome (SARS) coronavirus NTPase/helicase belongs to a distinct class of 5' to 3' viral helicases. J Biol Chem 278:39578–39582

Tong L (2002) Viral proteases. Chem Rev 102:4609–4626

Wei P, Fan K, Chen H, Ma L, Huang C, Tan L, Xi D, Li C, Liu Y, Cao A, Lai L (2006) The N-terminal octapeptide acts as a dimerization inhibitor of SARS coronavirus 3C-like proteinase. Biochem Biophys Res Commun 339:865–872

Yang H, Yang M, Ding Y, Liu Y, Lou Z, Zhou Z, Sun L, Mo L, Ye S, Pang H, Gao GF, Anand K, Bartlam M, Hilgenfeld R, Rao Z (2003) The crystal structures of severe acute respiratory syndrome virus main protease and its complex with an inhibitor. Proc Natl Acad Sci USA 100:13190–13195

Yang H, Xie W, Xue X, Yang K, Ma J, Liang W, Zhao Q, Zhou Z, Pei D, Ziebuhr J, Hilgenfeld R, Yuen KY, Wong L, Gao G, Chen S, Chen Z, Ma D, Bartlam M, Rao Z (2005) Design of wide-spectrum inhibitors targeting coronavirus main proteases. PLoS Biol 3:e324

Zheng K, Ma G, Zhou J, Zen M, Zhao W, Jiang Y, Yu Q, Feng J (2007) Insight into the activity of SARS main protease: Molecular dynamics study of dimeric and monomeric form of enzyme. Proteins 66:467–479

Zhong N, Zhang S, Zou P, Chen J, Kang X, Li Z, Liang C, Jin C, Xia B (2008) Without its N-finger, SARS-CoV main protease can form a novel dimer through its C-terminal domain. J Virol 82:4227–4234

Ziebuhr J, Bayer S, Cowley JA, Gorbalenya AE (2003) The 3C-like proteinase of an invertebrate nidovirus links coronavirus and potyvirus homologs. J Virol 77:1415–1426

Chapter 9
The Nucleocapsid Protein of the SARS Coronavirus: Structure, Function and Therapeutic Potential

Milan Surjit and Sunil K. Lal

Abstract As in other coronaviruses, the nucleocapsid protein is one of the core components of the SARS coronavirus (CoV). It oligomerizes to form a closed capsule, inside which the genomic RNA is securely stored thus providing the SARS-CoV genome with its first line of defense from the harsh conditions of the host environment and aiding in replication and propagation of the virus. In addition to this function, several reports have suggested that the SARS-CoV nucleocapsid protein modulates various host cellular processes, so as to make the internal milieu of the host more conducive for survival of the virus. This article will analyze and discuss the available literature regarding these different properties of the nucleocapsid protein. Towards the end of the article, we will also discuss some recent reports regarding the possible clinically relevant use of the nucleocapsid protein, as a candidate diagnostic tool and vaccine against SARS-CoV infection.

9.1 Introduction

By definition, nucleocapsid is a viral protein coat that surrounds the genome (either DNA or RNA). Nucleocapsid protein is the major constituent of a viral nucleocapsid. It is capable of associating with itself and with the genome, thus packaging the genome inside a closed cavity. In some viruses, nucleocapsid protein may also be assisted by other viral cofactors to form the capsid. However, in coronaviruses (including SARS-CoV), the nucleocapsid protein alone is capable of forming the capsid. The primary advantage of the virus for encoding the nucleocapsid protein is that the latter encloses and protects the viral genome from coming into direct contact with the harsh environment in the host. In fact, in some simple viruses

S.K. Lal (✉)
Virology Group, ICGEB, P. O. Box: 10504, Aruna Asaf Ali Road, New Delhi 110067, India
e-mail: sunillal@icgeb.res.in

like hepatitis E virus and polio virus, the nucleocapsid protein is the only coat that protects the genome from the outside world. However, in complex viruses, like hepatitis B virus and coronaviruses (including SARS-CoV), the nucleocapsid is covered by an additional coat composed of other viral proteins (spike protein is a major component of this coat). Besides this property, nucleocapsid proteins of several viruses have been demonstrated to play multiple regulatory roles during viral pathogenesis. They are equipped with specific structural motifs and/or signature sequences, by which they associate with other viral/ host factors and skew the host cellular machinery in such a manner that it becomes more favorable for the survival of the virus. Nucleocapsid protein is also one of the most abundantly expressed viral proteins and it is the major antigen recognized by convalescent antisera. Hence, it is tempting to evaluate its potential as a candidate diagnostic tool or vaccine against the virus.

Therefore, understanding the properties of the nucleocapsid protein is of utmost importance to any virologist in order to understand the biology of the virus and develop effective tools to control the infection. Since the identification and isolation of SARS-CoV in 2003, several laboratories around the world have focussed their research on characterization of various properties of the nucleocapsid protein. An indirect measure of the curiosity among SARS-CoV researchers to study the nucleocapsid protein is revealed from the fact that in PubMed the number of SARS-CoV research publications focussed on nucleocapsid protein is second only to those on spike protein. Evidence accumulated from these articles has helped us gain substantial understanding of the properties of this protein. In this article, we will provide a comprehensive description of all the different properties of the nucleocapsid protein, as established by independent workers from several laboratories. We will conclude this article with the discussion of some of the remaining challenges in this field that need to be addressed in future.

9.2 N-Protein: Structure and Composition

The nucleocapsid (N) protein is encoded by the ninth ORF of SARS-CoV. The same ORF also codes for another unique accessory protein called ORF9b, though in a different reading frame, whose function is yet to be defined. The N-protein is a 46-kDa protein composed of 422 amino acids (Rota et al. 2003). Its N-terminal region consists mostly of positively charged amino acids, which are responsible for RNA binding. A lysine-rich region is present between amino acids 373 and 390 at the C-terminus, which is predicted to be the nuclear localization signal. Besides these, an SR-rich motif is present in the middle region encompassing amino acids 177–207. Biophysical studies done by Chang et al. (2006) have suggested that this protein is composed of two independent structural domains and a linker region. The first domain is present at the N-terminus, inside the putative RNA binding domain, and the second domain consists of the C-terminal region that is capable of self-association. Between these two structural domains, there lies a highly disordered

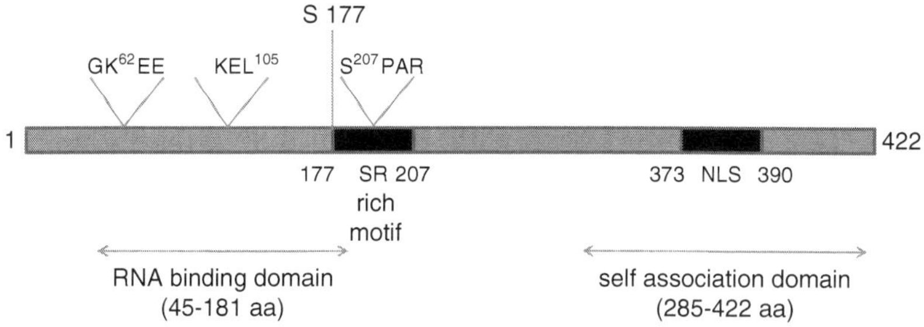

Fig. 9.1 Structure of the SARS-CoV nucleocapsid protein. A schematic diagram showing different domains identified to date. The numbers 1–422 correspond to the length in amino acids of the *N* gene. GKEE represents the sumoylation motif (lysine residue). KEL is the RXL motif, responsible for binding with cyclin D, and SPAR is the motif that gets phosphorylated by the cyclin–CDK complex (serine residue). S177 is the serine 177 residue that gets phosphorylated by GSK3

region, which serves as a linker. This region has been reported to interact with the membrane (M) protein and human cellular hnRNPA1 protein (Fang et al. 2006; Luo et al. 2005). Besides, this region is also predicted to be a hot spot for phosphorylation. Hence, in summary, the N-protein can be classified into three distinct regions (Fig. 9.1), which may serve completely different functions during different stages of the viral life-cycle. A similar mode of organization has been reported for other coronavirus nucleocapsid proteins.

9.3 Stability of the N-Protein

In-vitro thermodynamic studies done by Luo et al. (2004b) using purified recombinant N-protein have shown it to be stable between pH 7 and 10, with maximum conformational stability near pH 9. Further, it was observed to undergo irreversible thermal-induced denaturation. It starts to unfold at 35°C and is completely denatured at 55°C (Wang et al. 2004). However, denaturation of the N-protein induced by chemicals such as urea or guanidium chloride is a reversible process.

9.4 Posttranslational Modification

As in other coronavirus N-proteins, SARS-CoV N-protein has been predicted and later experimentally proven to undergo various posttranslational modifications such as acetylation, phosphorylation, and sumoylation.

Acetylation is the first modification of the N-protein to be experimentally proven. By mass spectrometric analysis of convalescent sera from several SARS patients, it has been shown that the N-terminal methionine of N is removed and all other methionines are oxidized and the resulting N-terminal serine is acetylated. However, the functional relevance of this modification, if any, remains to be elucidated (Krokhin et al. 2003).

Another unique modification of the N-protein is its ability to become sumoylated. Studies done by Li et al. (2005a) have clearly established that heterologously expressed N in mammalian cells is sumoylated. Using a site-directed mutagenesis approach, the sumoylation motif has been mapped to the 62nd lysine residue, which is present in a putative sumo-modification domain (GK^{62}EE). Their data further suggests that sumoylation may play a key role in modulating homo-oligomerization, nucleolar translocation and cell-cycle deregulatory property of the N-protein. Further experimental support regarding sumoylation of N-protein came from another independent study carried out by Fan et al. (2006) wherein they have demonstrated an association between the N-protein and Hubc9, which is a ubiquitin-conjugating enzyme of the sumoylation system. They have also mapped the interaction domain to the SR-rich motif, which is in agreement with the earlier report. However, they failed to detect the involvement of the GKEE motif in mediating this interaction (Fan et al. 2006).

Initially, the SARS-CoV N-protein was predicted to be heavily phosphorylated. Later on, from results obtained in our laboratory as well as by other researchers, it is now clear that the N-protein is a substrate of multiple cellular kinases. First experimental evidence for the phosphorylation status of the N-protein came from the study done by Zakhartchouk et al. (2005) in which, using [^{32}P]orthophosphate labelling, they were able to observe phosphorylation of adenovirus-vector-expressed N-protein in 293T cells. Further studies done in our laboratory clearly confirmed this observation. The majority of the N-protein was found to be phosphorylated at its serine residues (although the involvement of threonine and tyrosine residues could not be detected; they may be occurring in vivo). In addition, using a variety of biochemical assays, it was proved that, at least in vitro, the N-protein could become phosphorylated by mitogen-activated protein kinase (MAP kinase), cyclin-dependent kinase (CDK), glycogen synthase kinase 3 (GSK3), and casein kinase 2 (CK2). Also, this data provided preliminary indication regarding phosphorylation-dependent nucleo-cytoplasmic shuttling of the N-protein (Surjit et al. 2005). A recent report published by Wu et al. (2008) has further confirmed that N-protein is a substrate of GSK3 enzyme, both in vitro and in vivo. Using a variety of biochemical and genetic assays, it was clearly demonstrated that serine 177 residue of N-protein was phosphorylated by GSK3. An antibody specific to phospho 177 residue of the N-protein could efficiently detect the phospho N-protein both in vitro and in SARS-CoV infected cells. Interestingly, biochemically mediated inhibition of GSK3 activity in SARS-CoV infected cells also leads to around 80% reduction in viral titer and subsequent induction of a virus-induced cytopathic effect. The authors proposed that GSK3 may be a major regulator of SARS-CoV replication, possibly by virtue of its ability to phosphorylate the N-protein. However, phosphorylation of other viral and/or host proteins by GSK3 may also be a determinant of the observed cytopathic effect.

9.5 Localization of the N-Protein

In contrast to the N-protein of many other coronaviruses, the SARS-CoV N-protein is predominantly distributed in the cytoplasm, when expressed heterologously or in infected cells (Surjit et al. 2005; You et al. 2005; Rowland et al. 2005). In infected cells, a few cells exhibited nucleolar localization (You et al. 2005). As reported by You et al. (2005), the N-protein contains pat4, pat7 and bipartite-type nuclear localization signals. It has also been predicted to possess a potential CRM-1-dependent nuclear export signal. However, no clear experimental evidence could be obtained regarding the involvement of these signature sequences in regulating the localization of the N-protein. Interestingly, studies done in our laboratory revealed that the majority of N-protein localized to the nucleus in serum-starved cells. This phenomenon could be reproducibly observed both in biochemical fractionation as well as immunofluorescence studies. In addition, treatment of cells with specific inhibitors of different cellular kinases such as CK2 inhibitor and CDK inhibitor resulted in retention of a fraction of the N-protein in the nucleus, whereas GSK3 and MAPK inhibitor had very little effect. Further, N-protein was found to be efficiently phosphorylated by the cyclin–CDK complex, which is known to be active only in the nucleus. The N-protein was also found to associate with 14-3-3 protein in a phospho-specific manner and inhibition of the 14-3-3θ protein level by siRNA resulted in nuclear accumulation of the N-protein. Although these experiments are too preliminary to conclusively provide any answer regarding the intracellular localization of N-protein, nevertheless they do provide substantial clues regarding the physical presence of the N-protein in the nucleus, under certain circumstances, which may be a very dynamic phenomenon. Another study done by Timani et al. (2005) using different deletion mutants of the N-protein fused to EGFP showed that the N-terminal of N-protein, which contains the NLS 1 (aa 38–44), localizes to the nucleus, whereas the C-terminal region containing both NLS 2 (aa 257–265) and NLS 3 (aa 369–390) localizes to the cytoplasm and nucleolus. Using a combination of different deletion mutants, they concluded that the N-protein may act as a shuttle protein between cytoplasm–nucleus and nucleolus. Taken together, all these results further suggest that the N-protein per se has the physical ability to localize to the nucleus. Whether this localization is regulated through phosphorylation-mediated activation of a potential NLS or piggy-backing by association with another cellular nuclear protein or through any other mechanism remains to be established.

9.6 Genome Encapsidation: Primary Function of a Viral Capsid Protein

Being the capsid protein, the primary function of the N-protein is to package the genomic RNA in a protective covering. In order to achieve this structure, the N-protein must be equipped with two different characteristic properties; such as

(1) being able to recognize the genomic RNA and associate with it, and (2) self-associate into an oligomer to form the capsid. The N-protein of SARS-CoV has been experimentally proven to possess these properties in vitro, as discussed below.

9.6.1 Recognition and Binding with the Genomic RNA

The first experimental evidence regarding the RNA binding property of the N-protein came from the work of Huang et al. (2004), in which, by NMR studies, they proved the ability of the N-terminal domain to associate with several viral 3′ untranslated RNA sequences. Additionally, Chen et al. (2007) reported the presence of another RNA binding domain at the C-terminal region (residues 248–365) of the N-protein, which was proposed to be a stronger interaction than that at the N terminus. Based on structural analysis of the RNA–protein interaction, they have further suggested that the genomic RNA is packaged in a helical manner by the N-protein. In another report published by Luo et al. (2006), the RNA binding motif of the N-protein was mapped to amino acid residues 363–382. In summary, the RNA binding ability of the N-protein was attributed to its two distinct structural domains: the N-terminal domain (residues 45–181) and the C-terminal dimerization domain (residues 248–365). These two domains are spatially separated by long stretches of disordered region. A recent study done by Chang et al. (2008) has demonstrated RNA binding ability of these disordered regions. They have proposed that different RNA binding domains of the N-protein may cooperate to enhance the overall RNA binding efficiency of the N-protein and may also serve as interaction hubs for the association of N-protein with other viral and/or host nucleic acid and/or proteins.

Perhaps the most convincing proof to date regarding the ability of the N-protein to package the genomic RNA came from the work of Hsieh et al. (2005). They have established a system to produce SARS-CoV VLPs by cotransfection of spike, membrane, and envelope and nucleocapsid cDNAs into Vero E6 cells. While testing the packaging of an RNA-bearing GFP fused to SARS-CoV packaging signal into this particle, they observed that presence of the N-protein is an absolute requirement. However, the N-protein was not essential for the assembly of the empty particle per se. Further, by performing a filter binding assay using recombinant N-protein, they were able to identify two independent RNA binding domains in the N-protein; one at the N terminus (aa 1–235) and the other at the C terminus (aa 236–384). These results are in agreement with previous findings and further suggest that these two regions may be functional in vivo. Future experiments using a model infection system will confirm these observations.

9.6.2 Formation of the Capsid

One of the most crucial properties required by the N-protein for genome encapsidation is its ability to self-associate. Therefore, many laboratories have focused on

characterizing this phenomenon, with an eye on developing possible interference strategies that may help in limiting virus propagation.

Initial studies done in our laboratory using a yeast two-hybrid assay revealed that N-protein is able to self-associate through its C-terminal amino acid 209 residues (Surjit et al. 2004a). A parallel study done by He et al. (2004) using the mammalian two-hybrid system and sucrose gradient fractionation also proved the ability of the N-protein to self-associate to form an oligomer. They further mapped the interaction region to amino acid 184–196 residues, encompassing the SR-rich motif. However, there were some discrepancies regarding the interaction domain mapped in these two studies. Later on, extensive biophysical and biochemical analysis done by Chen's laboratory (Yu et al. 2005, 2006) and Jiang's laboratory (Luo et al. 2006, 2005) have enriched our understanding of the oligomerization process of the N-protein. In summary, the SR-rich motif does possess binding affinity, but this is specific for the central region (aa 211–290) of another molecule of N-protein, instead of the SR-rich motif itself. The C-terminal region (aa 283–422) possesses binding affinity for itself and to associate into a dimer, trimer, tetramer or hexamer, in a concentration-dependent manner. The essential sequence for oligomerization of the N-protein was identified to be residues 343–402. Interestingly, this region also encompasses the RNA binding motif of the N-protein, which prompts us to speculate that there might be mutual interplay between RNA binding and oligomerization activities of the N-protein. Further, the oligomerization was observed to be independent of electrostatic interactions and addition of single strand DNA to the reaction mixture containing tetramers of the N-protein promoted oligomerization. Thus, it has been proposed that once the tetramer is formed by protein–protein interaction between nucleocapsid molecules, binding with genomic RNA prompts further assembly of the complete nucleocapsid structure.

9.7 Perturbation of Host Cellular Process by the N-Protein

Besides being the capsid protein of the virus, the N-protein of many coronaviruses is known to double up as a regulatory protein. The N-protein of the SARS-CoV too has been shown to modulate the host cellular machinery in vitro, thereby indicating its possible regulatory role during its viral life-cycle. Some of the major cellular processes perturbed by heterologous expression of the N-protein are discussed below.

9.7.1 Deregulation of Host Cell Cycle

Three different groups have reported the ability of the N-protein to interfere with the host cell cycle in vitro. Work done by Li et al. (2005a, 2005b) proved that mutation of the sumoylation motif in the N-protein leads to cell cycle arrest.

Work done in our laboratory has shown the inhibition of S phase progression in cells expressing the N-protein (Surjit et al. 2006). Further, S-phase specific gene products like cyclin E and CDK2 were found to be downregulated in SARS-CoV infected cell lysate, which suggested that the observed phenomenon may be relevant in vivo. In an attempt to further characterize the mechanism of cell cycle blockage induced by the N-protein, several biochemical and mutational analysis were carried out. Results thus obtained demonstrated that the N-protein directly inhibits the activity of the cyclin–CDK complex, resulting in hypophosphorylation of retinoblastoma protein with a concomitant downregulation of E2F1-mediated transactivation. Analysis of RXL and CDK phosphorylation mutant N-protein identified the mechanisms of inhibition of CDK4 and CDK2 activity to be different. Whereas the N-protein could directly bind to cyclin D and inhibit the activity of the CDK4–cyclinD complex, inhibition of CDK2 activity appeared to be achieved in two different ways: indirectly by downregulation of protein levels of CDK2, cyclin E, and cyclin A, and by direct binding of N-protein to the CDK2–cyclin complex.

A third piee of evidence supporting the ability of N-protein to deregulate the host cycle came from the work of Zhou et al. (2008). They observed slower transition from S to G2/M phase and slower growth rate in N-protein-expressing 293T cells. They also observed a similar phenomenon in human peripheral blood lymphocyte and K 562 cells infected with a retrovirus expressing SARS-CoV N-protein.

9.7.2 Inhibition of Host Cell Cytokinesis

While searching for interaction partners for the C terminus of N-protein (aa 251–422) by following a yeast two-hybrid library screening approach, Zhou et al. (2008) discovered human elongation factor 1 alpha (EF1α) as a candidate partner. The specificity of the interaction was confirmed by various in-vitro and in-vivo assays. Further, expression of N-protein induced aggregation of EF1α. It is known that the majority of cellular EF1α is bound to F-actin and promotes F-actin bundling, which is a key event during cytokinesis (Kurasawa et al. 1996; Yang et al. 1990). Hence, the authors tested whether N-protein-induced aggregation of EF1α affected F-actin bundling and cytokinesis. As expected, they observed significantly fewer F-actin bundles in N-protein-expressing cells. In fact, a similar F-actin distribution pattern was also observed by Surjit et al. (2004b) in COS-1 cells. Further, the authors observed multinucleated cells in N-protein-expressing cells at a later time point (72 h post-transfection), indicating inhibition of cytokinesis in those cells. Specificity of the above data has been confirmed by the use of different deletion mutants of the N-protein, in which only the C-terminal domain of the N-protein (responsible for binding with EF1α) was able to reproduce the above results. Thus, it has been suggested that EF1α binding by the N-protein leads to its aggregation, resulting in inhibition of F-actin bundling and subsequent blocking of cytokinesis.

9.7.3 Inhibition of Host Cell Translation Machinery

EF1α is known to play a key role during the peptide elongation stage of translation. Therefore, it is an attractive candidate for pathogen proteins to manipulate its activity in order to skew the host translation machinery. For example, HIV-type 1 gag polyprotein has been shown to interact with EF1α and impair translation in vitro (Cimarelli and Luban 1999). Since Zhou et al. (2008) observed an interaction between EF1α and SARS-CoV N-protein, they further tested whether it interfered with the host translation machinery. Indeed, presence of the N-protein inhibited total cellular translation, both in vitro and in vivo, in a dose-dependent manner. Moreover, exogenous addition of excess EF1α could reverse the N-protein-induced translation inhibition, thus suggesting that N-protein exerts its effect by interfering with EF1α function. However, it remains to be confirmed whether a similar effect is recapitulated in vivo.

9.7.4 Inhibition of Interferon Production

Production of interferon (IFN) is one of the primary host defense mechanisms. However, SARS-CoV infection does not result in IFN production. Nevertheless, pretreatment of cells with IFN blocks SARS-CoV infection (Spiegel et al. 2005; Zheng et al. 2004). Based on this observation, Palese's laboratory has studied the IFN inhibitory property of different SARS-CoV proteins, which revealed that ORF3, ORF6 as well as the N-protein have the ability to independently inhibit IFN production through different mechanisms. The N-protein was found to inhibit the activity of IRF3 and NFkB in host cells, resulting in inhibition of IFN synthesis. IRF3 activity was also blocked by ORF3, ORF6 proteins, but inhibition of NFkB activity was a property unique to the N-protein. In addition, ORF3, ORF6 proteins were able to block STAT1 activity through different mechanisms (Kopecky-Bromberg et al. 2007). All these data suggest that SARS-CoV may employ multiple factors to check the activity of the host immune system and N-protein may be one of the major partners in this process. It may be possible that these different factors act independently during different stages of the viral life cycle. In that case, regulatory activity of the N-protein will be as indispensible as its structural activity.

9.7.5 Modulation of TGFβ Signaling Pathway

During the SARS outbreak, a large number of patients developed severe inflammation of the lungs, which subsequently led to acute respiratory distress syndrome (Ding et al. 2003; Nicholls et al. 2003). Acute respiratory distress syndrome is characterized by pulmonary fibrosis, which results in lung failure and subsequent death of the patient. The TGFβ signaling pathway plays a critical role in pulmonary fibrosis (Roberts et al. 2006; Border and Noble 1994). It enhances the expression of extracellular matrix (ECM) proteins, accelerates the secretion of protease inhibitors and reduces the secretion of proteases, thereby leading to deposition of ECM proteins. TGFβ may also induce pulmonary fibrosis directly by stimulating chemotactic migration and proliferation of fibroblasts as well as by fibroblast–myofibroblast transition. Hence, it is worth speculating that some of the SARS-CoV encoded factors may be modulating the TGFβ signaling pathway. In fact, proteins of several other viruses, such as hepatitis C virus core, NS3 and NS5 protein, adenovirus E1A, human papilloma virus E7, human T-lymphotropic virus Tax and Epstein–Barr virus LMP1, have been reported to modulate the TGFβ pathway. In general, these proteins directly bind with smad proteins and alter the innate signaling pathway.

Interestingly, a recent report published by Zhao et al. (2008) revealed that N-protein of SARS-CoV also interacts with smad3 and modulates the activity of the TGFβ pathway. By performing a smad binding element (SBE)-driven reporter assay, RT-PCR and immunohistological analysis of TGFβ target genes such as *PAI-1* (plasminogen activator inhibitor 1) and collagen in a variety of cell lines and SARS patients, the authors have clearly proved that N-protein indeed enhanced the activity of the TGFβ signaling pathway. Further, they observed that the effect of N-protein on TGFβ signaling was mediated through smad3 only (independent of the involvement of smad4). While trying to unravel the mechanism behind this phenomenon, they observed that N-protein specifically associated with the MH2 domain of smad3 (stronger binding affinity for phospho smad3) interrupted the interaction between smad3 and smad4, and enhanced the interaction between smad3 and transcriptional coactivator p300 in a dose-dependent manner. To further confirm the above data, they performed a chromatin immunoprecipitation assay at the SBE region of PAI-1 promoter in HPL1 cells and detected the presence of N-protein in the complex of smad3 and p300. Interestingly, however, N-protein inhibited TGFβ-induced apoptosis of HPL1 cells (it is a well established fact that smad3 activation induces apoptosis of HPL1 cells). Thus, N-protein appears to employ a clever mechanism whereby, on the one hand, it enhances the activity of the TGFβ signaling pathway, thus leading to enhanced expression of a subset of genes (such as ECM protein coding genes), and on the other hand, it blocks the programmed cell death of the host cell. It would be interesting to unravel the mechanism behind this unique property of the N-protein.

9.7.6 Upregulation of COX2 Production

Another major proinflammatory factor induced during viral infection is the cyclooxygenase-2 (COX2) protein. Using 293T cells expressing the N-protein, Yan et al. (2006) have shown that expression of the N-protein leads to upregulation of COX2 protein production in a transcriptional manner. They have further demonstrated that the N-protein directly binds to the NFkB response element present in the COX2 promoter through a 68 aa residue binding domain (aa 136–204) and activates its transcription.

Although the N-protein is known to associate with stretches of nucleic acids, to date there is no other documentation or prediction of its sequence-specific DNA binding activity (as a transcription factor). In such a scenario, the above observation, if reproducible in vivo, may really be a unique property of the N-protein and may further add to the established regulatory functions of the N-protein.

9.7.7 Upregulation of AP1 Activity

Exogenously expressed N-protein has been reported to enhance the DNA binding activity of c-fos, ATF-2, CREB-1, and fos B in an ELISA-based assay, thus suggesting an increase in AP1 activity in these cells (He et al. 2003). The mechanistic details and functional significance of this phenomenon remain to be elucidated.

9.7.8 Induction of Apoptosis

Earlier work done in our laboratory has shown that N-protein, when expressed in Cos-1 monkey kidney cells, induces apoptosis in the absence of growth factors. Attempts to understand the mechanism of programmed cell death revealed that the N-protein downmodulated the activity of prosurvival factors such as extracellular regulated kinase, Akt and bcl 2, and upregulated the activity of proapoptotic factors like caspase-3 and caspase-7 (Surjit et al. 2004b). However, this phenomenon was not observed in another cell line of epithelial lineage (huh7). The above observation was further confirmed by Zhang et al. (2007). They reported that serum starvation-induced apoptosis of N-protein-expressing COS-1 cells involved activation of mitochondrial pathway. Another elegant study done by Diemer et al. (2008) has further extended our understanding regarding the apoptotic property of the N-protein. Through a series of experiments involving both a model infection system of SARS-CoV and transient transfection of N-protein, the authors have confirmed that N-protein induces an intrinsic apoptotic pathway resulting in activation of caspase-9, which further leads to activation of caspase-3 and -6. Their data further revealed that these activated caspases cleave the N-protein at residues 400 and 403 and that nuclear localization of N-protein is an absolute requirement for cleavage. In addition, the authors have reported that the apoptosis-inducing ability of the N-protein is highly cell type specific. Only in cells where N-protein localizes to both nucleus and cytoplasm (Vero E6 and A549 cells), is it able to activate caspase and become cleaved; however, in cell lines where it localizes to the cytoplasm only (Caco2 and N-2a cells), no activation of caspase is observed. It remains to be studied whether this phenomenon is actually recapitulated in vivo.

9.7.9 Upregulation of Prothrombinase (hfgl2) Gene Transcription

A recent report by Han et al. (2008) revealed that, of all the SARS-CoV structural proteins, only N-protein specifically induced the transcription of prothrombinase gene in THP-1 and Vero cells. By performing luciferase reporter assay of *hfgl2* promoter in N-protein-expressing cells and electrophoretic mobility shift assay using N-protein-transfected cell lysate, they demonstrated that N-protein expression induced the binding of transcription factor C/EBPα to its cognate response element present in *hfgl2* promoter, leading to enhanced transcription of *hfgl2* gene. Since lungs of SARS patients have been shown to contain high amount of fibrin, the authors proposed that N-protein-mediated enhanced production of prothrombinase gene may contribute to the development of thrombosis in SARS patients.

9.7.10 Association with Host Cell Proteins

Luo et al. (2005) have reported the interaction between hnRNPA1 and N-protein by using a variety of biochemical and genetic assays. The interaction was found to be mediated through the middle region (aa 161–210) of N-protein. If relevant in vivo, this interaction may play a significant role in regulation of the viral RNA synthesis.

Another interesting study done by Luo et al. (2004a) has reported association between the N-protein and human cyclophylin A. By SPR (Surface Plasmn resonance) analysis they have shown it to be a high affinity interaction. Although the significance of this interaction is not known in vivo, they have proposed that this interaction might be crucial for viral infection. Notable is the fact that HIV-1 gag also binds with human cyclophylin A and this interaction is crucial for HIV infection (Gamble et al. 1996).

Recently, Zeng et al. (2008) have reported that N-protein associates with B23, a phosphoprotein in the nucleus. By performing in vivo coimmunoprecipitation in hela cells and GST pull-down assay using purified recombinant N-protein, the authors have demonstrated direct interaction between B23 and N-protein. The interaction domain has been mapped to amino acid residues 175–210 of N-protein, which include the SR-rich motif. B23 plays a key role in centrosome duplication during cell division. Phosphorylation of B23 at threonine-199 residue is known to regulate its function (Okuda et al. 2000, Tokuyama et al. 2001). In order to demonstrate the functional significance of N-protein interaction with B23 protein, the authors tested the phosphorylation status of threonine-199 residue of B23 in the presence of N-protein. Interestingly, N-protein was able to block threonine-199 phosphorylation. Based on this observation, the authors have proposed that N-protein exerts its effect on cell cycle deregulation by modulating the activity of B23 protein.

In summary, although several regulatory roles have been proposed for the SARS-CoV N-protein using a variety of in-vitro experimental systems, no clear evidence exists for their occurrence in vivo. In the absence of a suitable in-vivo experimental system, all these functions remain speculative.

9.8 N-Protein: An Efficient Diagnostic Tool

One of the most essential steps to limit the outbreak of any infectious disease is the ability to diagnose the causative agent at the earliest possible time, which can be achieved by detecting some of the markers that are specifically expressed by the pathogen or by identifying some of the host factors that are specifically produced during infection. N-protein, being one of the predominantly expressed proteins at the early stage of SARS-CoV infection, against which a strong antibody response is initiated by the host, has been proposed to be an attractive diagnostic tool.

In serum of SARS-CoV patients, the N-protein has been detected as early as day one of infection by ELISA using monoclonal antibodies against it (Che et al. 2004). Further, a comparative study to detect SARS-CoV-specific IgG, SARS-CoV RNA, and the N-protein during early stages of infection has demonstrated that the detection efficiency of the N-protein is significantly higher than the other two markers (Li et al. 2005b).

Researchers have been mainly focussing on two different strategies by which nucleocapsid can be used as a diagnostic tool: (1) development of efficient monoclonal antibodies against the N-protein, and (2) production of recombinantly expressed, highly purified N-protein for detection of N-protein-specific antibody in the host.

Using a phage display approach, Flego et al. (2005) have identified human antibody fragments that recognize distinct epitopes of the N-protein. These may help develop efficient reagents to detect N-protein in the infected host. Further, several laboratories have been trying to develop efficient monoclonal antibodies against the major immunodominant epitopes of the N-protein, that can be used in ELISA to detect SARS-CoV at an early stage of infection (Shang et al. 2005; Liu et al. 2003; He et al. 2005; Woo et al. 2005). In another interesting study, Liu et al. (2005) have developed an immunofluorescence assay using antirabbit N-protein antibody that can specifically detect N-protein from throat wash samples of SARS-CoV patients at day two of illness.

Several other workers have focussed on economical production of highly purified recombinant N-protein using a variety of heterologous expression systems that can be used in ELISA to detect N-protein-specific antibody in the patient sample. N-protein has been produced in abundant quantity using a codon-optimized gene in *E. coli* (Das and Suresh 2006). Saijo and coworkers have successfully expressed recombinant N-protein using a baculovirus expression system, which was found to be 92% efficient in neutralizing antibody assay (Saijo et al. 2005). In another study, Liu et al. have expressed full length N-protein using a yeast expression system

(Liu et al. 2004). However diagnostic use of recombinant N-protein has been a problematic issue because of several reasons as discussed below.

Bacterially expressed N-protein has been reported to produce false seropositivity owing to interference of bacterially derived antigens (Leung et al. 2006; Yip et al 2007). In addition, several studies have shown cross-reactivity between full-length N-protein of SARS and polyclonal antisera of group 1 animal coronaviruses, which may lead to faulty detection (Sun and Meng 2004). Another study done by Woo et al. also reported cross-reactivity of full-length recombinant N-protein with antisera of HCoV-OC43 and HCoV-229E infected patients, thus giving false positive results. They were able to minimize this false positivity by further verifying the ELISA results with Western blot assay using recombinant N and spike proteins of SARS-CoV (Woo et al. 2004).

Later, studies done by Qiu et al. and Bussmann et al. showed that the recombinantly expressed C-terminal of the N-protein acts more specifically in detecting SARS-CoV-specific antisera in comparison to full-length N-protein (Qiu et al. 2005; Bussmann et al. 2006). It is noteworthy that this region is predicted to encompass major antigenic sites of the N-protein.

In a recent report, Shin et al. (2007) demonstrated significantly higher efficacy of phosphorylated N-protein as a diagnostic antigen. They expressed the N-protein in insect cells, where it was phosphorylated by posttranslational modification. When the antigenicity of this protein was compared to that of a bacterially expressed N-protein (unphosphorylated) or to that of a dephosphorylated N-protein (by treatment with protein phosphatase 1) using SARS-positive or -negative patient serum, phosphorylated N-protein did not show any cross-reactivity with SARS-negative serum, thereby reducing the number of false positives. Also, the phosphorylated protein showed considerably stronger cross-reactivity with an N-protein-specific monoclonal antibody. Based on these observations, the authors have proposed the use of a phosphorylated N-protein as a better diagnostic agent.

Also, several reports have been published dealing with the detection of N-protein-specific IgM by ELISA or indirect immunofluorescent assay (Chang et al. 2004; Hsueh et al. 2004; Woo et al. 2004). However, in these studies, IgM antibodies became detectable later than IgG antibodies, which is in contrast to the phenomena observed in most other pathogens.

A recent report published by Yu et al. (2007) attempted to solve this problem by using a truncated N-protein (aa 122–422) as an antigen in IgM ELISA. They found the IgM response appeared three days before detection of the IgG response, which is in agreement with the results obtained from other known pathogens. Further, their results showed 100% specificity and sensitivity of the truncated protein in detecting N-protein-specific IgM from patients with laboratory confirmed SARS cases in comparison to healthy volunteers. The authors have suggested that the IgM capture ELISA using this truncated N-protein may be more effective in serodiagnosis of SARS-CoV at an earlier time.

In another interesting report, Woo et al. (2005) carried out comparative studies to evaluate the relative diagnostic efficacy of recombinantly expressed N and Spike proteins. They observed sensitivity of recombinant N-IgG ELISA to be significantly higher than that of recombinant S-IgG ELISA. The reverse was true in the case of IgM ELISA using recombinant N and S proteins. Based on this data, they have suggested the practise of ELISA for detection of IgM against both S and N proteins instead of N alone (Woo et al. 2005).

Taken together, all this data does support the notion that the N-protein may be used as an efficient diagnostic tool for detection of SARS-CoV infection. Nevertheless, production scale-up and further validation of specificity using patient samples will determine the possible clinical use of these reagents.

9.9 N-Protein: A Suitable Vaccine Candidate

One of the most clinically relevant uses of the N-protein can be its use as a protective vaccine against SARS-CoV infection. N-protein is one of the major antigens of the SARS-CoV. Also, N-protein analyzed from different patient samples shows least variation in the gene sequence (Tong et al. 2004), therefore indicating it to be a stable protein, which is a primary requirement for an efficient vaccine candidate.

Earlier studies carried in Collins', Rao's, and Li's laboratories have clearly shown that antiserum to the N-protein does not contain neutralizing antibodies against SARS-CoV (Buchholz et al. 2004; Pang et al. 2004; Liang et al. 2005). This may be attributed to the localization of N-protein inside the viral envelope, which will not be accessible to the antibody during infection. It is noteworthy that the most effective SARS-CoV structural protein that can induce neutralizing antibody production is the S-protein (Buchholz et al. 2004). The S-protein antibody could block viral infection with 100% efficiency. On the other hand, although unable to induce humoral immunity, expression of N-protein induced significant cytotoxic T-lymphocyte (CTL) response (Buchholz et al. 2004; Gao et al. 2003; Zhu et al. 2004). Induction of N-protein-specific CTLs will help limit the infection by lysing virus infected cells. This will also limit the spread of virus. Thus, N-protein-based vaccines may further augment the protection efficiency when coadministered with S-protein-based vaccine. Several laboratories have been exploring various strategies to evaluate the potential of N-protein as a vaccine candidate.

In an elegant work done by Kim et al. (2004), calreticulin-fused N-protein expressing vaccinia virus has been shown to generate potent N-protein-specific humoral and T-cell immune responses in mice. As reported by the authors, fusion with calreticulin specifically enhanced the efficiency and significantly reduced the titer of the challenging vector (vaccinia virus). The authors have proposed that N-protein may be the logical choice as a target antigen in the event of S-protein antibody-dependent enhancement (ADE) of infection. However, the ADE phenomenon has not been observed during spike-mediated vaccination (Buchholz et al. 2004). Another study done by Wang et al. (2005) has attempted to use plasmid DNA expressing S, M, and N proteins as an efficient vaccine candidate. Although they report the production of some B-cell and T-cell responses against N-protein, strong immune response was obtained for the S and M proteins, thus scaling down the choice of N-protein as a suitable candidate vaccine (Wang et al. 2005). A similar plasmid-mediated vaccination approach has also been reported by Zhao et al. (2004), in which they immunized mice with the DNA construct (pCI vector) expressing the N-protein. They too reported the generation of a robust B-cell and T-cell immune response in animals. Another group of workers has also reported successful use of the N-protein as a DNA vaccine. They immunized mice by intramucosal injection of the N-protein-expressing plasmid vector and were able to obtain specific humoral and T-cell responses (Zhu et al. 2004).

The N-protein has also been reported to be of potential interest as a peptide-based vaccine. A systematic study done by Liu et al. (2006) has revealed the immunodominant epitopes of the N-protein which could efficiently stimulate immune response. They have also deduced some conserved immunodominant epitopes in mouse, monkey, and humans, which may help in design of the vaccine.

A recent report published by Gao's laboratory provides further evidence regarding the efficiency of an N-protein-based vaccine (Zhao et al. 2007). By using overlapping synthetic peptides spanning the N-protein, they have identified dominant helper T-cell epitopes in the N-protein of SARS-CoV. Immunization of mice with peptides emcompassing these dominant TH cell epitopes resulted in strong cellular immunity in vivo. Priming with the helper peptides significantly accelerated the immune response induced by the N-protein. Further, by fusing with a conserved neutralizing epitope from the spike protein of SARS-CoV, two of the TH cell epitope-bearing peptides assisted in the production of higher titer neutralizing antibodies in vivo, in comparison to spike epitope alone or its mixture with TH epitope of N. Thus, it is practically possible to generate a better immune response by using a fusion of N and S protein. However, the TH epitopes identified in their report are specific to mouse, and will therefore not be useful for human. Nevertheless, their data provides useful information for the design of peptide-based anti-SARS-CoV vaccines.

Another interesting study conducted by Pei et al. (2005) reports the possible use of the N-protein as a mucosal vaccine candidate. They expressed the N-protein in *Lactobacillus lactis*, which is a food-grade bacteria, and challenged the mice either orally or intramucosally. As preliminary evidence, they were able to observe significant N-protein-specific IgG in the sera of orally challenged animals.

9.10 Future Perspective

It is a significant achievement for the research community that, within a short span of time, we have been able to obtain a more-or-less clear understanding regarding the structural and functional properties of the N-protein. However, it is a fact worth mentioning that all the studies done here were performed with in-vitro experiments, using recombinantly expressed N-protein, in isolation. So at present, all we can conclude is that the N-protein per se has the physical ability to perform the above described functions, in other words N-protein does bear the necessary signature sequence or motifs or conformation to perform these functions under suitable circumstances. Whether a similar event is recapitulated in vivo during viral infection will be dependent on several criteria: (1) the net effect of other viral factors on the activity of N-protein, (2) the net translation and turnover rate of N-protein, (3) a conducive intracellular milieu, and (4) the net modulation of an already skewed cellular pathway by other viral factors. Hence, it will be interesting to reevaluate the properties of N-protein in a SARS-CoV infection model. However, owing to the limited user-friendliness and accessibility of an infection system, we must probably still resort to in vitro systems for further analysis of the characteristics of N-protein. One of the better experimental systems has already been established by Chang's laboratory (Hsieh et al. 2005), in which all the structural proteins were coexpressed to form VLP in 293T cells. If this system can be further improved to optimize the rate of synthesis of these different proteins to a level near that in vivo, it will at least enable us to study the net effect of the N-protein with respect to other viral proteins. Further establishment of a replicon system may also be helpful. In addition, some of the interesting preliminary observations reported by several laboratories need to be analyzed in detail. To begin with, the reported interaction of the N-protein with the genomic RNA packaging signal needs to be further characterized and mapped. Since the oligomerization domain and the RNA binding regions of the N-protein overlap with each other, the suggested possibility of regulated genome incorporation and capsid assembly should be further characterized with the aid of a replicon system or a particle assembly system. In addition, the reported ability of the N-protein to modulate different cellular pathways should be further characterized in the particle assembly system or at least in the presence of other viral accessory proteins.

The most unique and significant property of the N-protein revealed by preliminary studies is its ability to act as a sequence-specific DNA binding factor. It has been shown to bind the NFkB response element of *COX2* promoter and to enhance *COX2* gene expression. This activity may be further empowering the N-protein to manipulate the entire gene expression programme of the infected cell. Therefore, studies should be initiated to analyze this phenomenon in detail. It seems to deserve so much attention because another study done by Palese's laboratory has proved the ability of the N-protein to inhibit NFkB activity, which results in inhibition of IFN synthesis. Further, Liao et al. (2005) have reported the activation of NFkB by N-protein in Vero E6 cells and He et al. (2005) failed to detect any change in NFkB activity in the same cells. Therefore it needs to be clarified whether N-protein enhances NFkB activity and, if so, whether upregulation of *COX2* transcription by direct DNA binding is a property specific to that promoter or whether it is a global phenomenon. In such a scenario, there may be complicated cross-talk between the ability of N-protein to deregulate the expression of COX2 and IFN in infected cells.

Lastly, the N-protein is known to be the most abundantly expressed protein of the SARS-CoV. Therefore, any information generated from the analysis of this protein, whether in vivo or ex vivo, will definitely help to increase our understanding of the biology of SARS-CoV and may someday help to design better protective tools against it.

Acknowledgments The authors wish to thank Ms. Alisha Lal for helping out in typing and formatting this review. We apologize to all those colleagues whose work we might have omitted to cite in this article.

References

Border WA, Noble NA (1994) Transforming growth factor beta in tissue fibrosis. N Engl J Med 331:1286–1292
Buchholz UJ, Bukreyev A, Yang L, Lamirande EW, Murphy BR, Subbarao K, Collins PL (2004) Contributions of the structural proteins of severe acute respiratory syndrome coronavirus to protective immunity. Proc Natl Acad Sci USA 101:9804–9809
Bussmann BM, Reiche S, Jacob LH, Braun JM, Jassoy C (2006) Antigenic and cellular localisation analysis of the severe acute respiratory syndrome coronavirus nucleocapsid protein using monoclonal antibodies. Virus Res 122:119–126
Chang WT, Kao CL, Chung MY, Chen SC, Lin SJ, Chiang WC, Chen SY, Su CP, Hsueh PR, Chen WJ, Chen PJ, Yang PC (2004) SARS exposure and emergency department workers. Emerg Infect Dis 10:1117–1119
Chang CK, Sue SC, Yu TH, Hsieh CM, Tsai CK, Chiang YC, Lee SJ, Hsiao HH, Wu WJ, Chang WL, Lin CH, Huang TH (2006) Modular organization of SARS coronavirus nucleocapsid protein. J Biomed Sci 13:59–72
Chang CK, Hsu YL, Chang YH, Chao FA, Wu MC, Huang YS, Hu CK, Huang TH (2008) Multiple nucleic acid binding sites and intrinsic disorder of SARS coronavirus nucleocapsid protein – implication for ribonucleocapsid protein packaging. J Virol 83(5):2255–2264
Che XY, Hao W, Wang Y, Di B, Yin K, Xu YC, Feng CS, Wan ZY, Cheng VC, Yuen KY (2004) Nucleocapsid protein as early diagnostic marker for SARS. Emerg Infect Dis 10:1947–1949
Chen CY, Chang CK, Chang YW, Sue SC, Bai HI, Riang L, Hsiao CD, Huang TH, (2007) Structure of the SARS Coronavirus nucleocapsid protein RNA-binding dimerization domain suggests a mechanism for helical packaging of viral RNA. J Mol Biol 368(4):1075–1086
Cimarelli A, Luban J (1999) Translation elongation factor 1-alpha interacts specifically with the HIV-1 Gag polyprotein. J Virol 73:5388–5401
Das D, Suresh MR (2006) Copious production of SARS-CoV nucleocapsid protein employing codon optimized synthetic gene. J Virol Methods 137:343–346
Diemer C, Schneider M, Seebach J, Quaas J, Frösner G, Schätzl HM, Gilch S (2008) Cell type-specific cleavage of nucleocapsid protein by effector caspases during SARS coronavirus infection. J Mol Biol 376:23–34
Ding Y, Wang H, Shen H, Li Z, Geng J, Han H, Cai J, Li X, Kang W, Weng D, Lu Y, Wu D, He L, Yao K (2003) The clinical pathology of severe acute respiratory syndrome (SARS): a report from China. J Pathol 200:282–289
Fan Z, Zhuo Y, Tan X, Zhou Z, Yuan J, Qiang B, Yan J, Peng X, Gao GF (2006) SARS-CoV nucleocapsid protein binds to hUbc9, a ubiquitin conjugating enzyme of the sumoylation system. J Med Virol 78:1365–1373
Fang X, Ye LB, Zhang Y, Li B, Li S, Kong L, Wang Y, Zheng H, Wang W, Wu Z (2006) Nucleocapsid amino acids 211 to 254, in particular, tetrad glutamines, are essential for the interaction between the nucleocapsid and membrane proteins of SARS-associated coronavirus. J Microbiol 44:577–580

Flego M, Di Bonito P, Ascione A, Zamboni S, Carattoli A, Grasso F, Cassone A, Cianfriglia M (2005) Generation of human antibody fragments recognizing distinct epitopes of the nucleocapsid (N) SARS-CoV protein using a phage display approach. BMC Infect Dis 5:73

Gamble TR, Vajdos FF, Yoo S, Worthylake DK, Houseweart M, Sundquist WI, Hill CP (1996) Crystal structure of human cyclophilin A bound to the amino-terminal domain of HIV-1 capsid. Cell 87:1285–1294

Gao W, Tamin A, Soloff A, D'Aiuto L, Nwanegbo E, Robbins PD, Bellini WJ, Barratt-Boyes S, Gambotto A (2003) Effects of a SARS-associated coronavirus vaccine in monkeys. Lancet 362:1895–1896

Han M, Yan W, Huang Y, Yao H, Wang Z, Xi D, Li W, Zhou Y, Hou J, Luo X, Ning Q (2008) The nucleocapsid protein of SARS-CoV induces transcription of hfgl2 prothrombinase gene dependent on C/EBP alpha. J Biochem 144:51–62

He R, Leeson A, Andonov A, Li Y, Bastien N, Cao J, Osiowy C, Dobie F, Cutts T et al (2003) Activation of AP-1 signal transduction pathway by SARS coronavirus nucleocapsid protein. Biochem Biophys Res Commun 311:870–876

He R, Dobie F, Ballantine M, Leeson A, Li Y, Bastien N, Cutts T, Andonov A, Cao J, Booth TF, Plummer FA, Tyler S, Baker L, Xm Li (2004) Analysis of multimerization of the SARS coronavirus nucleocapsid protein. Biochem Biophys Res Commun 316:476–483

He Q, Du Q, Lau S, Manopo I, Lu L, Chan SW, Fenner BJ, Kwang J (2005) Characterization of monoclonal antibody against SARS coronavirus nucleocapsid antigen and development of an antigen capture ELISA. J Virol Methods 127:46–53

Hsieh PK, Chang SC, Huang CC, Lee TT, Hsiao CW, Kou YH, Chen IY, Chang CK, Huang TH, Chang MF (2005) Assembly of severe acute respiratory syndrome coronavirus RNA packaging signal into virus-like particles is nucleocapsid dependent. J Virol 79:13848–13855

Hsueh PR, Huang LM, Chen PJ, Kao CL, Yang PV (2004) Chronological evolution of IgM, IgA, IgG and neutralisation antibodies after infection with SARS-associated coronavirus. Clin Microbiol Infect 10:1062–1066

Huang Q, Yu L, Petros AM, Gunasekera A, Liu Z, Xu N, Hajduk P, Mack J, Fesik SW, Olejniczak ET (2004) Structure of the N-terminal RNA-binding domain of the SARS CoV nucleocapsid protein. Biochemistry 43:6059–6063

Kim TW, Lee JH, Hung CF, Peng S, Roden R, Wang MC, Viscidi R, Tsai YC, He L, Chen PJ, Boyd DA, Wu TC (2004) Generation and characterization of DNA vaccines targeting the nucleocapsid protein of severe acute respiratory syndrome coronavirus. J Virol 78:4638–4645

Kopecky-Bromberg SA, Martinez-Sobrido L, Frieman M, Baric RA, Palese P (2007) Severe acute respiratory syndrome coronavirus open reading frame (ORF) 3b, ORF 6, and nucleocapsid proteins function as interferon antagonists. J Virol 81:548–557

Krokhin O, Li Y, Andonov A, Feldmann H, Flick R, Jones S, Stroeher U, Bastien N, Dasuri KV, Cheng K, Simonsen JN, Perreault H, Wilkins J, Ens W, Plummer F, Standing KG (2003) Mass spectrometric characterization of proteins from the SARS Virus: a preliminary report. Mol Cell Proteomics 2:346–356

Kurasawa Y, Watanabe Y, Numata O (1996) Characterization of F-actin bundling activity of Tetrahymena elongation factor 1 alpha investigated with rabbit skeletal muscle actin. Zool Sci 13:371–375

Leung DT, van Maren WW, Chan FK, Chan WS, Lo AW, Ma CH, Tam FC, To KF, Chan PK, Sung JJ, Lim PL (2006) Extremely low exposure of a community to severe acute respiratory syndrome coronavirus: false seropositivity due to use of bacterially derived antigens. J Virol 80:8920–8928

Li FQ, Xiao H, Tam JP, Liu DX (2005a) Sumoylation of the nucleocapsid protein of severe acute respiratory syndrome coronavirus. FEBS Lett 579:2387–2396

Li YH, Li J, Liu XE, Wang L, Li T, Zhou YH, Zhuang H (2005b) Detection of the nucleocapsid protein of severe acute respiratory syndrome coronavirus in serum: comparison with results of other viral markers. J Virol Methods 130:45–50

Liang MF, Du RL, Liu JZ, Li C, Zhang QF, Han LL, Yu JS, Duan SM, Wang XF, Wu KX, Xiong ZH, Jin Q, Li DX (2005) SARS patients-derived human recombinant antibodies to S and M proteins efficiently neutralize SARS-coronavirus infectivity. Biomed Environ Sci 18:363–374

Liao QJ, Ye LB, Timani KA et al (2005) Activation of NF-kappaB by the full-length nucleocapsid protein of the SARS coronavirus. Acta Biochim Biophys Sin 37:607–612

Liu G, Hu S, Hu Y, Chen P, Yin J, Wen J, Wang J, Lin L, Liu J, You B, Yin Y, Li S, Wang H, Ren Y, Ji J, Zhao X, Sun Y, Zhang X, Fang J, Wang J, Liu S, Yu J, Zhu H, Yang H (2003) The C-terminal portion of the nucleocapsid protein demonstrates SARS-CoV antigenicity. Genomics Proteomics Bioinformatics 1:193–197

Liu RS, Yang KY, Lin J, Lin YW, Zhang ZH, Zhang J, Xia NS (2004) High-yield expression of recombinant SARS coronavirus nucleocapsid protein in methylotrophic yeast Pichia pastoris. World J Gastroenterol 10:3602–3607

Liu IJ, Chen PJ, Yeh SH, Chiang YP, Huang LM, Chang MF, Chen SY, Yang PC, Chang SC, Wang WK (2005) Immunofluorescence assay for detection of the nucleocapsid antigen of the severe acute respiratory syndrome (SARS)-associated coronavirus in cells derived from throat wash samples of patients with SARS. J Clin Microbiol 43:2444–2448

Liu SJ, Leng CH, Lien SP, Chi HY, Huang CY, Lin CL, Lian WC, Chen CJ, Hsieh SL, Chong P (2006) Immunological characterizations of the nucleocapsid protein based SARS vaccine candidates. Vaccine 24:3100–3108

Luo C, Luo H, Zheng S, Gui C, Yue L, Yu C, Sun T, He P, Chen J, Shen J, Luo X, Li Y, Liu H, Bai D, Shen J, Yang Y, Li F, Zuo J, Hilgenfeld R, Pei G, Chen K, Shen X, Jiang H (2004a) Nucleocapsid protein of SARS coronavirus tightly binds to human cyclophilin A. Biochem Biophys Res Commun 321:557–565

Luo H, Ye F, Sun T, Yue L, Peng S, Chen J, Li G, Du Y, Xie Y, Yang Y, Shen J, Wang Y, Shen X, Jiang H (2004b) In vitro biochemical and thermodynamic characterization of nucleocapsid protein of SARS. Biophys Chem 112:15–25

Luo H, Chen Q, Chen J, Chen K, Shen X, Jiang H (2005) The nucleocapsid protein of SARS coronavirus has a high binding affinity to the human cellular heterogeneous nuclear ribonucleoprotein A1. FEBS Lett 579:2623–2628

Luo H, Chen J, Chen K, Shen X, Jiang H (2006) Carboxyl terminus of severe acute respiratory syndrome coronavirus nucleocapsid protein: self-association analysis and nucleic acid binding characterization. Biochemistry 45:11827–11835

Nicholls JM, Poon LL, Lee KC, Ng WF, Lai ST, Leung CY, Chu CM, Hui PK, Mak KL, Lim W, Yan KW, Chan KH, Tsang NC, Guan Y, Yuen KY, Peiris JS (2003) Lung pathology of fatal severe acute respiratory syndrome. Lancet 361:1773–1778

Okuda M, Horn HF, Tarapore P, Tokuyama Y, Smulian AG, Chan PK, Knudsen ES, Hofmann IA, Snyder JD, Bove KE, Fukasawa K (2000) Nucleophosmin/B23 is a target of CDK2/cyclin E in centrosome duplication. Cell 103:127–140

Pang H, Liu Y, Han X, Xu Y, Jiang F, Wu D, Kong X, Bartlam M, Rao Z (2004) Protective humoral responses to severe acute respiratory syndrome-associated coronavirus: implications for the design of an effective protein-based vaccine. J Gen Virol 85:3109–3113

Pei H, Liu J, Cheng Y, Sun C, Wang C, Lu Y, Ding J, Zhou J, Xiang H (2005) Expression of SARS-coronavirus nucleocapsid protein in Escherichia coli and Lactococcus lactis for serodiagnosis and mucosal vaccination. Appl Microbiol Biotechnol 68:220–227

Qiu M, Wang J, Wang H, Chen Z, Dai E, Guo Z, Wang X, Pang X, Fan B, Wen J, Wang J, Yang R (2005) Use of the COOH portion of the nucleocapsid protein in an antigen-capturing enzyme-linked immunosorbent assay for specific and sensitive detection of severe acute respiratory syndrome coronavirus. Clin Diagn Lab Immunol 12:474–476

Roberts AB, Tian F, Byfield SD, Stuelten C, Ooshima A, Saika S, Flanders KC (2006) Smad3 is key to TGF-beta-mediated epithelial-to-mesenchymal transition, fibrosis, tumor suppression and metastasis. Cytokine Growth Factor Rev 17:19–27

Rota PA, Oberste MS, Monroe SS, Nix WA, Campagnoli R, Icenogle JP, Penaranda S, Bankamp B, Maher K, Chen MH, Tong S, Tamin A, Lowe L, Frace M, DeRisi JL, Chen Q, Wang D, Erdman DD, Peret TC, Burns C, Ksiazek TG, Rollin PE, Sanchez A, Liffick S, Holloway B, Limor J, McCaustland K, Olsen-Rasmussen M, Fouchier R, Gunther S, Osterhaus AD, Drosten C, Pallansch MA, Anderson LJ, Bellini WJ (2003) Characterization of a novel coronavirus associated with severe acute respiratory syndrome. Science 300: 1394–1399

Rowland RR, Chauhan V, Fang Y, Pekosz A, Kerrigan M, Burton MD (2005) Intracellular localization of the severe acute respiratory syndrome coronavirus nucleocapsid protein: absence of nucleolar accumulation during infection and after expression as a recombinant protein in vero cells. J Virol 79:11507–11512

Saijo M, Ogino T, Taguchi F, Fukushi S, Mizutani T, Notomi T, Kanda H, Minekawa H, Matsuyama S, Long HT, Hanh NT, Kurane I, Tashiro M, Morikawa S (2005) Recombinant nucleocapsid protein-based IgG enzyme-linked immunosorbent assay for the serological diagnosis of SARS. J Virol Methods 125:181–186

Shang B, Wang XY, Yuan JW, Vabret A, Wu XD, Yang RF, Tian L, Ji YY, Deubel V, Sun B (2005) Characterization and application of monoclonal antibodies against N protein of SARS-coronavirus. Biochem Biophys Res Commun 336:110–117

Shin GC, Chung YS, Kim IS, Cho HW, Kang C (2007) Antigenic characterization of severe acute respiratory syndrome-coronavirus nucleocapsid protein expressed in insect cells: The effect of phosphorlation on immunoreactivity and specificity. Virus Res 127:71–80

Spiegel M, Pichlmair A, Martinez-Sobrido L, Cros J, Garcia-Sastre A, Haller O, Weber F (2005) Inhibition of beta interferon induction by severe acute respiratory syndrome coronavirus suggests a two-step model for activation of interferon regulatory factor 3. J Virol 79: 2079–2086

Sun ZF, Meng XJ (2004) Antigenic cross-reactivity between the nucleocapsid protein of severe acute respiratory syndrome (SARS) coronavirus and polyclonal antisera of antigenic group I animal coronaviruses: implication for SARS diagnosis. J Clin Microbiol. 42: 2351–2352

Surjit M, Liu B, Kumar P, Chow VT, Lal SK (2004a) The nucleocapsid protein of the SARS coronavirus is capable of self-association through a C-terminal 209 amino acid interaction domain. Biochem Biophys Res Commun 317:1030–1036

Surjit M, Liu B, Jameel S, Chow VT, Lal SK (2004b) The SARS coronavirus nucleocapsid protein induces actin reorganization and apoptosis in COS-1 cells in the absence of growth factors. Biochem J 383:13–18

Surjit M, Kumar R, Mishra RN, Reddy MK, Chow VT, Lal SK (2005) The severe acute respiratory syndrome coronavirus nucleocapsid protein is phosphorylated and localizes in the cytoplasm by 14-3-3-mediated translocation. J Virol 79:11476–11486

Surjit M, Liu B, Chow VT, Lal SK (2006) The nucleocapsid protein of severe acute respiratory syndrome-coronavirus inhibits the activity of cyclin-cyclin-dependent kinase complex and blocks S phase progression in mammalian cells. J Biol Chem 281:10669–10681

Timani KA, Liao Q, Ye L, Zeng Y, Liu J, Zheng Y, Ye L, Yang X, Lingbao K, Gao J, Zhu Y (2005) Nuclear/nucleolar localization properties of C-terminal nucleocapsid protein of SARS coronavirus. Virus Res 114:23–34

Tokuyama Y, Horn HF, Kawamura K, Tarapore P, Fukasawa K (2001) Specific phosphorylation of nucleophosmin on Thr(199) by cyclin-dependent kinase 2-cyclin E and its role in centrosome duplication. J Biol Chem 276:21529–21537

Tong S, Lingappa JR, Chen Q, Shu B, LaMonte AC, Cook BT, Birge C, Chern SW, Liu X, Galloway R, le Mai Q, Ng WF, Yang JY, Butany J, Comer JA, Monroe SS, Beard SR, Ksiazek TG, Erdman D, Rota PA, Pallansch MA, Anderson LJ (2004) Direct sequencing of SARS-coronavirus S and N genes from clinical specimens shows limited variation. J Infect Dis 190:1127–1131

Wang Y, Wu X, Wang Y, Li B, Zhou H, Yuan G, Fu Y, Luo Y (2004) Low stability of nucleocapsid protein in SARS virus. Biochemistry 43:11103–11108

Wang Z, Yuan Z, Matsumoto M, Hengge UR, Chang YF (2005) Immune responses with DNA vaccines encoded different gene fragments of severe acute respiratory syndrome coronavirus in BALB/c mice. Biochem Biophys Res Commun 327:130–135

Woo PCY, Lau SKP, Wong BHL, Chan KH, Chu CM, Tsoi HW, Huang Y, Peiris JSM, Yuen KY (2004) Longitudinal profile of immunoglobulin G (IgG), IgM, and IgA antibodies against the severe acute respiratory syndrome (SARS) coronavirus nucleocapsid protein in patients with pneumonia due to the SARS coronavirus. Clin Diagn Lab Immunol 11:665–668

Woo PC, Lau SK, Wong BH, Tsoi HW, Fung AM, Kao RY, Chan KH, Peiris JS, Yuen KY (2005) Differential sensitivities of severe acute respiratory syndrome (SARS) coronavirus spike polypeptide enzyme-linked immunosorbent assay (ELISA) and SARS coronavirus nucleocapsid protein ELISA for serodiagnosis of SARS coronavirus pneumonia. J Clin Microbiol 43:3054–3058

Wu CH, Yeh SH, Tsay YG, Shieh YH, Kao CL, Chen YS, Wang SH, Kuo TJ, Chen DS, Chen PJ (2008) Glycogen synthase kinase-3 regulates the phosphorylation of sars-coronavirus nucleocapsid protein and viral replication. J Biol Chem 284(8):5229–5239

Yan X, Hao Q, Mu Y, Timani KA, Ye L, Zhu Y, Wu J (2006) Nucleocapsid protein of SARS-CoV activates the expression of cyclooxygenase- 2 by binding directly to regulatory elements for nuclear factor-kappa B and CCAAT/enhancer binding protein. Int J Biochem Cell Biol 38:1417–1428

Yang F, Demma M, Warren V, Dharmawardhane S, Condeelis J (1990) Identification of an actin-binding protein from *Dictyostelium* as elongation factor 1a. Nature 347:494–496

Yip CW, Hon CC, Zeng F, Chow KY, Chan KH, Peiris JS, Leung FC (2007) Naturally occurring anti-Escherichia coli protein antibodies in the sera of healthy humans cause analytical interference in a recombinant nucleocapsid protein-based enzyme-linked immunosorbent assay for serodiagnosis of severe acute respiratory syndrome. Clin Vaccine Immunol 14:99–101

You J, Dove BK, Enjuanes L, DeDiego ML, Alvarez E, Howell G, Heinen P, Zambon M, Hiscox JA (2005) Subcellular localization of the severe acute respiratory syndrome coronavirus nucleocapsid protein. J Gen Virol 86:3303–3310

Yu IM, Gustafson CL, Diao J, Burgner JW 2nd, Li Z, Zhang J, Chen J (2005) Recombinant severe acute respiratory syndrome (SARS) coronavirus nucleocapsid protein forms a dimer through its C-terminal domain. J Biol Chem 280:23280–23286

Yu IM, Oldham ML, Zhang J, Chen J (2006) Crystal structure of the severe acute respiratory syndrome (SARS) coronavirus nucleocapsid protein dimerization domain reveals evolutionary linkage between corona- and arteriviridae. J Biol Chem 281:17134–17139

Yu F, Le MQ, Inoue S, Hasebe F, Parquet Mdel C, Morikawa S, Morita K (2007) Recombinant truncated nucleocapsid protein as antigen in a novel immunoglobulin M capture enzyme-linked immunosorbent assay for diagnosis of severe acute respiratory syndrome coronavirus infection. Clin Vaccine Immunol 14:146–149

Zakhartchouk AN, Viswanathan S, Mahony JB, Gauldie J, Babiuk LA (2005) Severe acute respiratory syndrome coronavirus nucleocapsid protein expressed by an adenovirus vector is phosphorylated and immunogenic in mice. J Gen Virol 86:211–215

Zeng Y, Ye L, Zhu S, Zheng H, Zhao P, Cai W, Su L, She Y, Wu Z (2008) The nucleocapsid protein of SARS-associated coronavirus inhibits B23 phosphorylation. Biochem Biophys Res Commun 369:287–291

Zhang L, Wei L, Jiang D, Wang J, Cong X, Fei R (2007) SARS-CoV nucleocapsid protein induced apoptosis of COS-1 mediated by the mitochondrial pathway. Artif Cells Blood Substit Immobil Biotechnol 35:237–253

Zhao P, Cao J, Zhao LJ, Qin ZL, Ke JS, Pan W, Ren H, Yu JG, Qi ZT (2004) Immune responses against SARS-coronavirus nucleocapsid protein induced by DNA vaccine. Virology 331:128–135

Zhao J, Huang Q, Wang W, Zhang Y, Lv P, Gao XM (2007) Identification and characterization of dominant helper T-cell epitopes in the nucleocapsid protein of severe acute respiratory syndrome coronavirus. J Virol 81:6079–6088

Zhao X, Nicholls JM, Chen YG (2008) Severe acute respiratory syndrome-associated coronavirus nucleocapsid protein interacts with Smad3 and modulates transforming growth factor-beta signaling. J Biol Chem 283:3272–3280

Zheng B, He ML, Wong KL, Lum CT, Poon LL, Peng Y, Guan Y, Lin MC, Kung HF (2004) Potent inhibition of SARS-associated coronavirus (SCOV) infection and replication by type I interferons (IFNalpha/ beta) but not by type II interferon (IFN-gamma). J Interferon Cytokine Res 24:388–390

Zhou B, Liu J, Wang Q, Liu X, Li X, Li P, Ma Q, Cao C (2008) The nucleocapsid protein of severe acute respiratory syndrome coronavirus inhibits cell cytokinesis and proliferation by interacting with translation elongation factor 1alpha. J Virol 82:6962–6971

Zhu MS, Pan Y, Chen HQ, Shen Y, Wang XC, Sun YJ, Tao KH (2004) Induction of SARS-nucleoprotein-specific immune response by use of DNA vaccine. Immunol Lett 92:237–243

Chapter 10
SARS Coronavirus Accessory Gene Expression and Function

Scott R. Schaecher and Andrew Pekosz

Abstract Coronavirus genomes are single-stranded positive-sense RNA that are transcribed into a nested set of 3′ coterminal subgenomic RNAs for gene expression. Members of the *Coronaviridae* express canonical polymerase genes, as well as structural genes, including S, E, M, and N, but also express a highly divergent set of accessory genes whose open reading frames are interspersed among the structural genes within the 3′ one-third of the viral genome. The accessory genes are thought to contain "luxury" functions that are often not required for in-vitro virus replication. The severe acute respiratory syndrome coronavirus (SARS-CoV) expresses eight such accessory genes (ORF3a, -3b, -6, -7a, -7b, -8a, -8b, and -9b), the most of any known coronavirus. This chapter will review our current knowledge of expression, structure, and function of each of the SARS-CoV accessory genes.

10.1 Introduction

Coronaviruses have large, positive-stranded, RNA genomes ranging from 27 to 31 kb in size (Fig. 10.1). The 3′ end of the genome contains open reading frames (ORFs) encoding for canonical structural proteins including envelope (E), membrane (M), spike (S), and the nucleocapsid protein (N). Interspersed among the structural genes are novel ORFs that encode additional virus-specific proteins. The novel ORFs have been aptly renamed "accessory genes." Coronavirus genomes differ greatly in number, location, and size of the accessory genes; as detailed in

A. Pekosz (✉)
W. Harry Feinstone Department of Molecular Microbiology and Immunology, Johns Hopkins University, Bloomberg School of Public Health, 615 North Wolfe St. Suite E5132, Baltimore, MD 21205, USA
e-mail: apekosz@jhsph.edu

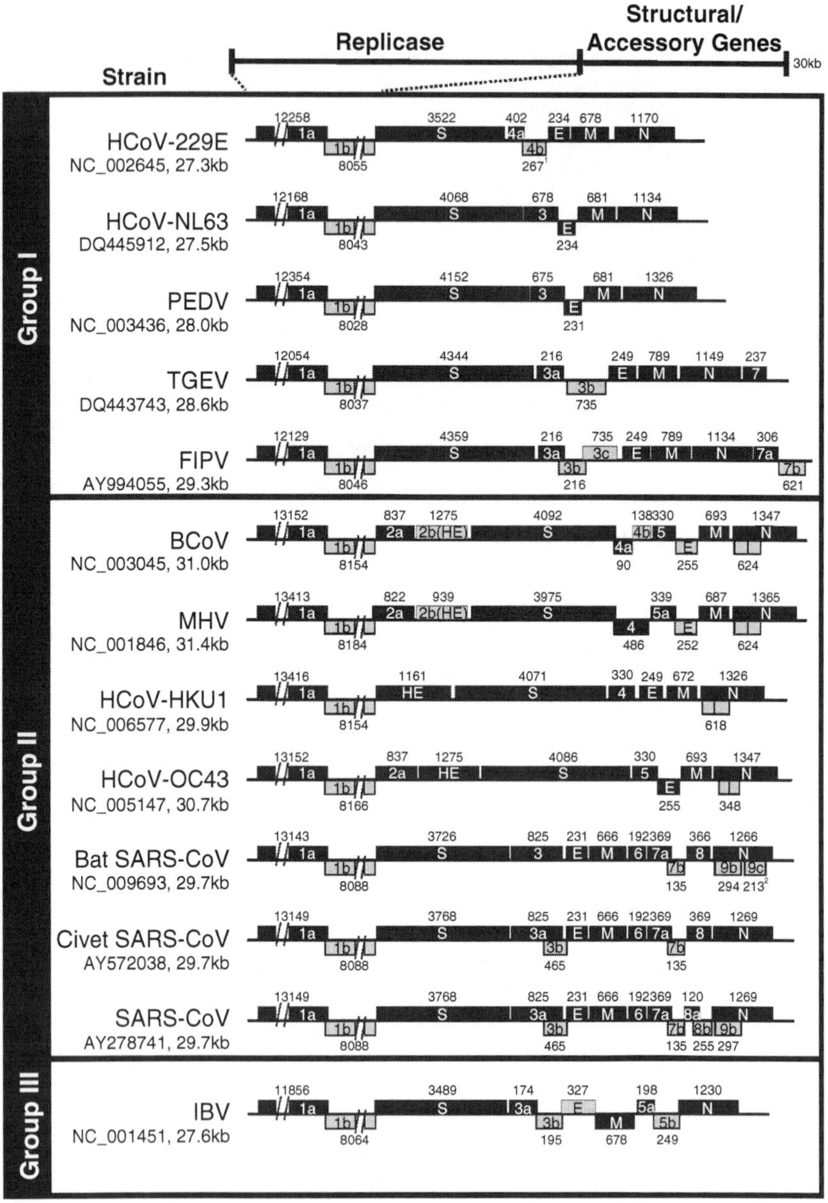

Fig. 10.1 Genomic diversity among coronaviruses. Genomic organization of representative virus strains from groups I, II, and III coronaviruses are detailed. Relative locations of each replicase, structural, and accessory ORF are highlighted for each virus, with primary ORFs on respective subgenomic RNAs in *black boxes* and downstream ORFs in multicistronic genes *shaded in gray*. The Genbank accession ID and genome size for each annotated strain is displayed and nucleotide length of each ORF is shown

Fig. 10.1, viruses such as hCoV-NL63 can express as few as one, or in the case of SARS-CoV and the recently identified whale coronavirus (Mihindukulasuriya et al. 2008), as many as eight accessory proteins.

Three independent reports simultaneously detailed the SARS-CoV genomic sequence and annotations (Marra et al. 2003; Rota et al. 2003; Thiel et al. 2003), which led to some confusion regarding the nomenclature of the accessory genes. Annotations by Thiel et al. were the least stringent and identified the eight currently accepted accessory genes; additionally, nomenclature consistent with other coronaviruses was given to them which will be used throughout this chapter. Numerous studies have reported on the expression of all eight SARS-CoV accessory genes. Few polymorphisms have been identified in the more than 100 sequenced SARS-CoV isolates, suggesting a selective pressure to maintain their expression. This chapter will review current data regarding the expression, structure, function, and molecular biology of the SARS-CoV accessory genes (Table 10.1).

Table 10.1 Expression and subcellular localization of SARS-CoV accessory proteins

Viral gene	Size	Subcellular localization	Structure	Coding strategy
ORF3a	37 kDa[a], 31 kDa, *274aa*	Golgi, plasma membrane	C-endo, triple membrane-spanning protein	Direct translation
ORF3b	18 kDa, *154aa*	Nucleolus[b], mitochondria	Unknown	Unknown
ORF6	7.5 kDa, *63aa*	ER, Golgi	N-endo:C-endo, amphipathic N-terminus membrane associated	Direct translation
ORF7a	17.5 kDa[c], 15 kDa, *122aa*	Golgi	Type I integral membrane protein, seven-stranded beta sandwich ectodomain	Direct translation
ORF7b	5.5 kDa, *44aa*	Golgi	C-endo, single transmembrane domain	Leaky scanning
ORF8ab[d]	14.0 kDa[e], *122aa*	ER	N-terminal signal sequence, membrane associated	Direct translation
ORF8a	5.3 kDa, *39aa*	Cytoplasm, mitochondria	N-terminal signal sequence, unknown	Direct translation
ORF8b	9.6 kDa, *84aa*	Cytoplasm	Unknown	Unknown or not translated
ORF9b	11 kDa, *98aa*	ER	Dimeric tent-like beta structure with central hydrophobic cavity	Unknown

[a] *O*-Glycosylated molecular weight
[b] Predominant subcellular localized site
[c] Prior to cleavage of N-terminal signal sequence
[d] ORF8ab exists as a single open reading frame in animal and early human SARS-CoV isolates
[e] ORF8ab is *N*-glycosylated. This is the predicted native, nonglycosylated molecular weight

10.2 SARS-CoV Accessory Gene Expression and Function

10.2.1 ORF3a and ORF3b

ORF3a is the largest of the SARS-CoV accessory proteins at 274 amino acids in length with a native molecular weight of 31 kDa (Tan et al. 2004c). ORF3a is expressed from the 5′-most ORF present in sgRNA3, which contains a minimal transcriptional regulatory sequence (TRS, 5′-ACGAAC-3′) (Snijder et al. 2003) immediately upstream of the translation initiation codon. Expression has been confirmed in virus-infected cells (Tan et al. 2004c; Yu et al. 2004; Zeng et al. 2004) and antibodies recognizing ORF3a have been detected in SARS-CoV patient convalescent sera (Guo et al. 2004; Tan et al. 2004b; Yu et al. 2004; Zeng et al. 2004; Qiu et al. 2005; Yount et al. 2005; Zhong et al. 2005).

ORF3a is a hydrophobic, triple membrane-spanning protein with an N-terminal ectodomain and a C-terminal intracellular domain (Tan et al. 2004c; Lu et al. 2006) and is predicted to contain an amino-terminal signal sequence spanning residues 1–16 (Law et al. 2005). ORF3a is O-glycosylated (Oostra et al. 2006) and localizes predominantly to the Golgi and plasma membrane (Tan et al. 2004c; Yu et al. 2004; Ito et al. 2005; Yuan et al. 2005a; Lu et al. 2006; von Brunn et al. 2007). Plasma membrane trafficking and endocytosis is mediated by a YxxΦ motif and a diacidic motif in the cytoplasmic tail that are juxtaposed to one another within a 14 amino acid region (Tan et al. 2004c). This region of the cytoplasmic tail has also been shown to mediate G1 phase cell cycle arrest in HEK-293T cells transfected with ORF3a cDNA (Yuan et al. 2007). No evidence has been presented regarding control of cell cycle progression in SARS-CoV infected cells, so the functional significance of this observation is unknown.

Analysis of virus-like particles and purified virions demonstrated that ORF3a is packaged into virions (Zeng et al. 2004; Ito et al. 2005; Shen et al. 2005). Potential interactions between ORF3a and the viral proteins S, E, M, and ORF7a have been demonstrated (Tan et al. 2004c; Zeng et al. 2004). Antibodies specific to ORF3a have shown the surprising ability to neutralize virus infection in multiple ways. Because some ORF3a is trafficked to the cell surface, antibodies recognizing the amino-terminal extracellular domain initiate complement-mediated lysis in the presence of functional complement proteins (Zhong et al. 2006). ORF3a N-terminus-specific antibodies are also capable of neutralizing infectious virus in microneutralization assays (Akerstrom et al. 2006). Although the precise mechanism of virus neutralization is unknown, these studies suggest that ORF3a may be a potential vaccine target.

ORF3a forms homodimers and homotetramers via disulfide linkages and the oligomeric complexes have been proposed to function as potassium-permeable ion channels in *Xenopus* oocytes (Lu et al. 2006). The authors propose a role for ORF3a in virus release; however, an ORF3a deletion virus is not defective in virus entry, replication, assembly, or budding (Yount et al. 2005). Additional evidence for ORF3a involvement in the virus lifecycle stems from the observation that ORF3a directly binds the 5′ untranslated region (UTR) of the viral genome, mediated by residues 125–200 of the cytoplasmic tail (Sharma et al. 2007). ORF3a is not required for virus replication in vitro or in a small animal model, as a recombinant SARS-CoV strain deleted of ORF3a replicates to wild-type levels in numerous cell lines and in BALB/c mice (Yount et al. 2005). Like many other SARS-CoV viral proteins, transient expression of ORF3a has been shown to induce apoptosis in Vero cells (Law et al. 2005). Expression of ORF3a in 293T cells can also augment NF-κB activity (Kanzawa et al. 2006; Narayanan et al. 2007) and expression in A549 cells upregulates the expression and secretion of fibrinogen (Tan et al. 2005).

ORF3b is also expressed from sgRNA3 with a coding sequence overlapping both the ORF3a and E sequences. Although present in the genomes of human and civet SARS-CoV isolates, the ORF3b open reading frame is not present in bat SARS-CoV isolates due to a stop codon early in the ORF3b sequence (Ren et al. 2006). The ORF3b protein is 154 amino acids in length and the translation initiation AUG codon lies 418 nucleotides downstream of the ORF3a AUG. No transcriptional regulatory sequence exists upstream of the ORF3b sequence and there are 11 AUG sequences between the ORF3a and ORF3b initiation codons which likely precludes leaky scanning as the mechanism of ORF3b translation. Although the ORF3b coding sequence lies in a translationally unfavorable context, the protein has been observed in virus-infected Vero cells (Chan et al. 2005) and antibodies recognizing ORF3b have also been found in patient convalescent sera (Guo et al. 2004; Chow et al. 2006).

Bioinformatics analysis predicted two nuclear localization signals (NLS) within the ORF3b protein and subsequent analysis has confirmed the nuclear and nucleolar localization of ORF3b (Pewe et al. 2005; Yuan et al. 2005c; Kopecky-Bromberg et al. 2007; von Brunn et al. 2007). A separate report has also suggested ORF3b localizes to mitochondria (Yuan et al. 2006a). No signal peptides or transmembrane domains (TMD) are predicted, suggesting ORF3b is an 18 kDa soluble protein.

ORF3b induces both apoptosis and necrosis in multiple cell lines (Yuan et al. 2005b; Khan et al. 2006) and prevents cell cycle transition from G0/G1 to S phase (Yuan et al. 2005b). Again, cell cycle regulation has not been identified in SARS-CoV infected cells, so the biological relevance of this finding is unknown. ORF3b has been shown to be an interferon antagonist (Kopecky-Bromberg et al. 2007). SARS-CoV has the ability to prevent interferon-β (IFN-β) production in infected cells (Spiegel et al. 2005) and ORF3b may be one of several viral proteins whose function is to prevent IFN-β induction. Like ORF6, the ORF3b protein prevents IFN-β expression by inhibiting the function of interferon regulatory factor 3 (IRF-3) and also prevents transcription from interferon stimulated response element (ISRE)-containing promoters.

10.2.2 ORF6

ORF6 is a 7.5 kDa protein of 63 amino acids and the only ORF translated from sgRNA6. ORF6 expression has been confirmed in lung and enteric tissue from infected patients (Chan et al. 2005; Geng et al. 2005). No signal peptide is predicted within ORF6 and the N-terminal 40 amino acids are predominantly hydrophobic, with the exception of six charged residues spaced approximately seven amino acids apart, suggesting an amphipathic alpha-helical structure (Netland et al. 2007). The C-terminal 23 amino acids are largely composed of hydrophilic, charged residues. The hydrophobic N-terminus of ORF6 mediates membrane association (Pewe et al. 2005); however, the evenly spaced charged residues suggest that the hydrophobic region may not form a genuine transmembrane domain. In agreement with this, Netland et al. identified that both the N- and C-termini are cytoplasmic (Netland et al. 2007).

ORF6 is packaged into virus particles and incorporated into virus-like particles; the protein was also released from HEK-293T cells into the supernatant when expressed independently of other viral proteins (Huang et al. 2007). The packaging of ORF6 into virions is likely due to specific interactions with other structural protein(s), as ORF6 is not present in purified mouse hepatitis virus (MHV) particles collected from cells coexpressing SARS-CoV ORF6 (Pewe et al. 2005; Tangudu et al. 2007). Direct interaction between ORF6 and the nonstructural protein nsp8 has been observed, suggesting that ORF6 may play a role in virus assembly or replication processes (Kumar et al. 2007). Given the subcellular localization of ORF6 to the approximate area of viral RNA replication and virus assembly, further studies analyzing contributions of ORF6 to virus replication are warranted.

Heterologous expression of ORF6 by an attenuated MHV variant resulted in increased virulence and higher viral titers in C57BL/6 mice (Pewe et al. 2005). Subsequently, Kopecky-Bromberg et al. (2007) identified that ORF6 antagonizes IFN-β production from cells infected with Sendai virus by inhibiting IRF-3 activation. ORF6 was also able to inhibit the nuclear translocation of STAT1 within IFN-β treated cells. Frieman et al. (2007) demonstrated that ORF6 interacts with the nuclear import factor karyopherin-α2 (KPNA2) at ER/Golgi membranes, sequestering KPNA2 and karyopherin-β1 complexes resulting in loss of STAT1 translocation into the nucleus and subsequent inhibition of STAT1-activated antiviral genes.

Deletion of ORF6 in a recombinant SARS-CoV was initially described as having little to no effect on virus replication in vitro or in the BALB/c mouse model of virus replication (Yount et al. 2005). In contrast, the deletion of ORF6 did result in subtle changes to SARS-CoV replication after low multiplicity of infection (MOI) infection of Vero cells (Zhao et al. 2009) and in virus virulence and replication in the human ACE2-transgenic C57BL/6 mouse model (Zhao et al. 2009).

10.2.3 ORF7a and ORF7b

Gene 7 contains two ORFs, ORF7a and ORF7b. The ORF7a translation initiation sequence lies immediately juxtaposed to the 5$'$ TRS and translation results in a 122 amino acid protein approximately 17.5 kDa in size prior to cleavage of an

N-terminal signal sequence (Fielding et al. 2004; Nelson et al. 2005). ORF7a is highly conserved among all sequenced human and animal SARS-CoV isolates including bat SARS-CoV and its expression has been detected in lung and enteric tissues from infected patients (Chan et al. 2005; Chen et al. 2005).

ORF7a is a type-I integral membrane protein containing a 15 amino acid N-terminal signal sequence, 81 residue luminal domain, 21 residue TMD, and a 5 amino acid cytoplasmic tail. The crystal structure of the luminal domain has been solved, revealing a seven-stranded beta sandwich that adopts a compact Ig-like fold (Nelson et al. 2005). The amino acid sequence contains little homology to any other known protein but the luminal domain structure does have some similarity to the D1 domain of human ICAM-1 (Nelson et al. 2005; Hanel and Willbold 2007).

ORF7a localizes predominantly to the Golgi region of transfected and infected cells (Nelson et al. 2005; Kopecky-Bromberg et al. 2006; Pekosz et al. 2006; Schaecher et al. 2007a,b). The short cytoplasmic tail contains a dibasic ER export motif (RK) × (RK) that mediates export from the ER through interactions with COPII machinery (Nelson et al. (2005) and our unpublished data) and the Golgi retention motif resides within the TMD and cytoplasmic tail (Nelson et al. 2005). ORF7a has also been observed in purified virus particles and virus-like particles (VLPs) (Huang et al. 2006).

The precise function of ORF7a in virus infected cells remains unclear. Numerous studies have reported proapoptotic effects of ORF7a when expressed via cDNA transfection (Tan et al. 2004a, 2007; Kopecky-Bromberg et al. 2006; Yuan et al. 2006b; Schaecher et al. 2007b). In-vitro expression of ORF7a results in apoptosis through a caspase-dependent pathway, inhibition of cellular protein synthesis, blockage of cell cycle progression at G0/G1 phase, activation of NF-κB, increased IL-8 promoter activity, and activation of p38 MAP kinase (Tan et al. 2004a, 2007; Kanzawa et al. 2006; Kopecky-Bromberg et al. 2006; Yuan et al. 2006b; Schaecher et al. 2007b). These varied results all suggest a role for ORF7a in altering the host cellular environment. A recombinant SARS-CoV lacking gene 7 induces early-stage apoptosis and cell death similar to wild-type virus, but latter stages of the apoptotic cascade are altered as oligonucleosomal DNA fragmentation is significantly reduced compared to cells infected with wild-type virus (Schaecher et al. 2007b); however, gene 7 deletion viruses have no discernable defect in virus replication, pathogenesis, or lethality in wild-type or immunodeficient Syrian golden hamsters (Schaecher et al. 2007b; Schaecher et al. 2008b) or in Balb/c mice (Yount et al. 2005).

Several studies have identified potential interactions between cellular proteins and ORF7a. Tan et al. have reported that the ORF7a transmembrane domain mediates interaction with the antiapoptotic cellular protein BCL-X_L (Tan et al. 2007). Although ORF7a does not localize to mitochondria, it is possible that direct interaction with BCL-X_L occurs at the ER membranes. This could tilt the balance of pro- and antiapoptotic regulators at the mitochondria resulting in activation of a cell death cascade. Our preliminary data has not revealed any detectable colocalization between ORF7a and BCL-X_L at any subcellular site (data not shown), so further studies are necessary to analyze the potential interactions of ORF7a with BCL-X_L.

Potential interactions between ORF7a and the small glutamine-rich tetraicopeptide repeat-containing protein hSGT were identified using a two-hybrid screen (Fielding et al. 2006); residues 1–96 of ORF7a were determined to be the interacting region. These residues include the signal peptide (which is not present in the mature protein), and the ORF7a ectodomain domain which is in the Golgi lumen (Nelson et al. 2005). Since hSGT is a cytosolic protein, it is unclear how this interaction can occur in mammalian cells.

Interactions between ORF7a and a third cellular protein, lymphocyte function-associated antigen 1 (LFA-1), have been identified (Hanel and Willbold 2007). LFA-1 is expressed exclusively on leukocytes and is involved in lymphocyte homing and intercellular interactions during inflammatory immune responses. Recombinant ORF7a binds to the extracellular domain of LFA-1; however, the significance of this is unknown as ORF7a is not secreted from expressing cells, nor is it present on the plasma membrane of infected cells (Nelson et al. 2005; Huang et al. 2006).

The ORF7b protein is encoded by an ORF beginning 365 nucleotides from the sgRNA7 TRS. The start codon for ORF7b overlaps but is out of frame with the stop codon for ORF7a and translation occurs via ribosome leaky scanning (Schaecher et al. 2007a). ORF7b localizes to the Golgi region of cDNA transfected and SARS-CoV infected cells (Pekosz et al. 2006; Schaecher et al. 2007a,b). The ORF7b protein is a highly hydrophobic protein of 44 amino acids and a molecular weight of 5.5 kDa. ORF7b has no predicted signal sequence and is an integral membrane protein containing a 22 amino acid transmembrane domain and a cytoplasmic C-terminus (Schaecher et al. 2007b), suggesting it is a type-III integral membrane protein. The Golgi localization motif is present within the ORF7b TMD and can confer Golgi localization to a protein that is normally found on the plasma membrane (Schaecher et al. 2008a).

The ORF7b protein is associated with intracellular virus particles and is present in purified virions (Schaecher et al. 2007a). Expression of ORF7b induces apoptosis in transfected cells, but to a lesser extent than ORF7a (Kopecky-Bromberg et al. 2006; Schaecher et al. 2007b). Like the other SARS-CoV accessory proteins, however, it is not required for replication in vitro, in mice, or in Syrian golden hamsters (Sims et al. 2005; Yount et al. 2005; Schaecher et al. 2007b; Schaecher et al. 2008b).

10.2.4 ORF8a and ORF8b

SARS-CoV strains isolated from civet cats, raccoon dogs, and bats in China as well as early human isolates contain a single, continuous ORF8 sequence referred to as ORF8ab. Virus isolated from middle and late stages of the human outbreak, comprising the majority of human isolates, was found to have a 29-nucleotide deletion within the ORF creating two overlapping reading frames designated ORF8a and ORF8b (Fig. 10.1) (Guan et al. 2003; Chinese 2004; Lau et al. 2005). The ORF8a/8ab translation initiation codon lies proximal to the strong sgRNA8 TRS, and translation of full length ORF8ab results in a 122 amino acid protein with an N-terminus identical to that of ORF8a, and C-terminus identical to that of ORF8b. Translation of ORF8a and ORF8b, however, results in products of 39 and 84 residues, respectively. It has been suggested that the ORF8ab protein is a functional protein that was lost upon transmission to humans, and the resultant ORF8a and/or ORF8b proteins are likely nonfunctional (Snijder et al. 2003; Oostra et al. 2007).

The N-terminus of both the large ORF8ab protein and truncated ORF8a contains a predicted hydrophobic signal sequence. ORF8ab colocalizes with ER markers and is membrane-associated; it is presumed that ORF8ab is not an integral membrane protein and exists either as a soluble protein in the lumen ER lumen or is peripherally associated with the luminal face of ER membranes (Oostra et al. 2007).

The ORF8a translation product is a small protein of 5.3 kDa and antibodies recognizing ORF8a have been detected in patient convalescent sera (Chen et al. 2007). An ORF8a–GFP fusion protein colocalizes with the ER marker calnexin (Oostra et al. 2007); however, the 39 amino acid ORF8a is too small to cotranslationally interact with the signal recognition particle (SRP) and is likely released into the cytosol prior to translocation. The cytosolic accumulation of ORF8a in SARS-CoV infected Vero cells supports this theory (Keng et al. 2006), although ORF8a localization to mitochondria has also been described where it may impart a proapoptotic effect (Chen et al. 2007). Little is known regarding ORF8a function in the virus lifecycle; however, a dose-dependent increase in virus replication has been observed in stable cell lines expressing varying levels of ORF8a (Keng et al. 2006).

No TRS exists upstream of ORF8b and translation initiation most likely does not involve a conventional ribosomal scanning mechanism, as scanning ribosomes would have to pass two AUG sequences prior to reaching the predicted ORF8b initiation codon. Contradictory evidence regarding the expression of ORF8b exists. Keng et al. (2006) have demonstrated that ORF8b is expressed in SARS-CoV infected Vero cells. However, two independent reports have suggested that ORF8b is not expressed from the ORF8a/b message either in SARS-CoV infected cells or ORF8a/b cDNA-transfected cells (Le et al. 2007; Oostra et al. 2007). When expressed independently, ORF8b localization is distributed evenly throughout the cytoplasm (Keng et al. 2006; Le et al. 2007).

Little is known regarding the function of ORF8b in virus-infected cells. Cotransfection of ORF8b and SARS-CoV E protein results in a posttranslational downregulation of E, a phenomenon not observed in cells coexpressing E and ORF8ab (Keng et al. 2006). Direct interactions between ORF8b and E were detected suggesting a potential role for ORF8b in modulating degradation or stability of E. Further analysis of ORF8b expression and function is required to confirm whether ORF8b is in fact translated and functional in SARS-CoV infected cells.

10.2.5 ORF9b

Some group 2 coronaviruses encode a protein in an alternative reading frame contained entirely within the *N* gene, termed I or "internal" (Fig. 10.1). SARS-CoV also encodes an internal ORF, termed ORF9b. The predicted initiation codon lies ten nucleotides from the nucleocapsid initiation AUG codon with no intervening AUG sequences, providing a prime opportunity for ribosomal leaky scanning mediated translation initiation. Antibodies specific to ORF9b have been detected in patient sera (Guo et al. 2004; Qiu et al. 2005; Zhong et al. 2005) and protein was detected in tissues from SARS-CoV infected patients (Chan et al. 2005), suggesting that ORF9b is expressed during virus infection.

The ORF9b protein is 98 amino acids in length and has a molecular weight of approximately 11 kDa. The crystal structure of the protein reveals an intertwined symmetrical dimer whose β-sheets form a tent-like structure (Meier et al. 2006). Mass spectroscopy analysis revealed a long fatty acid or fatty acid ester molecule present in the structure's hydrophobic tunnel; it has been speculated that the ORF9b protein may anchor to intracellular membranes by internalizing one or more lipidic tails from membrane lipids (Meier et al. 2006). ORF9b localizes to the ER region of transfected cells, suggesting that the protein may interact with intracellular membranes although no transmembrane regions are predicted within ORF9b. Yeast-two-hybrid experiments have suggested that ORF9b interacts with viral proteins nsp8, nsp14, and ORF7b, and provide further evidence of homotypic interactions (von Brunn et al. 2007); however, no functional analysis has been described to date regarding the role of ORF9b in the context of the virus replication cycle.

10.3 Conclusions

The large genetic diversity among coronaviruses raises questions regarding the origins of the accessory genes. Have they been acquired through horizontal transfer from host cellular genetic material or heterologous viruses? Have they been acquired through gene duplication and subsequent mutation within the virus's own genome? Evidence has been presented pointing to both possibilities. For example, the group 2 coronavirus hemagglutinin esterase (*HE*) genes have approximately 30% amino acid identity with the influenza C virus (ICV) HE-1 subunit of the *HEF* gene; it has been suggested that an early group 2 coronavirus captured HE from ICV during a mixed infection (Luytjes et al. 1988). For the latter theory, several of the SARS-CoV accessory genes may have originated from internal duplication, fusion, or shifting events (Inberg and Linial 2004). ORF3a may have arisen through gene duplication and mutation of the M ORF (Inberg and Linial 2004; Masters 2006; Oostra et al. 2006). Furthermore, ORF7b and ORF8a have some amino acid sequence similarity to ORF3a, suggesting that the larger ORF3a may have given rise to ORF7b and ORF8 (Inberg and Linial 2004). Regardless of the origins of the SARS-CoV accessory genes, the severe disease caused by the virus highlights the need to study the functions of its eight accessory genes.

The observation that the SARS-CoV accessory proteins are dispensable for virus replication in vitro is not uncommon among coronaviruses (Yount et al. 2005). It is striking that a virus lacking each of the accessory proteins, either individually or in combination, replicates efficiently in BALB/c mice. This finding may indicate that other animal models for the in-vivo analysis of accessory gene function are required. It is also possible that some or all of the accessory genes play a more important role during infection of humans, bats or other animal reservoirs of the virus. Identifying the functions of these accessory proteins may prove beneficial not only for further understanding of SARS-CoV pathogenesis, but also for better understanding and preparedness for future viral infectious diseases.

References

Akerstrom S, Tan YJ et al (2006) Amino acids 15–28 in the ectodomain of SARS coronavirus 3a protein induces neutralizing antibodies. FEBS Lett 580(16):3799–3803

Chan WS, Wu C et al (2005) Coronaviral hypothetical and structural proteins were found in the intestinal surface enterocytes and pneumocytes of severe acute respiratory syndrome (SARS). Mod Pathol 18:1432–1439

Chen YY, Shuang B et al (2005) The protein X4 of severe acute respiratory syndrome-associated coronavirus is expressed on both virus-infected cells and lung tissue of severe acute respiratory syndrome patients and inhibits growth of Balb/c 3 T3 cell line. Chin Med J 118(4):267–274

Chen CY, Ping YH et al (2007) Open reading frame 8a of the human severe acute respiratory syndrome coronavirus not only promotes viral replication but also induces apoptosis. J Infect Dis 196(3):405–415

Chinese SMEC (2004) Molecular evolution of the SARS Coronavirus during the course of the SARS epidemic in China. Science 303(5664):1666–1669

Chow SC, Ho CY et al (2006) Specific epitopes of the structural and hypothetical proteins elicit variable humoral responses in SARS patients. J Clin Pathol 59(5):468–476

Fielding BC, Tan Y-J et al (2004) Characterization of a unique group-specific protein (U122) of the severe acute respiratory syndrome Coronavirus. J Virol 78(14):7311–7318

Fielding BC, Gunalan V et al (2006) Severe acute respiratory syndrome coronavirus protein 7a interacts with hSGT. Biochem Biophys Res Commun 343(4):1201–1208

Frieman M, Yount B et al (2007) Severe acute respiratory syndrome coronavirus ORF6 antagonizes STAT1 function by sequestering nuclear import factors on the rough endoplasmic reticulum/Golgi membrane. J Virol 81(18):9812–9824

Geng H, Liu Y-M et al (2005) The putative protein 6 of the severe acute respiratory syndrome-associated coronavirus: expression and functional characterization. FEBS Lett 579(30):6763–6768

Guan Y, Zheng BJ et al (2003) Isolation and characterization of viruses related to the SARS Coronavirus from animals in Southern China. Science 302(5643):276–278

Guo JP, Petric M et al (2004) SARS corona virus peptides recognized by antibodies in the sera of convalescent cases. Virology 324(2):251–256

Hanel K, Willbold D (2007) SARS-CoV accessory protein 7a directly interacts with human LFA-1. Biol Chem 388(12):1325–1332

Huang C, Ito N et al (2006) Severe acute respiratory syndrome coronavirus 7a accessory protein is a viral structural protein. J Virol 80(15):7287–7294

Huang C, Peters CJ et al (2007) Severe acute respiratory syndrome coronavirus accessory protein 6 is a virion-associated protein and is released from 6 protein-expressing cells. J Virol 81 (10):5423–5426

Inberg A, Linial M (2004) Evolutional insights on uncharacterized SARS coronavirus genes. FEBS Lett 577(1–2):159–164

Ito N, Mossel EC et al (2005) Severe acute respiratory syndrome Coronavirus 3a protein is a viral structural protein. J Virol 79(5):3182–3186

Kanzawa N, Nishigaki K et al (2006) Augmentation of chemokine production by severe acute respiratory syndrome coronavirus 3a/X1 and 7a/X4 proteins through NF-[kappa]B activation. FEBS Lett 580(30):6807–6812

Keng C-T, Choi Y-W et al (2006) The human severe acute respiratory syndrome coronavirus (SARS-CoV) 8b protein is distinct from its counterpart in animal SARS-CoV and down-regulates the expression of the envelope protein in infected cells. Virology 354(1):132–142

Khan S, Fielding BC et al (2006) Over-expression of severe acute respiratory syndrome coronavirus 3b protein induces both apoptosis and necrosis in Vero E6 cells. Virus Res 122(1–2):20–27

Kopecky-Bromberg SA, Martinez-Sobrido L et al (2006) 7a protein of severe acute respiratory syndrome Coronavirus inhibits cellular protein synthesis and activates p38 mitogen-activated protein kinase. J Virol 80(2):785–793

Kopecky-Bromberg SA, Martinez-Sobrido L et al (2007) Severe acute respiratory syndrome Coronavirus open reading frame (ORF) 3b, ORF 6, and nucleocapsid proteins function as interferon antagonists. J Virol 81(2):548–557

Kumar P, Gunalan V et al (2007) The nonstructural protein 8 (nsp8) of the SARS coronavirus interacts with its ORF6 accessory protein. Virology 366(2):293–303

Lau SKP, Woo PCY et al (2005) Severe acute respiratory syndrome coronavirus-like virus in Chinese horseshoe bats. PNAS 102(39):14040–14045

Law PTW, Wong C-H et al (2005) The 3a protein of severe acute respiratory syndrome-associated coronavirus induces apoptosis in Vero E6 cells. J Gen Virol 86(7):1921–1930

Le TM, Wong HH et al (2007) Expression, post-translational modification and biochemical characterization of proteins encoded by subgenomic mRNA8 of the severe acute respiratory syndrome coronavirus. FEBS J 274(16):4211–4222

Lu W, Zheng B-J et al (2006) Severe acute respiratory syndrome-associated coronavirus 3a protein forms an ion channel and modulates virus release. PNAS 103(33):12540–12545

Luytjes W, Bredenbeek PJ et al (1988) Sequence of mouse hepatitis virus A59 mRNA 2: indications for RNA recombination between coronaviruses and influenza C virus. Virology 166 (2):415–422

Marra MA, Jones SJM et al (2003) The genome sequence of the SARS-associated Coronavirus. Science 300(5624):1399–1404

Masters PS (2006) The molecular biology of coronaviruses. Adv Virus Res 66:193–292

Meier C, Aricescu AR et al (2006) The crystal structure of ORF-9b, a lipid binding protein from the SARS Coronavirus. Structure 14(7):1157–1165

Mihindukulasuriya KA, Wu G, et al (2008) Identification of a novel Coronavirus from a Beluga Whale using a pan-viral microarray. J Virol 82(10):5084–5088

Narayanan K, Huang C, et al (2007) SARS coronavirus accessory proteins. Virus Res 133(1): 113–121

Nelson CA, Pekosz A et al (2005) Structure and intracellular targeting of the SARS-Coronavirus Orf7a accessory protein. Structure 13(1):75–85

Netland J, Ferraro D et al (2007) Enhancement of murine coronavirus replication by severe acute respiratory syndrome coronavirus protein 6 requires the N-terminal hydrophobic region but not C-terminal sorting motifs. J Virol 81(20):11520–11525

Oostra M, de Haan CAM et al (2006) Glycosylation of the severe acute respiratory syndrome coronavirus triple-spanning membrane proteins 3a and M. J Virol 80(5):2326–2336

Oostra M, de Haan CA et al (2007) The 29-nucleotide deletion present in human but not in animal severe acute respiratory syndrome coronaviruses disrupts the functional expression of open reading frame 8. J Virol 81(24):13876–13888

Pekosz A, Schaecher SR et al (2006) Structure, expression, and intracellular localization of the SARS-CoV accessory proteins 7a and 7b. Adv Exp Med Biol 581:115–120

Pewe L, Zhou H et al (2005) A severe acute respiratory syndrome-associated coronavirus-specific protein enhances virulence of an attenuated murine coronavirus. J Virol 79(17):11335–11342

Qiu M, Shi Y et al (2005) Antibody responses to individual proteins of SARS coronavirus and their neutralization activities. Microbes Infect 7(5–6):882–889

Ren W, Li W et al (2006) Full-length genome sequences of two SARS-like coronaviruses in horseshoe bats and genetic variation analysis. J Gen Virol 87(11):3355–3359

Rota PA, Oberste MS et al (2003) Characterization of a novel coronavirus associated with severe acute respiratory syndrome. Science 300(5624):1394–1399

Schaecher SR, Mackenzie JM et al (2007a) The ORF7b protein of severe acute respiratory syndrome coronavirus (SARS-CoV) is expressed in virus-infected cells and incorporated into SARS-CoV particles. J Virol 81(2):718–731

Schaecher SR, Touchette E et al (2007b) Severe acute respiratory syndrome coronavirus gene 7 products contribute to virus-induced apoptosis. J Virol 81(20):11054–11068

Schaecher SR, Diamond MS et al (2008a) The transmembrane domain of the severe acute respiratory syndrome coronavirus ORF7b protein is necessary and sufficient for its retention in the Golgi complex. J Virol 82(19):9477–9491

Schaecher SR, Stabenow J et al (2008b) An immunosuppressed Syrian golden hamster model for SARS-CoV infection. Virology 380(2):312–321

Sharma K, Surjit M et al (2007) The 3a accessory protein of SARS coronavirus specifically interacts with the 5'UTR of its genomic RNA, using a unique 75 amino acid interaction domain. Biochemistry 46(22):6488–6499

Shen S, Lin PS et al (2005) The severe acute respiratory syndrome coronavirus 3a is a novel structural protein. Biochem Biophys Res Commun 330(1):286–292

Sims AC, Baric RS et al (2005) Severe acute respiratory syndrome coronavirus infection of human ciliated airway epithelia: role of ciliated cells in viral spread in the conducting airways of the lungs. J Virol 79(24):15511–15524

Snijder EJ, Bredenbeek PJ et al (2003) Unique and conserved features of genome and proteome of SARS-coronavirus, an early split-off from the coronavirus group 2 lineage. J Mol Biol 331(5):991–1004

Spiegel M, Pichlmair A et al (2005) Inhibition of beta interferon induction by severe acute respiratory syndrome coronavirus suggests a two-step model for activation of interferon regulatory factor 3. J Virol 79(4):2079–2086

Tan Y-J, Fielding BC et al (2004a) Overexpression of 7a, a protein specifically encoded by the severe acute respiratory syndrome coronavirus, induces apoptosis via a caspase-dependent pathway. J Virol 78(24):14043–14047

Tan Y-J, Goh P-Y et al (2004b) Profiles of antibody responses against severe acute respiratory syndrome coronavirus recombinant proteins and their potential use as diagnostic markers. Clin Diagn Lab Immunol 11(2):362–371

Tan Y-J, Teng E et al (2004c) A novel severe acute respiratory syndrome coronavirus protein, U274, is transported to the cell surface and undergoes endocytosis. J Virol 78(13):6723–6734

Tan Y-J, Tham P-Y et al (2005) The severe acute respiratory syndrome coronavirus 3a protein up-regulates expression of fibrinogen in lung epithelial cells. J Virol 79(15):10083–10087

Tan YX, Tan TH et al (2007) The induction of apoptosis by the severe acute respiratory syndrome (SARS)-Coronavirus 7a protein is dependent on its interaction with the Bcl-XL protein. J Virol 81(12):6346–6355

Tangudu C, Olivares H et al (2007) Severe acute respiratory syndrome coronavirus protein 6 accelerates murine coronavirus infections. J Virol 81(3):1220–1229

Thiel V, Ivanov KA et al (2003) Mechanisms and enzymes involved in SARS coronavirus genome expression. J Gen Virol 84(9):2305–2315
von Brunn A, Teepe C et al (2007) Analysis of intraviral protein-protein interactions of the SARS coronavirus ORFeome. PLoS ONE 2(5):e459
Yount B, Roberts RS et al (2005) Severe acute respiratory syndrome coronavirus group-specific open reading frames encode nonessential functions for replication in cell cultures and mice. J Virol 79(23):14909–14922
Yu CJ, Chen YC et al (2004) Identification of a novel protein 3a from severe acute respiratory syndrome coronavirus. FEBS Lett 565(1–3):111–116
Yuan X, Li J et al (2005a) Subcellular localization and membrane association of SARS-CoV 3a protein. Virus Res 109(2):191–202
Yuan X, Shan Y et al (2005b) G0/G1 arrest and apoptosis induced by SARS-CoV 3b protein in transfected cells. Virol J 2(66):66
Yuan X, Yao Z et al (2005c) Nucleolar localization of non-structural protein 3b, a protein specifically encoded by the severe acute respiratory syndrome coronavirus. Virus Res 114 (1–2):70–79
Yuan X, Shan Y et al (2006a) Mitochondrial location of severe acute respiratory syndrome coronavirus 3b protein. Mol Cells 21(2):186–191
Yuan X, Wu J et al (2006b) SARS coronavirus 7a protein blocks cell cycle progression at G0/G1 phase via the cyclin D3/pRb pathway. Virology 346(1):74–85
Yuan X, Yao Z et al (2007) G1 phase cell cycle arrest induced by SARS-CoV 3a protein via the cyclin D3/pRb pathway. Am J Respir Cell Mol Biol 37(1):9–19
Zeng R, Yang RF et al (2004) Characterization of the 3a protein of SARS-associated coronavirus in infected vero E6 cells and SARS patients. J Mol Biol 341(1):271–279
Zhao J, Falcon A et al (2009) Severe acute respiratory syndrome coronavirus protein 6 is required for optimal replication. J Virol 83(5):2368–2373
Zhong X, Yang H et al (2005) B-cell responses in patients who have recovered from severe acute respiratory syndrome target a dominant site in the S2 domain of the surface spike glycoprotein. J Virol 79(6):3401–3408
Zhong X, Guo Z et al (2006) Amino terminus of the SARS coronavirus protein 3a elicits strong, potentially protective humoral responses in infected patients. J Gen Virol 87(2):369–373

Chapter 11
SARS Accessory Proteins ORF3a and 9b and Their Functional Analysis

Wei Lu, Ke Xu, and Bing Sun

Abstract The SARS coronavirus (CoV) positive-stranded RNA viral genome encodes 14 open reading frames (ORFs), eight of which encode proteins termed as "accessory proteins." These proteins help the virus infect the host and promote virulence. In this chapter we describe some of our latest investigations into the structure and function of two such accessory proteins: ORF3a and 9b. The ORF3a accessory protein is the largest accessory protein in SARS-CoV and is a unique membrane protein consisting of three transmembrane domains. It colocalizes on the cell membrane and host Golgi networks and may be involved in ion channel formation during infection. Similarly the ORF9b accessory protein is 98 amino acids, associates with the spike and nucleocapsid proteins and has unusual membrane binding properties. In this chapter we have suggested possible new roles for these two accessory proteins which may in the long run contain answers to many unanswered questions and also give us new ideas for drugs and vaccine design.

11.1 The SARS-CoV ORF3a Protein

The SARS-CoV genome contains more than 14 open reading frames (ORFs). These ORFs encode various viral proteins, which take part in different virus infection steps (Narayanan et al. 2007). These viral proteins can be classified into three groups: structural proteins, nonstructural proteins and accessory proteins. Generally, researchers take Spike protein, Membrane protein, Envelope protein, and Nucleocapsid protein as the four major structure proteins, since they are all located on the mature virus particles and construct the virus framework. Nonstructural proteins generally refer to viral enzymes, for example the virus RNA dependent

B. Sun (✉)
Laboratory of Molecular Virology, Institut Pasteur of Shanghai, Chinese Academy of Sciences, 225 South Chongqing Road, Shanghai 200025, China
e-mail: bsun@sibs.ac.cn

RNA polymerase and other viral proteinases. Besides these structural proteins and nonstructural proteins, there are many other viral proteins encoded by many other ORFs. The existence and functions of these proteins are all still debatable.

Coronavirus structural proteins and nonstructural proteins have been studied for quite a long time, and their biochemical characteristics and biological functions have been clearly demonstrated. However these studies are not enough for us to understand and conquer the coronavirus. One major problem with the SARS-CoV is that it contains a great number of accessory proteins which help the virus infect the host and promote virulence, and contain answers to many unanswered questions and ideas for drugs and vaccine design.

At the beginning of the SARS-CoV genome are two large ORFs, 1a and 1b, the major source of viral nonstructural proteins. After that, four structural proteins genes are located on the genome in the following order: spike, envelope, membrane and nucleocapsid. Interestingly, there are many small and separated ORFs randomly located among these structural protein genes (Narayanan et al. 2007); these are the SARS-CoV accessory proteins. Researchers in previous years have devoted great efforts to these proteins to understand their purpose and biological significance.

ORF3a is located between the spike and envelope genes on the SARS-CoV genome. It has also been named as SNE protein, X1 protein and U274 protein (Zeng et al. 2004; Kanzawa et al. 2006; Tan et al. 2004a). Protein sequence analysis on ORF3a suggests that it is a membrane protein which has three transmembrane domains at the N terminus. Upon analysis of the full-length sequence of ORF3a, we found that it lacks similarities to any known proteins. However, some researchers divided the protein sequence into several parts and then BLAST-searched them separately. Results clearly showed that the N-terminal domain of ORF3a had 29% identity with the putative cytochrome B-561 transmembrane protein from a bacterium (*Ralstonia solanacearum*) and 24% identity with the opsin fragments from a hawkmoth (*Manduca sexta*) and the rhodopsin from a butterfly (*Popilio glaucus*). The C terminus of ORF3a had moderate similarity to the calcium-transporting ATPase (calcium pump) from a parasite (*Plasmodium falciparum*; 41% identity) and the outer-membrane porin from a bacterium (*Shewanella oneidensis*; 27% identity) (Yu et al. 2004).

At nearly the same time, several research groups reported more new findings on ORF3a using different methods. Using confocal methods, they found ORF3a expressed in SARS-CoV-infected Vero-E6 cells, located mainly on the cell membrane and host cell Golgi networks. They also detected ORF3a in lung sections of SARS-CoV-infected patients (Yu et al. 2004; Yuan et al. 2005; Ito et al. 2005).

Zeng et al. reported that when carrying out proteomics analysis on SARS-CoV-infected Vero-E6 cells, two specific peptides were identified which corresponded to the virus ORF3a sequences. They also detected the presence of anti-ORF3a antibody in the SARS-CoV-infected patient sera by using an antigen peptide derived from the ORF3a sequence (Zeng et al. 2004). Some other groups also demonstrated the existence of ORF3a in purified SARS-CoV virus particles – so that some researchers even proposed that ORF3a may be classified as a new member of the viral structural proteins. However, no further evidence to date supports ORF3a's function as a structural protein. A gold-labeled immuno-electron microscopy test on purified virus particles is still needed to prove the above hypothesis (Shen et al. 2005).

As ORF3a is the largest protein in the SARS-CoV accessory protein family, it was considered the most important one, and is postulated to have crucial biological roles. However ORF3a is a very unusual protein as well; compared with other coronavirus proteins, no homologs can be identified with this protein, and hence little functional hint can be obtained using bioinformatics.

11.2 Functional Analysis of ORF3a

More and more evidence has confirmed the presence of ORF3a in SARS-CoV infection. The function of ORF3a becomes the most urgent unanswered question. We know that the viruses in general lead a simple life-style. A limited number of proteins are expressed during their life-cycle and they usually do not carry abundant or unnecessary protein ORFs in their genome.

ORF3a is a membrane-associated viral protein. At first, researchers hypothesized that it must be associated with virus entry into the host cells due to the fact that immunoprecipitation tests with ORF3a showed interaction with the S protein. Thus it is postulated that ORF3a can promote S protein function, and then facilitate virus entry. But little further study has focused on this possibility. One research group proposed that the N-terminal ectodomain of ORF3a can elicit a strong humoral response, which may provide a protective humoral response in SARS patients (Akerström et al. 2006). However, another group tested the neutralization ability of the anti-ORF3a antibody (Qiu et al. 2005), and the results suggested that anti-ORF3a antibody did not show neutralization ability compared with some anti-S antibodies that were tested. We therefore cannot say definitely that ORF3a is a participant in SARS-CoV entry, and many critical experiments still need to be done to answer this question.

Another possible function of ORF3a that researchers have recently uncovered by over-expressing it in several mammalian cell lines is cell apoptosis, using DNA ladder assay and TUNEL assay (Law et al. 2005). This observation is also somewhat debatable because another group compared the apoptosis induction ability of different SARS-CoV proteins, for example N, M, E, ORF7a and ORF3a, and found that only the over-expression of viral ORF7a can induce cell apoptosis (Tan et al. 2004b). However it is noteworthy that both experimental results used two different kinds of cell lines, which may very well be the reason for the differing results.

In 2006, our group found that ORF3a can form an ion channel (Lu et al. 2006). Our previous work demonstrated the presence of ORF3a in virus infection, but we were more interested in its function in the host during SARS-CoV infection. When performing ORF3a expression in mammalian cells, we found that it could interact with itself by disulfide bonds to form homo-dimers and -tetramers. Both immunoprecipitation and FRET tests confirmed these findings. We also screened the ORF3a sequence and identified that the sixth cystiene is in charge of this homopolymer formation.

ORF3a is a membrane-associated protein and can form a homo-tetramer. All ion channel proteins form homo- or hetero-polymers and associate with the membrane. Hence we proposed that ORF3a may form an ion channel.

In order to check our hypothesis, we used an electrophysiological method, which combines biology, physiology and physics. There are two fundamental electrophysiological test methods: two-electrode voltage clamp (TEVC), and patch clamp. TEVC is based on the *Xenopus* oocyte cell system. Initially we obtained cDNA from selected samples, transcribed them in vitro into complementary mRNA, and then microinjected these mRNAs into oocyte cells, which are nearly one millimeter in diameter and easily used in microinjection. Several days later, using a TEVC amplifier on these injected oocytes, we analyzed their membrane potential and conductance to test their channel activity. When using ORF3a in this TEVC test, we found a dramatic membrane current through ORF3a-expressing oocytes compared with control oocytes. This result clearly indicated that ORF3a may form an ion channel. We also found this current to be partially inhibited by barium, which is also a specific potassium channel inhibitor. We also tested the channel permeability of ORF3a to potassium ions and found that potassium ions can pass through its channel pore.

What is the possible biological role of the ORF3a channel during the virus infection? To answer this question, we suppressed ORF3a expression by siRNA during SARS-CoV infection. We found that after knocking down ORF3a in the virus-infected cells, virus release into the cell culture is apparently decreased. We proposed that the SARS-CoV ORF3a channel is closely related to virus release. It is interesting to note that some other viral ion channels are also thought to be associated with virus release: for example, HIV encoded ion channel Vpu protein can promote the release of HIV virus particles specifically in human cell lines. p7 protein from HCV has also been proved to form a ion channel, and its absence will cause reduction of HCV virus assembly and release. However, in-depth studies on these viral ion channels have not been reported to date. How could SARS-CoV ORF3a regulate virus release? And is this function dependent on ORF3a ion channel activity?

11.3 Ion-Channel Activity of the ORF3a Homologue in Other Human Coronavirus

When we initiated our in-depth studies on the relationship between ORF3a channel activity and virus release, several SARS-CoV laboratory infection cases were reported, and regulatory authorities forbid the handling of SARS-CoV in research labs indefinitely. However we found a new way to conduct our research on this subject. We found that in nearly all the human coronavirus genomes, there was an ORF3a homolog gene located between the viral spike gene and envelope gene. For convenience, we named it *SNE* gene, S-neighbor-E (Zeng et al. 2004).

We used HCoV-229E and OC43 as tools and models to analyze the SNE protein functions. Meanwhile, in order to improve the accuracy of our experiments in electrophysiology, we used another technology, patch clamp analysis, on all these SNE proteins.

Fortunately, we found that all these SNE proteins can form ion channels. Both the TEVC results and patch clamp results confirmed our findings, and we also found that all these ion channels can be permeable to both sodium ions and potassium ions, whereas calcium ions cannot pass through this channel pore. Since SNE channels are nonspecific monovalent cation channels, they result in cell membrane depolarization. Membrane depolarization is an important biological stimulus. There are many cascades downstream of the membrane depolarization, and the most important one is the activation of voltage-dependent calcium channels (VDCC). There are several types of VDCC; L-type is the most common one, found on nearly all types of cells (Lehmann-Horn and Jurkat-Rott 1999). At physiologic or resting membrane potential, L-type calcium channels are normally closed. They are activated at depolarized membrane potentials (Lehmann-Horn and Jurkat-Rott 1999; Friis et al. 2004; Lipscombe et al. 2004; Pringle 2004; Røttingen and Iversen 2000; Catterall et al. 2005).

Using a fluorescence calcium probe, we found that SNE-mediated channel activity causes a dramatic membrane depolarization, which exceeds the threshold of L-type calcium channels, and consequently leads to a calcium influx into the host cells.

Under normal conditions, the intracellular calcium concentration in the cell cytoplasm is very low. There are two pathways to increase cell intracellular calcium after specific stimulation: one is calcium release from the cell organelles' calcium store, such as ER; the other is calcium influx through cell voltage-dependent calcium channels on the cell membrane to allow the calcium ions to be transported from the outside to the inside of the cells (Pringle 2004; Røttingen and Iversen 2000). We found that the latter pathway was responsible for SNE-mediated intracellular increase. When we blocked this calcium influx through L-type calcium channels with the specific inhibitor nimodipine, we found that the virus assembly inside the virus-infected cells was arrested. Virus budding seemed to be inhibited in the absence of calcium influx, and consequently virus release was also interrupted.

It has already been reported that coronavirus assembly begins with the viral genome RNAs interacting with nucleoproteins to form ribonucleoproteins in the host-cell cytoplasm. These ribonucleoproteins move to the budding sites at the ER–Golgi intermediate compartment and the Golgi region (Garoff et al. 1998; Kuo and Masters 2002; Stertz et al. 2007). From the Golgi compartments, infected cells synthesize many vesicles, called multivesicular bodies (MVB). Virus core particles can bud into these vesicles and cause the phospholipid bilayer to form mature virions. Vesicles containing a large amount of mature viruses will then be transported to the cell membrane, where viruses are released after fusion of vesicles with the cell membrane (Weiss and Navas-Martin 2005a; Qinfen et al. 2004). One or more steps of coronavirus assembly may be influenced by calcium influx through L-type calcium channels. We thus propose that one possibility is that the formation of MVBs is calcium-dependent during coronavirus infection. Calcium influx can lead to an increase in the number or size of MVBs, and subsequently promote efficient virus assembly and release. Viral SNE proteins simply trigger this calcium requirement via activation of L-type calcium channels.

Coronavirus envelope proteins have recently also been shown to have monovalent cation channel activity (Wilson et al. 2004, 2006). It is well documented that coronavirus envelope protein can modulate virus assembly (Ye and Hogue 2007). Expression of coronavirus membrane protein and envelope protein together can result in abundant virus-like particle formation (Vennema et al. 1996). Since envelope protein can form ion channels, perhaps it also employs some pathways to increase intracellular calcium and then influence virus assembly and release.

In conclusion, our findings on ORF3a illustrate the function of human coronavirus SNE proteins and show that they are related to host-cell L-type calcium channels. We also propose that both the viral ion channel and the L-type calcium channel may be appropriate drug targets in therapeutic approaches against coronavirus infection.

11.4 The SARS-CoV ORF9b Protein

There are nine subgenomic mRNAs in SARS-CoV, four of which (mRNA3, 7, 8, 9) are bicistronic producing two ORFs starting at the first or additional downstream start codon separately (Thiel et al. 2003; Weiss and Navas-Martin 2005b). On bicistronic mRNA9, a 98 amino acid viral accessory protein 9b was encoded from a complete internal ORF within the *N* gene.

We expressed ORF9b protein in *E. coli*, and raised specific antibodies against it. In order to test the presence of ORF9b, we produced SARS-CoV infected cell supernatant and by sucrose density gradient centrifugation, the released viral particles in this supernatant were collected. These viral particles were then Western blot analyzed, using anti-ORF9b antibodies. A specific band was detected, which clearly proved the existence of PRF9b in the mature virus particles. We also found that ORF9b was present in the virus-like-particles (VLP), using the same density gradient centrifugation technique.

ORF9b was shown to be associated with viral S and N proteins. S protein is known to be located on the membrane of the virion envelope, and N protein is located inside the particles. The interactions among S, N and ORF9b may thus help to localize ORF9b inside the particles but closer to the envelope. We also investigated the intracellular localization of ORF9b in both transfected and infected cells. Cytoplasmic ORF9b was diffused in the cytoplasm, but it was difficult to determine whether the signal really represented ORF9b localization.

We detected the expression of ORF9b in infected cells and in clinical specimens; however, further details of the properties and function of ORF9b and its role in viral pathogenesis remain a mystery.

11.5 Functional Role of ORF9b

ORF9b is even less studied than ORF3a. The crystal structure was reported recently; it indicated that ORF9b is an unusual membrane binding protein with a long hydrophobic lipid-binding tunnel. It was therefore proposed to be associated with intracellular vesicles and may have some function in SARS-CoV assembly (Meier et al. 2006). However, further analysis of the properties and function of ORF9b are still necessary to understand its contribution to virus pathogenesis.

We found there was a similar protein, called I protein, present in the mouse hepatitis virus (MHV). There are also corresponding proteins similar to SARS-CoV ORF9b, so-called "internal" or "I" proteins, in other group II coronaviruses (Fischer et al. 1997; Lapps et al. 1987; Senanayake and Brian 1997). In MHV, I protein is an accessory viral structural protein which can contribute to plaque morphology (Fischer et al. 1997). In our study, the I protein in SARS-CoV (ORF9b) was also shown to be a structural component of virions. It is reasonable to propose that ORF9b in SARS-CoV may contribute to viral pathogenesis as I protein in MHV does. However, a detailed functional analysis of ORF9b is still needed to understand its role in the viral life-cycle.

In order to investigate the function of ORF9b, we used the *ORF9b* gene as bait in a yeast two-hybrid screening experiment. We found several interesting genes that can strongly interact with ORF9b, some of which are virus infection related host factors. Further work on this is still underway.

11.6 SARS-CoV Accessory Proteins: An Important Part of the Virus Genome

SARS-CoV possesses a long positive-strand RNA as genome. It is a single-stranded positive-sense RNA 29,727 nucleotides in length, excluding the polyadenylation tract at the 3' end (Rota et al. 2003). Fourteen ORFs have been identified, encoding two replicases (1a and 1b), four structural proteins (S, E, M, N) and eight accessory proteins (ORFs 3a, 3b, 6, 7a, 7b, 8a, 8b, and 9b) (Rota et al. 2003; Marra et al. 2003). All current studies on accessory proteins of coronaviruses including SARS-CoV suggest that they are not essential for virus replication (de Haan et al. 2002; Yount et al. 2005), but do affect viral release, stability, and pathogenesis, and finally contribute to virulence (Weiss and Navas-Martin 2005c).

Our study on ORF3a also shows similar results where it functions as an ion channel in SARS-CoV to facilitate virus release (Lu et al. 2006). In addition, ORF7a has also been characterized as a virion-associated protein of SARS-CoV (Huang et al. 2006), which induces apoptosis (Tan et al. 2004b; Kopecky-Bromberg et al. 2006) and arrests the cell cycle (Yuan et al. 2006) when over-expressed. Most recently, SARS-CoV ORF6 and ORF7b proteins have been identified as being incorporated into virus particles, although the functions of these two proteins are still unknown (Huang et al. 2007; Schaecher et al. 2007).

Understanding the properties and functions of SARS-CoV specific accessory proteins may help explain the differences in pathogenicity between SARS-CoV and other known coronaviruses.

References

Akerström S, Tan YJ, Mirazimi A (2006) FEBS Lett 580(16):3799–3803
Catterall WA, Perez-Reyes E, Snutch TP, Striessnig J (2005) Pharmacol Rev 57(4):411–425
de Haan CA, Masters PS, Shen X, Weiss S, Rottier PJ (2002) Virology 296:177–189
Fischer F, Peng D, Hingley ST, Weiss SR, Masters PS (1997) J Virol 71:996–1003
Friis UG, Jørgensen F, Andreasen D, Jensen BL, Skøtt O (2004) Acta Physiol Scand 181 (4):391–396
Garoff H, Hewson R, Opstelten DJ (1998) Microbiol Mol Biol Rev 62(4):1171–1190
Huang C, Ito N, Tseng CT, Makino S (2006) J Virol 80:7287–7294
Huang C, Peters CJ, Makino S (2007) J Virol 81:5423–5426
Ito N, Mossel EC, Narayanan K, Popov VL, Huang C, Inoue T, Peters CJ, Makino S (2005) J Virol 79(5):3182–3186
Kanzawa N, Nishigaki K, Hayashi T, Ishii Y, Furukawa S, Niiro A, Yasui F, Kohara M, Morita K, Matsushima K, Le MQ, Masuda T, Kannagi M (2006) FEBS Lett 580(30):6807–6812
Kopecky-Bromberg SA, Martinez-Sobrido L, Palese P (2006) J Virol 80:785–793
Kuo L, Masters PS (2002) J Virol 76(10):4987–4999
Lapps W, Hogue BG, Brian DA (1987) Virology 157:47–57
Law PT, Wong CH, Au TC, Chuck CP, Kong SK, Chan PK, To KF, Lo AW, Chan JY, Suen YK, Chan HY, Fung KP, Waye MM, Sung JJ, Lo YM, Tsui SK (2005) J Gen Virol 86:1921–1930
Lehmann-Horn F, Jurkat-Rott K (1999) Physiol Rev 79(4):1317–1372
Lipscombe D, Helton TD, Xu W (2004) J Neurophysiol 92(5):2633–2641
Lu W, Zheng BJ, Xu K, Schwarz W, Du L, Wong CK, Chen J, Duan S, Deubel V, Sun B (2006) Proc Natl Acad Sci U S A 103(33):12540–12545
Marra MA, Jones SJ, Astell CR, Holt RA, Brooks-Wilson A, Butterfield YS, Khattra J, Asano JK, Barber SA, Chan SY, Cloutier A, Coughlin SM, Freeman D, Girn N, Griffith OL, Leach SR, Mayo M, McDonald H, Montgomery SB, Pandoh PK, Petrescu AS, Robertson AG, Schein JE, Siddiqui A, Smailus DE, Stott JM, Yang GS, Plummer F, Andonov A, Artsob H, Bastien N, Bernard K, Booth TF, Bowness D, Czub M, Drebot M, Fernando L, Flick R, Garbutt M, Gray M, Grolla A, Jones S, Feldmann H, Meyers A, Kabani A, Li Y, Normand S, Stroher U, Tipples GA, Tyler S, Vogrig R, Ward D, Watson B, Brunham RC, Krajden M, Petric M, Skowronski DM, Upton C, Roper RL (2003) Science 300:1399–1404
Meier C, Aricescu AR, Assenberg R, Aplin RT, Gilbert RJ, Grimes JM, Stuart DI (2006) Structure 14:1157–1165
Narayanan K, Huang C, Makino S (2007) Virus Res 133(1):113–121
Pringle AK (2004) Cell Calcium 36(3–4):235–245

Qinfen Z, Jinming C, Xiaojun H, Huanying Z, Jicheng H, Ling F, Kunpeng L, Jingqiang Z (2004) J Med Virol 73(3):332–337
Qiu M, Shi Y, Guo Z, Chen Z, He R, Chen R, Zhou D, Dai E, Wang X, Si B, Song Y, Li J, Yang L, Wang J, Wang H, Pang X, Zhai J, Du Z, Liu Y, Zhang Y, Li L, Wang J, Sun B, Yang R (2005) Microbes Infect 7(5–6):882–889
Rota PA, Oberste MS, Monroe SS, Nix WA, Campagnoli R, Icenogle JP, Penaranda S, Bankamp B, Maher K, Chen MH, Tong S, Tamin A, Lowe L, Frace M, DeRisi JL, Chen Q, Wang D, Erdman DD, Peret TC, Burns C, Ksiazek TG, Rollin PE, Sanchez A, Liffick S, Holloway B, Limor J, McCaustland K, Olsen-Rasmussen M, Fouchier R, Gunther S, Osterhaus AD, Drosten C, Pallansch MA, Anderson LJ, Bellini WJ (2003) Science 300:1394–1399
Røttingen J, Iversen JG (2000) Acta Physiol Scand 169:203–219
Schaecher SR, Mackenzie JM, Pekosz A (2007) J Virol 81:718–731
Senanayake SD, Brian DA (1997) Virus Res 48:101–105
Shen S, Lin PS, Chao YC, Zhang A, Yang X, Lim SG, Hong W, Tan YJ (2005) Biochem Biophys Res Commun 330(1):286–292
Stertz S, Reichelt M, Spiegel M, Kuri T, Martínez-Sobrido L, García-Sastre A, Weber F, Kochs G (2007) Virology 361(2):304–315
Tan YJ, Teng E, Shen S, Tan TH, Goh PY, Fielding BC, Ooi EE, Tan HC, Lim SG, Hong W (2004a) J Virol 78(13):6723–6734
Tan YJ, Fielding BC, Goh PY, Shen S, Tan TH, Lim SG, Hong W (2004b) J Virol 78(24):14043–14047
Thiel V, Ivanov KA, Putics A, Hertzig T, Schelle B, Bayer S, Weissbrich B, Snijder EJ, Rabenau H, Doerr HW, Gorbalenya AE, Ziebuhr J (2003) J Gen Virol 84:2305–2315
Vennema H, Godeke GJ, Rossen JW, Voorhout WF, Horzinek MC, Opstelten DJ, Rottier PJ (1996) EMBO J 15(8):2020–2028
Weiss SR, Navas-Martin S (2005a) Microbiol Mol Biol Rev 69(4):635–664
Weiss SR, Navas-Martin S (2005b) Microbiol Mol Biol Rev 69:635–664
Weiss SR, Navas-Martin S (2005c) Microbiol Mol Biol Rev 69:635–664
Wilson L, McKinlay C, Gage P, Ewart G (2004) Virology 330(1):322–331
Wilson L, Gage P, Ewart G (2006) Virology 353(2):294–306
Ye Y, Hogue BG (2007) J Virol 81(7):3597–3607
Yount B, Roberts RS, Sims AC, Deming D, Frieman MB, Sparks J, Denison MR, Davis N, Baric RS (2005) J Virol 79:14909–14922
Yu CJ, Chen YC, Hsiao CH, Kuo TC, Chang SC, Lu CY, Wei WC, Lee CH, Huang LM, Chang MF et al (2004) FEBS Lett 565:111–116
Yuan X, Li J, Shan Y, Yang Z, Zhao Z, Chen B, Yao Z, Dong B, Wang S, Chen J, Cong Y (2005) Virus Res 109(2):191–202
Yuan X, Wu J, Shan Y, Yao Z, Dong B, Chen B, Zhao Z, Wang S, Chen J, Cong Y (2006) Virology 346:74–85
Zeng R, Yang RF, Shi MD, Jiang MR, Xie YH, Ruan HQ, Jiang XS, Shi L, Zhou H, Zhang L, Wu XD, Lin Y, Ji YY, Xiong L, Jin Y, Dai EH, Wang XY, Si BY, Wang J, Wang HX, Wang CE, Gan YH, Li YC, Cao JT, Zuo JP, Shan SF, Xie E, Chen SH, Jiang ZQ, Zhang X, Wang Y, Pei G, Sun B, Wu JR (2004) J Mol Biol 341(1):271–279

Chapter 12
Molecular and Biochemical Characterization of the SARS-CoV Accessory Proteins ORF8a, ORF8b and ORF8ab

Choong-Tat Keng and Yee-Joo Tan

Abstract A novel coronavirus was identified as the aetiological agent for the global outbreak of severe acute respiratory syndrome (SARS) at the beginning of the twenty-first century. The SARS coronavirus genome encodes for proteins that are common to all members of the coronavirus, i.e. replicase polyproteins (pp1a and pp1b) and structural proteins (spike, membrane, nucleocapsid and envelope), as well as eight accessory proteins. The accessory proteins have been designated as open reading frames (ORF) 3a, 3b, 6, 7a, 7b, 8a, 8b and 9b, and they do not show significant homology to viral proteins of other known coronaviruses. Epidemiological studies have revealed that the part of the viral genome that encodes for ORF8a and ORF8b showed major variations and the animal isolates contain an additional 29-nucleotide sequence which is absent in most of the human isolates. As a result, ORF8a and ORF8b in the human isolates become one ORF, termed ORF8ab. In this chapter, we will discuss the genetic variation in the ORF8 region, expression of ORF8a, ORF8b and ORF8ab during infection, cellular localization and posttranslational modification of ORF8a, ORF8b and ORF8ab, participation of ORF8a, ORF8b and ORF8ab in viral–viral interactions, their effects on other viral proteins and impact on viral replication and/or pathogenesis.

12.1 Introduction

A novel coronavirus was identified as the aetiological agent for the severe acute respiratory syndrome (SARS) epidemic in 2003 (Drosten et al. 2003; Poon et al. 2004). In addition to the replicase polyproteins (pp1a and pp1ab) and structural proteins (spike (S), membrane (M), nucleocapsid (N) and envelope (E)) which are

Y.-J. Tan (✉)
Collaborative Anti-Viral Research Group, Institute of Molecular and Cell Biology, 61 Biopolis Drive, Proteos, Singapore 138673
e-mail: mcbtanyj@imcb.a-star.edu.sg

common to all members of the coronavirus genus, the SARS coronavirus (SARS-CoV) genome also encodes eight accessory proteins, i.e. open reading frames (ORFs) 3a, 3b, 6, 7a, 7b, 8a, 8b and 9b, varying in length from 39 to 274 amino acids (Marra et al. 2003; Snijder et al. 2003). While the SARS-CoV replicase gene products and structural proteins share some degree of sequence homology with those of other coronaviruses, the accessory proteins do not show significant homology to viral proteins of known coronaviruses. Indeed, many laboratories have performed molecular and biochemical characterization of these accessory proteins (Tan et al. 2006; Narayanan et al. 2008), as a better understanding of these unique SARS-CoV proteins may offer clues as to why the SARS-CoV causes such a severe and rapid attack in humans while other coronaviruses that infect humans seem to be more forgiving.

Interestingly, epidemiological studies have revealed that the part of the viral genome that encodes for two of these accessory proteins, ORF8a and ORF8b, shows major variations. In one of these studies, Guan and coworkers analysed SARS-CoV isolates obtained from animals in a live-market in Guangdong and found that all the animal isolates contained a 29-nucleotide (nt) sequence which is absent in most of the human isolates (Guan et al. 2003) (Fig. 12.1a). As a result of this, the ORF8a and ORF8b (also termed as ORF10 and ORF11 respectively) in the

Fig. 12.1 (a) Schematic diagram showing the genetic difference in the ORF8 region of the SARS-CoV isolated from animals and humans infected during the middle phase of the SARS epidemic in 2003. The animal isolates have an extra 29-nucleotide insertion such that the subgenomic RNA8 encodes for a single protein, termed ORF8ab, while that of the human isolates (from the middle phase) encodes two proteins, ORF8a and ORF8b. Human isolates from early phase of the epidemic also have the 29-nucleotide insertion found in the animal SARS-CoV. (b) Alignment of the protein sequences of ORF8a, ORF8b and ORF8ab. Mismatches between ORF8a and ORF8ab or ORF8b and ORF8ab are boxed. The predicted signal peptides in ORF8a and ORF8ab are *underlined* and the predicted *N*-glycosylation sites in these proteins are showed in *bold*. The ORF8ab is reconstructed from a human isolate from the middle phase (SIN2774) by insertion of the 29-nucleotides found in a human isolate from the early phase (GZ02). *Dot* represents the stop codon

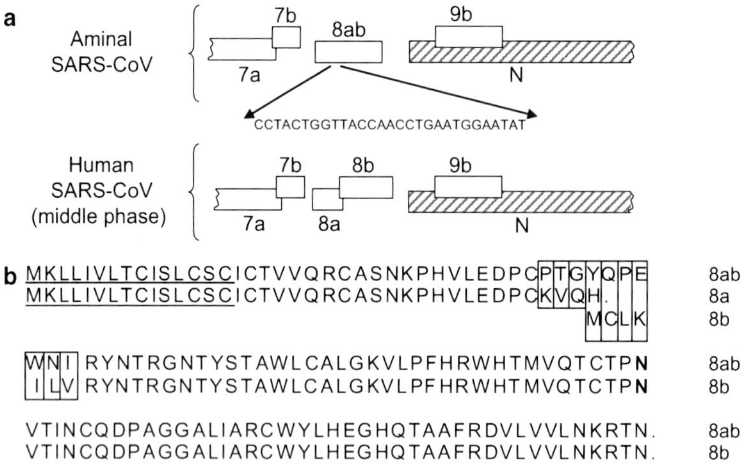

human isolates become one ORF, termed as ORF8ab. ORF8ab encodes a protein of 122 amino acids, whose N terminus is identical to ORF8a and C terminus is identical to ORF8b (Fig. 12.1b). For simplification, we shall term the isolates obtained from wet-market animals as animal SARS-CoVs and those without the additional 29-nt in the ORF8 region as human SARS-CoVs.

In this chapter, we summarize present knowledge of the two accessory proteins, ORF8a and ORF8b, encoded by subgenomic mRNA8 of the human SARS-CoV, as well as their counterpart in animal SARS-CoV, ORF8ab. Although the mutations in the ORF8 region do not appear to have any adverse effect on the survival of the virus, it is conceivable that the ORF8a, ORF8b and ORF8ab proteins have different stabilities and/or functions, and hence would contribute differently to viral replication and/or pathogenesis in vivo. In order to understand how the variations in the ORF8 region of the viral genome may influence SARS-CoV infection, we will discuss the genetic variations in the ORF8 region, expression of ORF8a, ORF8b and ORF8ab during infection, cellular localization and posttranslational modification of ORF8a, ORF8b and ORF8ab, participation of ORF8a, ORF8b and ORF8ab in viral–viral interactions, their effects on other viral proteins and impact on viral replication and/or pathogenesis.

12.2 Genetic Variations in the ORF8 Region

When a virus is first introduced into the human population from an animal source, it has to undergo evolution in order to optimize the entry, replication and budding processes as well as to evade immune responses. Thus, genetic and epidemiological studies yield valuable insights on how viruses cross the species barrier and evolve to cause disease in humans. Indeed, such studies conducted on the SARS-CoV have revealed that this virus crossed the animal–human barrier recently (Donnelly et al. 2004; Guan et al. 2003; Lau et al. 2005; Song et al. 2005). Large genetic variations have also been found in the ORF8 region. Interestingly, the animal strains of SARS-CoV, isolated from a raccoon dog and palm civets in markets/restaurants and from wild bats, contain an extra 29-nt in the ORF8 region (Guan et al. 2003; Lau et al. 2005).

Another extensive study of 63 SARS-CoV isolates obtained from the SARS epidemic in China also showed that there are major variations in this region of the viral genome (The Chinese SARS Molecular Epidemiology Consortium 2004). In this study, the course of the epidemic was divided into the early, middle and late phases with the early phase defined as the period of first emergence of SARS in Guangdong Province, China, 2002. The middle phase referred to all events up to the first cluster of SARS cases in the Metropole hotel in Hong Kong and the late phase referred to all cases following this cluster. Interestingly, the clustering of patients with different patterns of variation in ORF8 region was correlated with the different phases of the epidemic. The 29-nt sequence is not found in all the human strains that were isolated in the middle phase of the epidemic but is present in most of the human isolates from the early phase. Indeed, human infection in the early phase probably represents the first breach of the animal–human barrier as these isolates share the closest phylogenetic relationship with the animal isolates. These findings were subsequently verified by researchers who studied the SARS-CoVs isolated in different countries (Lan et al. 2005; Qin et al. 2003; Wang et al. 2005).

After the WHO's declaration of the end of the SARS epidemic, there were four confirmed SARS patients in Guangzhou, China, in late 2003 to early 2004 (Liang et al. 2004; Song et al. 2005). These patients did not have any contact history with previously documented SARS cases. Sequence analysis of viruses isolated from these patients showed that they were not derived from the preceding epidemic in 2003 but rather suggested that these cases represented new zoonotic transmissions (Song et al. 2005). Like the animal isolates, these viruses also contain the additional 29-nt in the ORF8 region.

These findings clearly indicate that the deletion of the extra 29-nt in ORF8 is not necessary for animal–human transmission, as human SARS-CoVs isolated from early phase still contain these nucleotides. However, during infection in humans, the deletion of the 29-nt seems to occur quickly. Besides this deletion, other types of deletion close to or within the ORF8 region have also been reported. For example, the viruses isolated from some patients in Hong Kong were found to contain a deletion of 386-nt that flanks the 29-nt site (The Chinese SARS Molecular Epidemiology Consortium 2004; Chiu et al. 2005). This causes the ORF7b to become smaller and eliminates both ORF8a and ORF8b. In addition, a frame-shift mutation in the *ORF8a* gene was also reported for two isolates that had been passaged in Vero E6 or FRhK4 cells (Poon et al. 2005). This mutation was not found in the original clinical samples, suggesting that this mutation was acquired as a result of cell culture adaptation. Interestingly, deletion mutations that lead to premature stop of the ORF8a protein have also been found in three other human isolates (Lan et al. 2005).

Analysis of the variation in the sequences of S protein show that SARS-CoV has rapidly evolved during the SARS epidemic and that the virus was undergoing adaptation in the human host (Song et al. 2005; The Chinese SARS Molecular Epidemiology Consortium 2004). While it is clear that the *S* gene was undergoing positive selection (Holmes 2005; Zhao 2007), whether the genetic variation in the ORF8 region is a result of viral adaptation or genomic instability remains to be determined.

12.3 Expression of ORF8a, ORF8b and ORF8ab During Infection

12.3.1 Expression During Infection In Vivo

During viral infection, the host's immune system is stimulated to defend it against the invading virus. Subsequently, antibody responses against viral proteins can be detected in the sera of infected patients. Serological assays have been used to determine if SARS-CoV infected patients have antibodies against the SARS-CoV accessory proteins (Tan et al. 2005). In one study, two out of 37 patients were found to have anti-ORF8a antibodies suggesting that ORF8a was expressed during infection in vivo (Chen et al. 2007). Since only 5.4% of the patients had anti-ORF8a antibodies, this suggests that ORF8a may be expressed at low levels or that it is not highly immunogenic. While this study used full-length ORF8a expressed in bacteria as well as mammalian cells to determine the presence of anti-ORF8a antibodies, another study used overlapping peptides of ORF8a instead (Guo et al. 2004). Here, they examined four pairs of acute and convalescent sera and found that two of the convalescent sera contained IgA antibodies against ORF8a while none of the acute sera had such antibodies. In the same study, no anti-ORF8b antibody was found.

Thus, these studies show that ORF8a can be expressed during infection in vivo but there is no evidence for the expression of ORF8b. Due to the high similarity in the amino acid sequences of ORF8a and ORF8ab, it is difficult to distinguish between the antibodies against ORF8a and ORF8ab in these patients. As no sequencing was performed for the samples from these patients, it is also not known if the 29-nt insertion is present in the viral genomes.

12.3.2 Expression During Infection In Vitro

Since SARS-CoV grows well in cell culture, different groups also generated ORF8a and ORF8b specific antibodies in order to determine their expression during infection in vitro. In our laboratory, we have been successful in detecting the expression of ORF8a and ORF8b in SARS-CoV infected Vero E6 cells that were infected with an isolate from a SARS patient in Singapore (SARS CoV 2003VA2774) (Keng et al. 2006). ORF8a and ORF8b were found to be localized in the cytoplasm of the infected cells at a late time point of infection. Similar results were obtained when the experiment was repeated with the HKU39849 Hong Kong isolate (unpublished data). However, another laboratory could not detect ORF8b in SARS-CoV (strain 5688) infected Vero E6 cells, indicating that ORF8b is not, or only very inefficiently, expressed during infection (Oostra et al. 2007). The difference between these two studies may be due to the specificity and affinity of the antibodies used or the different time-point of infection studied. Thus far, the expression of ORF8ab during infection in vitro has not been documented.

ORF8a and ORF8b are encoded by the bicistronic subgenomic RNA8 produced in SARS-CoV infected cells (Marra et al. 2003; Snijder et al. 2003). Since its translation initiation codon is not the first AUG in the subgenomic RNA8, ORF8b is likely to be expressed via an internal ribosomal entry mechanism or by a leaky ribosomal scanning mode of translation, as have been described for viral proteins encoded by other bicistonic or tricistronic coronaviral mRNAs (Liu and Inglis 1992; Senanayake and Brian 1997; Thiel and Siddell 1994). Thus, the recombinant vaccinia virus bacteriophage T7 RNA polymerase expression system was further used to generate mRNA that mimics the subgenomic RNA8 generated in infected cells (Oostra et al. 2007). Again, ORF8b could not be expressed using this system, indicating that the human SARS-CoV may not contain an internal ribosomal entry site for the expression of ORF8b. Similar results were independently reported by another laboratory (Le et al. 2007).

12.4 Cellular Localization and Posttranslational Modification of ORF8a, ORF8b and ORF8ab

12.4.1 Cellular Localization

Immunofluorescence experiments performed on Vero E6 transfected with a cDNA construct containing the *ORF8a* gene show that untagged ORF8a is localized in the mitochondria (Chen et al. 2007). In contrast, ORF8a tagged at the C terminus with the enhanced green fluorescent protein (EGFP) was reported to colocalize with calreticulin, an endoplasmic reticulum (ER) marker (Oostra et al. 2007). In the latter study, ORF8ab-EGFP was also found in the ER while ORF8b-EGFP was distributed throughout the entire cell.

In a separate study, the cellular localization of untagged ORF8a, ORF8b and ORF8ab, also expressed using cDNA constructs, were compared by using specific antibodies against ORF8a and ORF8b (Keng et al. 2006). ORF8a and ORF8b were found in punctuate vesicle-like structures throughout the cytoplasm but no cellular marker was used to determine the nature of these structures. On the other hand, ORF8ab was found to be diffused in the cytoplasm. Hence, there appear to be significant differences in the conformations of ORF8a and ORF8ab, although 35 out of 39 amino acids of ORF8a are present in ORF8ab (Fig. 12.1b). Similarly, 77 out of 84 amino acids of ORF8b are present in ORF8ab, but the cellular localization of ORF8b is distinct from ORF8ab. The vesicular staining for the ORF8b protein was also observed when the protein is tagged at either the N or C terminus with the FLAG tag (von Brunn et al. 2007).

The discrepancies between the different studies may be due to the use of different antibodies. In addition, some laboratories expressed the native proteins while other expressed them fused with different tags (for detection purposes). Given that ORF8a, ORF8b and ORF8ab are rather small proteins, it is probable that fusion with some tags can affect their cellular localization. As all these studies were performed by over-expressing the viral proteins, the exact cellular localization of these proteins in the SARS-CoV infected cells still needs to be examined carefully.

12.4.2 Glycosylation

ORF8ab is predicted to contain a cleavable signal sequence which directs the precursor to the ER and mediates its translocation into the lumen (Oostra et al.

2007). Indeed, when ORF8ab was expressed using the vaccinia virus T7 expression system, the precursor was observed to be cleaved efficiently to yield a cleaved protein that became N-glycosylated, assembled into disulfide-linked homomultimeric complexes, and remained stably in the ER. This was confirmed by an independent study which also used site-directed mutagenesis to show that ORF8ab is N-glycosylated on the Asn81 residue (Le et al. 2007).

ORF8a contains the same signal peptide as ORF8ab but this signal cannot function due to the shortness of the ORF8a polypeptide (Oostra et al. 2007). However, if ORF8a is tagged at its C terminus with EGFP, then the glycosylation of the fusion protein can be observed. Similar experiments performed on ORF8b showed that it is not glycosylated.

12.4.3 Ubiquitination

Another possible explanation for the difficulties encountered by some laboratories in detecting ORF8b in SARS-CoV infected cells (see Sect. 12.3.2) is that the protein may be unstable. In order to determine if ORF8b and ORF8ab are posttranslationally modified by ubiquitination, co-immunoprecipitation experiments were performed after the proteins and myc-tagged ubiquitin were over-expressed in HeLa cells (Le et al. 2007). The results showed that ORF8b and ORF8ab can interact with mono-ubiquitin and polyubiquitin and this was also confirmed by using ORF8b and ORF8ab expressed as glutathione S-transferase-fusion proteins in bacteria.

Although both ORF8b and ORF8ab can become ubiquitinated, ORF8b, but not ORF8ab, seems to undergo rapid degradation when expressed using an infectious clone system derived from the coronavirus infectious bronchitis virus (IBV) (Le et al. 2007). The degradation of ORF8b can be blocked with proteasome inhibitors, suggesting that it is degraded via the proteasome pathway. Interestingly, when the glycosylation of ORF8ab is abolished through the substitution of Asn81 with Asp (see Sect. 12.4.2), it also becomes more susceptible to degradation. Thus, it appears that glycosylation may assist in the folding of ORF8ab, and the lack of glycosylation of ORF8b results in a highly unstable protein.

12.5 Participation of ORF8a, ORF8b and ORF8ab in Viral–Viral Interactions and Their Effects on Other SARS-CoV Proteins

12.5.1 Interaction of ORF8a, ORF8b and ORF8ab with Other Viral Proteins

Co-immunoprecipitation experiments have been performed to determine if ORF8a, ORF8b and ORF8ab can interact with the SARS-CoV structural proteins, S, M, E

Table 12.1 The abilities of ORF8a, ORF8b and ORF8ab to interact with other viral proteins were determined by yeast-two-hybrid and co-immunoprecipitation assays

Interacting partners of:	ORF8a	ORF8b	ORF8ab
nsp3N	No[a]	Yes[a]	N.D.
nsp8	Yes[a,b]	Yes[a,b]	N.D.
nsp15	Yes[a]	No[a]	N.D.
S	No[a]/yes[c]	Yes[a]/no[c]	Yes[c]
E	No[a]/yes[c]	No[a]/yes[c]	No[c]
M	No[a,c]	No[a]/yes[c]	Yes[c]
ORF3a	No[a,c]	No[a]/yes[c]	Yes[c]
ORF7a	No[a,c]	No[a]/yes[c]	Yes[c]
ORF7b	No[a]	Yes[a]	N.D.
ORF8a	No[a]	Yes[a]	N.D.
ORF8b	Yes[a]	No[a]	N.D.
ORF9b	Yes[a]	Yes[a]	N.D.
ORF14	No[a]	Yes[a]	N.D.

[a]Results taken from yeast-two-hybrid assay in von Brunn et al. (2007)
[b]Results taken from co-immunoprecipitation assay in von Brunn et al. (2007)
[c]Results taken from co-immunoprecipitation assay in Keng et al. (2006)
N.D.:not determined

and N, as well as two SARS-CoV accessory proteins, ORF3a and ORF7a (Keng et al. 2006). These results show that the binding profiles of ORF8a, ORF8b and ORF8ab are clearly distinct (Table 12.1), suggesting that the conformations of the ORF8a and ORF8b may be quite different from ORF8ab. Another group also performed a comprehensive analysis of intraviral protein–protein interactions of the SARS-CoV (von Brunn et al. 2007). The results from the yeast-two-hybrid system show that ORF8a and ORF8b are involved in numerous interactions with other viral proteins (Table 12.1). ORF8ab was not examined in this study. However, only the interactions of nsp8 with ORF8a and ORF8b were validated by co-immunoprecipitation experiments. There are some differences between these two studies (Table 12.1) but this is not surprising since protein–protein interactions observed in yeast are frequently not replicated in mammalian cells.

Although these viral–viral protein interactions need to be verified in infected cells, the findings of these studies suggest that ORF8a, ORF8b and ORF8ab may modulate the activities of other viral proteins by interacting with them. It is also interesting to note that ORF8a and ORF8b can interact with each other (von Brunn et al. 2007), but further studies are required to determine if this complex has the same function as ORF8ab.

12.5.2 Downregulation of the E Protein by ORF8b

The expression of ORF8b was observed to downregulate the expression of E but the expression of E was not affected by either ORF8a or ORF8ab (Keng et al. 2006).

Northern blot analysis showed that the mRNA level of E was not decreased in the presence of ORF8b, suggesting that the effect of ORF8b on the expression of the E protein is likely to be posttranslational. This finding was further confirmed by immunofluorescence studies which showed that the expressions of ORF8b and E in SARS-CoV infected cells were mutually exclusive (Keng et al. 2006). The effect of ORF8b on E can be blocked by proteasome inhibitors, suggesting that it involves the proteasome pathway (Le et al. 2007).

Although the coexpression of SARS-CoV E and M is sufficient for the assembly of viral-like particles in the baculovirus system (Ho et al. 2004; Mortola and Roy 2004), it was demonstrated by reverse genetic techniques that E is not essential for the replication of SARS-CoV in Vero E6, HuH-7 and CaCo-2 cells (DeDiego et al. 2007). However, the recombinant SARS-CoV lacking the *E* gene (rSARS-CoV-DeltaE) grew to lower titers than the recombinant wild-type virus, indicating that the E protein has an effect on virus growth. In addition, the rSARS-CoV-DeltaE virus was found to be attenuated in both hamsters and transgenic mice expressing the SARS-CoV receptor, human angiotensin converting enzyme-2 (hACE-2) (DeDiego et al. 2007, 2008), indicating that E is an important virulence factor. Based on this, it is attempting to speculate that the downregulation of E by ORF8b may have an effect on the virulence of SARS-CoV. However, this effect is likely to be only modulative as only a fraction of the infected cells expresses detectable levels of ORF8b (Keng et al. 2006). In order for ORF8b to be expressed via an internal ribosomal entry mechanism or by a leaky ribosomal scanning mode of translation (see Sect. 12.3.2), activation of certain host translational machineries may be necessary (Komar and Hatzoglou 2005; Stoneley and Willis 2004). Such activation may happen only in the fraction of the cells that is in a certain phase of the cell cycle or subjected to certain stimuli. This is probably why ORF8b can only be detected in a fraction of the SARS-CoV infected cells (Keng et al. 2006). Further experiments are needed to address this possibility and determine the temporal expression of ORF8b during the viral replication cycle. In addition, it was reported that over-expression of SARS-CoV E can induce apoptosis in T-cells (Yang et al. 2005), and thus the downregulation of E by ORF8b may also have an impact on viral pathogenesis.

12.6 Impact of ORF8a, ORF8b and ORF8ab on Viral Replication and/or Pathogenesis

12.6.1 Contribution to Viral Replication

It has been demonstrated that palm civets are equally susceptible to the human SARS-CoV isolate BJ01 from the middle phase (containing ORF8a and ORF8b) and the isolate GZ01 from the early phase (containing ORF8ab) (Wu et al. 2005). Although infection of the animals with BJ01 seems to result in higher average body temperature and stronger antibody responses, further experiments are required to confirm these observations. Using reverse genetic methods, Yount and coworkers created an Urbani strain with the 29-nt insertion so that the ORF8 of this human SARS-CoV became like the animal SARS-CoV (Yount et al. 2005). Consistent with Wu et al. 2005, their results show that this mutant virus replicates as well as the wild-type virus in mice, suggesting the 29-nt insertion does not enhance the replication of SARS-CoV in young mice, which cleared both viruses by day 7 postinfection.

These results suggest that ORF8a, ORF8b and ORF8ab are not essential for viral replication or pathogenesis in the mice and palm civet models. However, Chen and coworkers demonstrated that the over-expression of ORF8a can enhance viral replication in HuH-7 cells (Chen et al. 2007). When HuH-7 cell lines stably expressing ORF8a were infected with SARS-CoV, higher viral loads and greater cytopathic effects were observed when compared to the original HuH-7 cells. In order to resolve the discrepancies between these observations, it is important to determine how much higher is the level of ORF8a in these stable cell lines compared to that expressed during SARS-CoV infection. In addition, it will be interesting to investigate if the differences are due to the types of cells used.

12.6.2 Abilities to Modulate Cellular Events

Both apoptosis and necrosis have been observed in various infected tissues obtained during autopsy studies on SARS casualties (Chau et al. 2004; Chong et al. 2004; Ding et al. 2003; Lang et al. 2003; Wei et al. 2007), indicating the occurrence of cell death during SARS infection in vivo. In addition, one of the most common abnormalities in SARS-CoV infected patients is lymphopenia (Peiris et al. 2003; Chng et al. 2005; Chen et al. 2006), which could be caused by the depletion of T lymphocytes by apoptosis. In one study, extensive apoptosis was observed in the hepatocytes of three SARS patients who had liver impairment, suggesting that liver damage in these patients may be mediated by apoptosis (Chau et al. 2004). Apoptosis was also observed in thyroid glands obtained from five fatal SARS cases (Wei et al. 2007).

The occurrence of apoptosis during SARS-CoV infection in vitro has also been studied by various groups (Mizutani et al. 2004; Tan et al. 2004; Yan et al. 2004; Ren et al. 2005; Bordi et al. 2006). The induction of apoptosis was dependent on viral replication and could be inhibited by caspase inhibitors or the over-expression of the prosurvival protein, Bcl-2 (Ren et al. 2005; Bordi et al. 2006). Furthermore, the over-expression of some of the SARS-CoV proteins can induce apoptosis and/or necrosis (Tan et al. 2007). Five of the accessory proteins, ORF3a, ORF3b, ORF6, ORF7a and ORF8a, have also been shown to induce apoptosis when they were over-expressed (Law et al. 2005; Tan et al. 2004; Yuan et al. 2005; Khan et al. 2006; Chen et al. 2007; Ye et al. 2008). In addition, ORF8a was found to be localized in the mitochondria (Chen et al. 2007). As a result, there is a high level of reactive oxygen species production, suggesting that the over-expression of ORF8a causes mitochondrial dysfunction, which results in apoptosis.

The over-expression of ORF8b has been shown to induce DNA synthesis, suggesting that it may have the ability to increase cell proliferation (Law et al. 2006). Another SARS-CoV accessory protein, ORF6, also has the same property (Geng et al. 2005). Coexpression of ORF8b and ORF6 did not yield any synergistic effects, suggesting that they may function independently (Law et al. 2006). However, the abilities of these proteins to modulate DNA synthesis have yet to be examined in the context of SARS-CoV infection.

12.7 Conclusion

Since the identification of the SARS-CoV in the year 2003, extensive research on the SARS-CoV has yielded significant understanding of this newly emerged virus. Interestingly, the SARS-CoV isolated from humans infected during the peak of epidemic, encodes two accessory proteins termed as ORF8a and ORF8b while the SARS-CoV isolated from animals contains an extra 29-nt in this region such that these proteins are fused to become a single protein, ORF8ab. As described above, the accessory proteins ORF8a and ORF8b are expressed during infection in vitro, but replacing them with ORF8ab does not affect SARS-CoV replication in cell culture or small animal models. However, it is still not known if these proteins contribute to viral replication and/or pathogenesis in the natural host. Significant differences in the molecular and biochemical characteristics of ORF8a, ORF8b and ORF8ab have been observed but the impact on the function of these proteins remains unclear. Another unanswered question is why large genetic variations have occurred most frequently in this particular region of the viral genome.

References

Bordi L, Castilletti C, Falasca L, Ciccosanti F, Calcaterra S, Rozera G, Di Caro A, Zaniratti S, Rinaldi A, Ippolito G, Piacentini M, Capobianchi MR (2006) Bcl-2 inhibits the caspase-dependent apoptosis induced by SARS-CoV without affecting virus replication kinetics. Arch Virol 151(2):369–377

Chau TN, Lee KC, Yao H, Tsang TY, Chow TC, Yeung YC, Choi KW, Tso YK, Lau T, Lai ST, Lai CL (2004) SARS-associated viral hepatitis caused by a novel coronavirus: report of three cases. Hepatology 39(2):302–310

Chen RF, Chang JC, Yeh WT, Lee CH, Liu JW, Eng HL, Yang KD (2006) Role of vascular cell adhesion molecules and leukocyte apoptosis in the lymphopenia and thrombocytopenia of patients with severe acute respiratory syndrome (SARS). Microbes Infect 8(1):122–127

Chen CY, Ping YH, Lee HC, Chen KH, Lee YM, Chan YJ, Lien TC, Jap TS, Lin CH, Kao LS, Chen YM (2007) Open reading frame 8a of the human severe acute respiratory syndrome coronavirus not only promotes viral replication but also induces apoptosis. J Infect Dis 196 (3):405–415

Chiu RW, Chim SS, Tong YK, Fung KS, Chan PK, Zhao GP, Lo YM (2005) Tracing SARS-coronavirus variant with large genomic deletion. Emerg Infect Dis 11(1):168–170

Chng WJ, Lai HC, Earnest A, Kuperan P (2005) Haematological parameters in severe acute respiratory syndrome. Clin Lab Haematol 27(1):15–20

Chong PY, Chui P, Ling AE, Franks TJ, Tai DY, Leo YS, Kaw GJ, Wansaicheong G, Chan KP, Ean Oon LL, Teo ES, Tan KB, Nakajima N, Sata T, Travis WD (2004) Analysis of deaths during the severe acute respiratory syndrome (SARS) epidemic in Singapore: challenges in determining a SARS diagnosis. Arch Pathol Lab Med 128(2):195–204

DeDiego ML, Alvarez E, Almazan F, Rejas MT, Lamirande E, Roberts A, Shieh WJ, Zaki SR, Subbarao K, Enjuanes L (2007) A severe acute respiratory syndrome coronavirus that lacks the E gene is attenuated in vitro and in vivo. J Virol 81(4):1701–1713

Dediego ML, Pewe L, Alvarez E, Rejas MT, Perlman S, Enjuanes L (2008) Pathogenicity of severe acute respiratory coronavirus deletion mutants in hACE-2 transgenic mice. Virology 376(2):379–389

Ding Y, Wang H, Shen H, Li Z, Geng J, Han H, Cai J, Li X, Kang W, Weng D, Lu Y, Wu D, He L, Yao K (2003) The clinical pathology of severe acute respiratory syndrome (SARS): a report from China. J Pathol 200(3):282–289

Donnelly CA, Fisher MC, Fraser C, Ghani AC, Riley S, Ferguson NM, Anderson RM (2004) Epidemiological and genetic analysis of severe acute respiratory syndrome. Lancet Infect Dis 4(11):672–683

Drosten C, Preiser W, Gunther S, Schmitz H, Doerr HW (2003) Severe acute respiratory syndrome: identification of the etiological agent. Trends Mol Med 9(8):325–327

Geng H, Liu YM, Chan WS, Lo AW, Au DM, Waye MM, Ho YY (2005) The putative protein 6 of the severe acute respiratory syndrome-associated coronavirus: expression and functional characterization. FEBS Lett 579(30):6763–6768

Guan Y, Zheng BJ, He YQ, Liu XL, Zhuang ZX, Cheung CL, Luo SW, Li PH, Zhang LJ, Guan YJ, Butt KM, Wong KL, Chan KW, Lim W, Shortridge KF, Yuen KY, Peiris JS, Poon LL (2003) Isolation and characterization of viruses related to the SARS coronavirus from animals in southern China. Science 302(5643):276–278

Guo JP, Petric M, Campbell W, McGeer PL (2004) SARS corona virus peptides recognized by antibodies in the sera of convalescent cases. Virology 324(2):251–256

Ho Y, Lin PH, Liu CY, Lee SP, Chao YC (2004) Assembly of human severe acute respiratory syndrome coronavirus-like particles. Biochem Biophys Res Commun 318(4):833–838

Holmes KV (2005) Structural biology. Adaptation of SARS coronavirus to humans. Science 309 (5742):1822–1823

Keng CT, Choi YW, Welkers MR, Chan DZ, Shen S, Gee Lim S, Hong W, Tan YJ (2006) The human severe acute respiratory syndrome coronavirus (SARS-CoV) 8b protein is distinct from its counterpart in animal SARS-CoV and down-regulates the expression of the envelope protein in infected cells. Virology 354(1):132–142

Khan S, Fielding BC, Tan TH, Chou CF, Shen S, Lim SG, Hong W, Tan YJ (2006) Over-expression of severe acute respiratory syndrome coronavirus 3b protein induces both apoptosis and necrosis in Vero E6 cells. Virus Res 122(1–2):20–27

Komar AA, Hatzoglou M (2005) Internal ribosome entry sites in cellular mRNAs: mystery of their existence. J Biol Chem 280(25):23425–23428

Lan YC, Liu HF, Shih YP, Yang JY, Chen HY, Chen YM (2005) Phylogenetic analysis and sequence comparisons of structural and non-structural SARS coronavirus proteins in Taiwan. Infect Genet Evol 5(3):261–269

Lang ZW, Zhang LJ, Zhang SJ, Meng X, Li JQ, Song CZ, Sun L, Zhou YS, Dwyer DE (2003) A clinicopathological study of three cases of severe acute respiratory syndrome (SARS). Pathology 35(6):526–531

Lau SK, Woo PC, Li KS, Huang Y, Tsoi HW, Wong BH, Wong SS, Leung SY, Chan KH, Yuen KY (2005) Severe acute respiratory syndrome coronavirus-like virus in Chinese horseshoe bats. Proc Natl Acad Sci USA 102(39):14040–14045

Law PT, Wong CH, Au TC, Chuck CP, Kong SK, Chan PK, To KF, Lo AW, Chan JY, Suen YK, Chan HY, Fung KP, Waye MM, Sung JJ, Lo YM, Tsui SK (2005) The 3a protein of severe acute respiratory syndrome-associated coronavirus induces apoptosis in Vero E6 cells. J Gen Virol 86(Pt 7):1921–1930

Law PY, Liu YM, Geng H, Kwan KH, Waye MM, Ho YY (2006) Expression and functional characterization of the putative protein 8b of the severe acute respiratory syndrome-associated coronavirus. FEBS Lett 580(15):3643–3648

Le TM, Wong HH, Tay FP, Fang S, Keng CT, Tan YJ, Liu DX (2007) Expression, post-translational modification and biochemical characterization of proteins encoded by subgenomic mRNA8 of the severe acute respiratory syndrome coronavirus. FEBS J 274(16):4211–4222

Liang G, Chen Q, Xu J, Liu Y, Lim W, Peiris JS, Anderson LJ, Ruan L, Li H, Kan B, Di B, Cheng P, Chan KH, Erdman DD, Gu S, Yan X, Liang W, Zhou D, Haynes L, Duan S, Zhang X, Zheng H, Gao Y, Tong S, Li D, Fang L, Qin P, Xu W (2004) Laboratory diagnosis of four recent sporadic cases of community-acquired SARS, Guangdong Province, China. Emerg Infect Dis 10(10):1774–1781

Liu DX, Inglis SC (1992) Internal entry of ribosomes on a tricistronic mRNA encoded by infectious bronchitis virus. J Virol 66(10):6143–6154

Marra MA, Jones SJ, Astell CR, Holt RA, Brooks-Wilson A, Butterfield YS, Khattra J, Asano JK, Barber SA, Chan SY, Cloutier A, Coughlin SM, Freeman D, Girn N, Griffith OL, Leach SR, Mayo M, McDonald H, Montgomery SB, Pandoh PK, Petrescu AS, Robertson AG, Schein JE, Siddiqui A, Smailus DE, Stott JM, Yang GS, Plummer F, Andonov A, Artsob H, Bastien N, Bernard K, Booth TF, Bowness D, Czub M, Drebot M, Fernando L, Flick R, Garbutt M, Gray M, Grolla A, Jones S, Feldmann H, Meyers A, Kabani A, Li Y, Normand S, Stroher U, Tipples GA, Tyler S, Vogrig R, Ward D, Watson B, Brunham RC, Krajden M, Petric M, Skowronski DM, Upton C, Roper RL (2003) The genome sequence of the SARS-associated coronavirus. Science 300(5624):1399–1404

Mizutani T, Fukushi S, Saijo M, Kurane I, Morikawa S (2004) Phosphorylation of p38 MAPK and its downstream targets in SARS coronavirus-infected cells. Biochem Biophys Res Commun 319(4):1228–1234

Mortola E, Roy P (2004) Efficient assembly and release of SARS coronavirus-like particles by a heterologous expression system. FEBS Lett 576(1–2):174–178

Narayanan K, Huang C, Makino S (2008) SARS coronavirus accessory proteins. Virus Res 133(1):113–121

Oostra M, de Haan CA, Rottier PJ (2007) The 29-nucleotide deletion present in human but not in animal severe acute respiratory syndrome coronaviruses disrupts the functional expression of open reading frame 8. J Virol 81(24):13876–13888

Peiris JS, Yuen KY, Osterhaus AD, Stohr K (2003) The severe acute respiratory syndrome. N Engl J Med 349(25):2431–2441

Poon LL, Guan Y, Nicholls JM, Yuen KY, Peiris JS (2004) The aetiology, origins, and diagnosis of severe acute respiratory syndrome. Lancet Infect Dis 4(11):663–671

Poon LL, Leung CS, Chan KH, Yuen KY, Guan Y, Peiris JS (2005) Recurrent mutations associated with isolation and passage of SARS coronavirus in cells from non-human primates. J Med Virol 76(4):435–440

Qin E, He X, Tian W, Liu Y, Li W, Wen J, Wang J, Fan B, Wu Q, Chang G, Cao W, Xu Z, Yang R, Yu M, Li Y, Xu J, Si B, Hu Y, Peng W, Tang L, Jiang T, Shi J, Ji J, Zhang Y, Ye J, Wang C, Han Y, Zhou J, Deng Y, Li X, Hu J, Yan C, Zhang Q, Bao J, Li G, Chen W, Fang L, Li C, Lei M, Li D, Tong W, Tian X, Zhang B, Zhang H, Zhao H, Zhang X, Li S, Cheng X, Liu B, Zeng C, Tan X, Liu S, Dong W, Wong GK, Yu J, Zhu Q, Yang H (2003) A genome sequence of novel SARS-CoV isolates: the genotype, GD-Ins29, leads to a hypothesis of viral transmission in South China. Genomics Proteomics Bioinformatics 1(2):101–107

Ren L, Yang R, Guo L, Qu J, Wang J, Hung T (2005) Apoptosis induced by the SARS-associated coronavirus in Vero cells is replication-dependent and involves caspase. DNA Cell Biol 24(8):496–502

Senanayake SD, Brian DA (1997) Bovine coronavirus I protein synthesis follows ribosomal scanning on the bicistronic N mRNA. Virus Res 48(1):101–105

Snijder EJ, Bredenbeek PJ, Dobbe JC, Thiel V, Ziebuhr J, Poon LL, Guan Y, Rozanov M, Spaan WJ, Gorbalenya AE (2003) Unique and conserved features of genome and proteome of SARS-coronavirus, an early split-off from the coronavirus group 2 lineage. J Mol Biol 331(5):991–1004

Song HD, Tu CC, Zhang GW, Wang SY, Zheng K, Lei LC, Chen QX, Gao YW, Zhou HQ, Xiang H, Zheng HJ, Chern SW, Cheng F, Pan CM, Xuan H, Chen SJ, Luo HM, Zhou DH, Liu YF, He JF, Qin PZ, Li LH, Ren YQ, Liang WJ, Yu YD, Anderson L, Wang M, Xu RH, Wu XW, Zheng HY, Chen JD, Liang G, Gao Y, Liao M, Fang L, Jiang LY, Li H, Chen F, Di B, He LJ, Lin JY, Tong S, Kong X, Du L, Hao P, Tang H, Bernini A, Yu XJ, Spiga O, Guo ZM, Pan HY, He WZ, Manuguerra JC, Fontanet A, Danchin A, Niccolai N, Li YX, Wu CI, Zhao GP (2005) Cross-host evolution of severe acute respiratory syndrome coronavirus in palm civet and human. Proc Natl Acad Sci USA 102(7):2430–2435

Stoneley M, Willis AE (2004) Cellular internal ribosome entry segments: structures, trans-acting factors and regulation of gene expression. Oncogene 23(18):3200–3207

Tan YJ, Fielding BC, Goh PY, Shen S, Tan TH, Lim SG, Hong W (2004) Overexpression of 7a, a protein specifically encoded by the severe acute respiratory syndrome coronavirus, induces apoptosis via a caspase-dependent pathway. J Virol 78(24):14043–14047

Tan YJ, Lim SG, Hong W (2005) Characterization of viral proteins encoded by the SARS-coronavirus genome. Antiviral Res 65(2):69–78

Tan YJ, Lim SG, Hong W (2006) Understanding the accessory viral proteins unique to the severe acute respiratory syndrome (SARS) coronavirus. Antiviral Res 72(2):78–88

Tan YJ, Lim SG, Hong W (2007) Regulation of cell death during infection by the severe acute respiratory syndrome coronavirus and other coronaviruses. Cell Microbiol 9(11):2552–2561

The Chinese SARS Molecular Epidemiology Consortium (2004) Molecular evolution of the SARS coronavirus during the course of the SARS epidemic in China. Science 303(5664):1666–1669

Thiel V, Siddell SG (1994) Internal ribosome entry in the coding region of murine hepatitis virus mRNA 5. J Gen Virol 75(Pt 11):3041–3046

von Brunn A, Teepe C, Simpson JC, Pepperkok R, Friedel CC, Zimmer R, Roberts R, Baric R, Haas J (2007) Analysis of intraviral protein-protein interactions of the SARS coronavirus ORFeome. PLoS ONE 2(5):e459

Wang ZG, Zheng ZH, Shang L, Li LJ, Cong LM, Feng MG, Luo Y, Cheng SY, Zhang YJ, Ru MG, Wang ZX, Bao QY (2005) Molecular evolution and multilocus sequence typing of 145 strains of SARS-CoV. FEBS Lett 579(22):4928–4936

Wei L, Sun S, Xu CH, Zhang J, Xu Y, Zhu H, Peh SC, Korteweg C, McNutt MA, Gu J (2007) Pathology of the thyroid in severe acute respiratory syndrome. Hum Pathol 38(1):95–102

Wu D, Tu C, Xin C, Xuan H, Meng Q, Liu Y, Yu Y, Guan Y, Jiang Y, Yin X, Crameri G, Wang M, Li C, Liu S, Liao M, Feng L, Xiang H, Sun J, Chen J, Sun Y, Gu S, Liu N, Fu D, Eaton BT, Wang LF, Kong X (2005) Civets are equally susceptible to experimental infection by two different severe acute respiratory syndrome coronavirus isolates. J Virol 79(4):2620–2625

Yan H, Xiao G, Zhang J, Hu Y, Yuan F, Cole DK, Zheng C, Gao GF (2004) SARS coronavirus induces apoptosis in Vero E6 cells. J Med Virol 73(3):323–331

Yang Y, Xiong Z, Zhang S, Yan Y, Nguyen J, Ng B, Lu H, Brendese J, Yang F, Wang H, Yang XF (2005) Bcl-xL inhibits T-cell apoptosis induced by expression of SARS coronavirus E protein in the absence of growth factors. Biochem J 392(Pt 1):135–143

Ye Z, Wong CK, Li P, Xie Y (2008) A SARS-CoV protein, ORF-6, induces Caspase-3 mediated, ER stress and JNK dependent apoptosis. Biochim Biophys Acta 1780(12):1383–1387

Part IV
Viral Pathogenesis and Host Immune Response

Chapter 13
SARS Coronavirus Pathogenesis and Therapeutic Treatment Design

Timothy P. Sheahan and Ralph S. Baric

Abstract Emerging pathogens are either new or newly recognized or those that are increasing in incidence and spread. Since the identity of emerging pathogens from animal reservoirs is difficult to predict, the development for pathogen-specific therapeutics and vaccines is problematic. The highly pathogenic SARS coronavirus (SARS-CoV) emerged from zoonotic pools in 2002 to cause a global epidemic of severe acute respiratory syndrome (SARS). Many patients with SARS-CoV experienced an exacerbated form of disease called acute respiratory distress syndrome (ARDS) requiring mechanical ventilation and supplemental oxygen and half of these patients died. Similar to other viral pathogens like influenza and West Nile Virus, the severity of SARS-CoV disease increased with age. Unfortunately, successful vaccination in the most vulnerable populations is a difficult task because of immunological deficiencies associated with aging (immune senescence). Due to the rapidity of virus emergence, technologies like synthetic biology can be harnessed to facilitate rapid recombinant virus construction for studying the novel virus biology, pathogenesis and the evaluation of therapeutic interventions. Since predicting the antigenic identity of future emergence is difficult, candidate vaccines and therapeutics should have a maximal breadth of cross-protection, and panels of antigenically divergent synthetically reconstructed viruses can be used as tools for this evaluation. We discuss how synthetic reconstruction of many animal and human SARS-CoV has provided a model to study the molecular mechanisms governing emergence and pathogenesis of viral diseases. In addition, we review the evolution, epidemiology, and pathogenesis of epidemic and zoonotic SARS-CoV with focus on the development of broadly reactive therapeutics and vaccines that protect aged populations from the zoonotic pool.

R.S. Baric (✉)
Department of Microbiology and Immunology, University of North Carolina at Chapel Hill, Chapel Hill, NC, USA
e-mail: rbaric@email.unc.edu

13.1 Introduction

Many diseases of great human historical significance in the ancient and recent past like plague, smallpox, HIV, and influenza A emerged from wild or domestic animal populations to cause devastating human disease (Achtman et al. 1999; Li et al. 2007; Nguyen et al. 2005; Parrish and Kawaoka 2005; Tumpey et al. 2005; Wolfe et al. 2007; Woo et al. 2006). In 2002, a novel coronavirus (SARS-CoV) emerged suddenly as the causative agent of severe acute respiratory syndrome (SARS) and spread worldwide causing about 8,000 cases and >700 deaths (Christian et al. 2004; Ksiazek et al. 2003; Rota et al. 2003). Viruses similar to the epidemic strain were isolated from civets for sale within wet markets in China during the epidemic in 2003 and the reemergence of 2004 (Chinese 2004; Guan et al. 2003). Genome sequences of viruses isolated from bats, civets and humans suggest that viruses circulating in bats crossed the species barrier to infect civets who then served as an amplification host for yet another host-range shift to generate human tropic virus (Chinese 2004; Guan et al. 2003; Lau et al. 2005; Li et al. 2005b). Since viruses similar to the epidemic strain of SARS-CoV are currently circulating in zoonotic pools, the future emergence of a SARS-CoV-like virus may occur, as has occurred with Ebola, influenza H5N1, Marburg, and chikungunya virus (Gonzalez et al. 2007; Kaur et al. 2008; Leroy et al. 2005; Towner et al. 2007; Woo et al. 2006). Therefore, it is imperative that we understand the pathogenic mechanisms of coronavirus lung diseases and that current vaccination and passive sero-therapies be effective in protecting humans from infection by zoonotic SARS-CoV. Though a considerable amount of work has enhanced our knowledge of SARS-CoV pathogenesis and therapeutic treatment design, many questions remain unanswered. What host factors have contributed to the protection or prevention of severe disease? Will SARS-CoV therapeutics be effective against future emergence? How do we rationally design an antiviral therapy against future emergence of unknown antigenic identity? Will SARS-CoV therapeutics protect the most vulnerable human populations? The current research aimed at answering these questions is the focus of this review.

13.2 Human SARS-CoV Pathogenesis

13.2.1 The Clinical Course of Human SARS-CoV Infection

SARS-CoV is thought to have emerged suddenly from zoonotic pools of virus. Molecular epidemiology suggests that the epidemic strain evolved from bat-associated SARS-CoV-like viruses by way of a intermediate civet host (Fig. 13.1). Without SARS-CoV evolution promoting efficient infection of human cells and person-to-person transmission, the emergent SARS-CoV epidemic would not have occurred. Thus, zoonotic SARS-CoV adaptation was a necessary initial step in

Fig. 13.1 Schematic of SARS-CoV emergence from zoonotic pools. Viruses similar to the epidemic strain of SARS-CoV, SARS Urbani, were found within Chinese horseshoe bats, which are believed to be the animal reservoir for SARS-CoV like viruses. Within live animal markets in China, it is postulated that interaction between bat and Himalayan palm civets helped the virus to jump species to infect civets. Viruses more closely related to SARS Urbani were found in civets in markets during the epidemic. Over time, civet viruses were transmitted to humans, ultimately resulting in a virus that could be transmitted from person to person. During the early middle and late phases of the epidemic, SARS-CoV evolved further until the virus stabilized as SARS Urbani

SARS-CoV human pathogenesis. Sequence analysis of zoonotic, early, middle and late-stage epidemic strains, coupled with in vitro evolution experimentation, has demonstrated that zoonotic isolates can rapidly adapt to efficient growth in human airway cells by multiple genetic pathways (Chinese 2004; Guan et al. 2003; Li et al. 2005a, 2005b; Sheahan et al. 2008a, 2008b). The plasticity of the SARS-CoV spike (S) glycoprotein and receptor interaction is a particularly troubling harbinger for the ease and potential of future cross-species transmission. SARS-CoV is thought to be transmitted by direct patient contact, airborne droplet nuclei, contact with fomites or urine/fecal contact with mucous membranes (Peiris et al. 2003a, 2003b). After a brief incubation period of approximately 6 days, the patient enters the acute phase of infection characterized by fever (>100°F), chills, malaise, and myalgia (Booth et al. 2003; Liang et al. 2004; Peiris et al. 2003a). During the acute phase of the infection patients develop a nonproductive cough/shortness of breath (dyspnea), and bilateral pulmonary infiltrates are seen by chest radiography. Pulmonary lesions visible by radiography continue to worsen until 7 days after the onset of symptoms (AOS), after which most patients begin to improve. Approximately 30% of patients show clinical improvement after the first week of illness while the remaining 70% present with recurring fever and shortness of breath (Peiris et al. 2003a, 2003b). A case study of health care workers in Toronto ($n = 14$, age mean $= 42$ years, age range 27–63 years) provides a typical example of the course of SARS-CoV convalescence, where a week after hospital discharge, all patients complained of dyspnea, weakness, and lethargy and all suffered from significant weight loss (anorexia) from SARS-CoV disease (Avendano et al. 2003). Three weeks after discharge, the Toronto healthcare worker cohort were no longer weak and continued to gain weight but still suffered from dyspnea (14/14 patients) and a

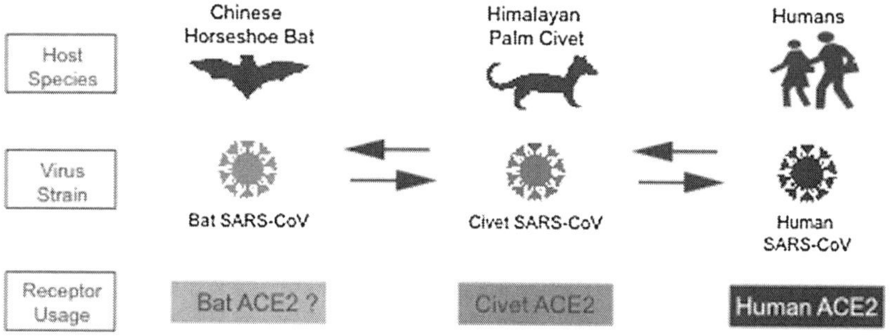

few still presented with an abnormal chest X-ray (5/14 patients) (Avendano et al. 2003). A case study in Hong Kong provides a detailed example of a nonconvalescent cohort where lung damage continues to progress in a minority (20–30%) of patients where "diffuse ground glass" changes are seen in the chest X-ray typical of acute lung injury (ALI) and acute respiratory distress syndrome (ARDS) (Peiris et al. 2003a). ALI can progress to ARDS, which is characterized primarily by an acute onset, bilateral infiltrates on chest radiograph, hypoxemia, fever, and leukopenia (Bauer et al. 2006a).

ALI and ARDS are inflammatory lung diseases characterized by diffuse alveolar infiltration, hypoxia, respiratory failure, and death due to the failure of multiple organs. The most severe form of ALI, ARDS, is fatal in almost 50% of patients and affects ~1,000,000 people/year worldwide, ranking it among the most difficult challenges in critical-care medicine (Kuba et al. 2006a). ARDS is characterized by diffuse alveolar damage (DAD), which includes a protein-rich edema, an exudative phase with hyaline membranes, and inflammation leading to surfactant dysfunction and severe hypoxia. Neutrophils dominate in bronchoalveolar lavage (BAL) fluid, and cytokines TGF-β1, TNFα, IL-1β, IL-6, IL8 are often elevated (Bauer et al. 2006b; Dahlem et al. 2007; Guidot et al. 2006). The exudative phase lasts about a week, and a progressive proliferative phase can progress and organize, resulting in fibrosis (fibrotic phase). Approximately 20% of SARS patients required intensive care unit (ICU) treatment due to ARDS symptoms and a majority of those admitted to the ICU require mechanical ventilation and oxygen support (Booth et al. 2003; Peiris et al. 2003a). Though overall mortality rates of the global SARS-CoV epidemic approached 8%, the mortality rates for patients over the age of 65 ranged from approximately 25 to 55% as a result of comorbidities and immune senescence (Booth et al. 2003; Leung et al. 2004; Liang et al. 2004; Peiris et al. 2003a).

In the lung, the primary target cells of SARS-CoV infection of humans remains controversial. Several thorough pathological studies of post-mortem tissues from SARS-CoV infection have identified the ciliated epithelial cells, and type I and II pneumocytes within the lung as the primary target of virus infection, though virus antigen has also been found in macrophages (Mø), dendritic cells (DCs), T cells, B cells, NK cells, and putative lung stem/progenitor CD34 + Oct-4+ cells (Chen et al. 2007; Gu et al. 2005; Hwang et al. 2005; Nicholls et al. 2006; Tse et al. 2004; Ye et al. 2007). Unfortunately, the in-situ hybridization utilized in these pathological studies was unable to differentiate between active viral infection and uptake of virus by passive cellular means like phagocytosis. Several in vitro studies have demonstrated that human Mø and DCs were unable to support active virus replication and instead instigated cell activation leading to upregulation of MHC II and secretion of inflammatory cytokines (Cheung et al. 2005; Law et al. 2005; Tseng et al. 2005). In support of gross lung pathology seen by X-ray, microscopic evaluation of SARS-CoV lung pathology has repeatedly been described in various cohorts as showing various phases of exudative and proliferative acute lung injury (ALI) (Gu et al. 2005; Hwang et al. 2005; Nicholls et al. 2006; Tse et al. 2004). Typical SARS-CoV lung pathology is characterized by inflammatory cell infiltration, pulmonary edema, hyaline membrane formation, mild to moderate fibrosis, alveolar epithelial hyperplasia, and alveolar/epithelial cell desquamination (i.e., sloughing) (Gu et al. 2005; Hwang et al. 2005; Tse et al. 2004). In two postdischarge cohorts, the percentages of patients with continued symptoms and pulmonary fibrosis ranged from 21 to 62% and the development of fibrosis was associated with increasing age and admission to intensive care (Antonio et al. 2003; Wang et al. 2008). Pulmonary fibrosis (PF) is a devastating disease with an almost universally fatal outcome characterized by inflammation of the alveoli, damage to lung tissues, and progressive interstitial fibrosis (hardening of tissues). There are five million people worldwide that are affected by PF, including some 200,000 cases culminating in 40,000 deaths/year in the US (http://www.pulmonaryfibrosis.org/home.htm). Models of virus-induced PF are essential for understanding and managing these devastating end-stage clinical diseases. The ultimate clinical course of ARDS is often determined by the ability of the injured lung to repopulate the alveolar epithelium with functional cells. Death may occur when fibrosis predominates during the healing response, worsening lung compliance and oxygenation.

13.2.2 The Human Adaptive Immune Response to SARS-CoV

Chest X-ray, serology and virological data suggest direct involvement of the adaptive immune response in viral clearance. A detailed virological/immunological longitudinal analysis was performed on a cohort of SARS patients in Hong Kong (Peiris et al. 2003a). Five days AOS, nasopharyngeal aspirates contained between 3 and 7 $\log_{10}$ genomes/ml of SARS-CoV and by day 10 the range tightened to 5–7 $\log_{10}$ genomes/ml. In most patients SARS-CoV-specific IgG seroconversion begins near day 10 AOS (mean = 20 days AOS) after which virus titers begin to fall. Interestingly, 40% of the cohort ($n = 75$) did not seroconvert until 24 days AOS but 93% had seroconverted by day 29 AOS. Even after seroconversion, viral genomes were detected in nasopharyngeal aspirate (47%), stool (67%) and urine (21%) as far as 21 days AOS. Of note, many patients in the Hong Kong cohort were treated with corticosteroids and these drugs may have delayed the onset of seroconversion. In an animal model of coronavirus lung infection (PRCV) with similar pathologies to SARS-CoV, the administration of corticosteroids alleviated signs of PRCV pneumonia early (2 dpi) though exacerbated later stages of disease (4, 10, 21 dpi) probably due to the lack of a cell-mediated response creating an environment for extended virus lung replication (Jung et al. 2007). Together, these data suggest that the administration of corticosteroids during SARS-CoV acute infection may exacerbate disease due to dampening of the cell-mediated immune response. In another Hong Kong SARS cohort, seroconversion was detected as early as 4 days AOS with a median seroconversion occurring at 15 days AOS (Lee et al. 2006). Lastly, Cameron et al. correlated elevated levels of anti-SARS-CoV S-specific antibody in patient sera with less-severe disease (Cameron et al. 2007). Nevertheless, these data suggest that as the adaptive immune response mounts, viral load is depressed paving the way for convalescence. Unfortunately, all current animal models fail to recapitulate the adaptive immune response to SARS-CoV.

13.2.3 The Human Innate Immune Response to SARS-CoV

Important insights into the mechanisms of the innate inflammatory response of SARS-CoV infection has been gleaned from several clinical, in vitro and in vivo animal model studies, yet a clear and concise model of this innate response and its relation to viral pathogenesis remains to be elucidated. Inconsistencies in experimental protocol, clinical treatment, cell type infected in vitro, and possible species-specific effects within various animal models have created a confusing and complicated body of data making the generation of a comprehensive model of SARS-CoV pathogenesis difficult. Nevertheless, concordant data between experimental systems has provided the coronavirus field with a great body of useful information and those data are summarized below.

A thorough evaluation of inflammatory gene expression in SARS-CoV patient peripheral blood mononuclear cell (PBMC) was performed by both Cameron et al. and Regunathan et al. but the investigators arrived at disparate conclusions. Cameron et al. concluded that a robust type I INF response was observed early in the progression of disease and was predicted to be essential for viral clearance and convalescence (Cameron et al. 2007). In addition to a strong INF response, genes typically induced by INF like CCL2 (MCP-1) and CXCL10 (IP-10) were also upregulated in patients early during SARS-CoV disease (Cameron et al. 2007). Interestingly, Cameron and collaborators present data that suggest that the cytokine storm that serves to protect some SARS-CoV patients can progress in an unchecked manner and contribute to the development of an inadequate adaptive immune response and more severe disease (Cameron et al. 2007). Reghunathan and collaborators also sampled SARS patient PBMCs for microarray analysis but did not find type I INF induced in their samples and instead found upregulation of several other tissue inflammation/remodeling, homeostasis, and cell-cycle genes (Reghunathan et al. 2005). Compounding the fact that Reghunathan did not discuss virological data and failed to report the stage of SARS-CoV disease at which these samples were extracted (acute, convalescent, etc.), the sample size ($n = 10$) was small in comparison to the Cameron et al. cohort ($n = 50$) (Cameron et al. 2007; Reghunathan et al. 2005). Of note, many cohorts of SARS patients, including the Cameron cohort discussed above, were treated with immunosuppressive corticosteroids making the resulting immunological and virological data more difficult to interpret and evaluate (Stockman et al. 2006).

13.2.4 Animal Models of SARS-CoV Pathogenesis

Since human clinical SARS data is complicated by host genetic variation, disease-exacerbating comorbidities, age variation, and variable drug-treatment regimens, animal models provide a more homogeneous and controlled environment within which to ask questions related to the involvement of the host immune response to SARS-CoV infection. In nonhuman primate (NHP, cynomolgus macaque) infection with SARS-CoV, Haagmans et al. demonstrated the prophylactic administration of pegylated INFα controlled virus replication and lessened disease pathology (Haagmans et al. 2004). Concordantly, transcriptional profiles of NHP-infected lung tissue suggest that INFα, β, λ, and γ and also IP-10, MCP-1, IL-6, and IL-8 genes were all upregulated in SARS-CoV-infected NHPs (de Lang et al. 2007). Since NHPs do not succumb to infection, it is proposed that the inflammatory response to virus infection described above aids in the control and clearance of this acute infection.

The development of mouse models that recapitulate components of human disease have been invaluable in viral pathogenesis research. Young BALB/c and C57BL/6 strains of mice support SARS Urbani replication in the lung and infection causes lung pathology similar to that seen in human cases, but these models lack both morbidity and mortality and the more severe lung pathologies noted in human patients (Roberts and Subbarao 2006; Rockx et al. 2007). Since replication models provide little utility regarding disease pathogenesis, several models of severe SARS-CoV disease were developed in 2007 (Table 13.1). Subbarao et al. created a mouse-adapted SARS-CoV (MA15) through repeated passage of SARS Urbani in BALB/c mice (Roberts et al. 2007). Our laboratory created a molecular clone of MA15 (rMA15) through the introduction of the six mouse-adapting amino acid changes into our infections clone for SARS Urbani (icSARS) (Roberts et al. 2007) (Fig. 13.2). When administered intranasally to young adult BALB/c mice (6–10 weeks), rMA15 causes significant weight loss (~20% of starting weight) resulting from a severe acute infection of the lung, resulting in almost 100% mortality by 4 dpi (Roberts et al. 2007). rMA15 mortality ensues more rapidly with increasing age in BALB/c mice, where year-old mice succumb to infection beginning on 3 dpi (unpublished observation, Sheahan and Baric). Similarly, Rockx et al. developed an age-related mouse model of severe SARS-CoV disease where infection of old BALB/c mice with SARS-CoV bearing the civet HCSZ61/03 or early human GZ02 S glycoprotein genes caused uniform mortality by 4 dpi and lung pathologies quite similar to those seen in human cases (Rockx et al. 2007). Due to the acute and severe nature of the infections within these lethal models (100% mortality by 4 dpi), many aspects of natural SARS-CoV infection such as development of the adaptive immune response, death after virus clearance due to immunopathology, and the development of pulmonary fibrosis cannot be assessed. Since a majority of the human SARS-CoV cases resulted in patient morbidity and survival, models with morbidity but without mortality should be developed in order to study the mechanisms of host protection reflecting the prevailing course of SARS-CoV disease.

Though current animal models are not able to assess the importance of adaptive immunity in SARS-CoV pathogenesis, multiple studies have implicated the importance of the innate response in viral clearance. Glass et al. demonstrated that SARS-CoV was cleared with similar kinetics in WT C57BL/6 mice or strains that lacked that T, B and NK cells, suggesting that the innate immune response alone is sufficient for viral clearance (Glass et al. 2004). Like gene expression data from human samples, MCP-1, MIP1-α, and IP-10 are all upregulated in the lung during SARS-CoV infection of C57BL/6 mice, suggesting a role for these chemokines in

Table 13.1 Models of lethal disease for studying SARS-CoV pathogenesis and passive immunization/vaccine efficacy

Virus name	Description	Total AA Δs from Urbani	Spike AA Δs from Urbani		Pathogenic phenotype			
					Young BALB/c	Aged BALB/c	Young C57BL/6	Aged C57BL/6
icSARS	Molecular clone of the epidemic strain SARS Urbani	0	0	Weight loss	+	++	No	No
				Replication in lung	++	+++	++	++
				Mortality	No	+/−	No	No
icGZ02	icSARS bearing an early epidemic phase virus spike	6	6	Weight loss	+	+++	N/A	N/A
				Virus replication in lung	++	+++	N/A	N/A
				Mortality	No	Yes	N/A	N/A
icHC/SZ/ 61/03	icSARS bearing a zoonotic civet SARS-CoV spike	21	21	Weight loss	+	+++	N/A	N/A
				Virus replication in lung	++	++	N/A	N/A
				Mortality	No	Yes	N/A	N/A
rMA15	Molecular clone of the mouse adapted SARS-CoV	6	1	Weight loss	+++	+++	++	+++
				Virus replication in lung	+++	+++	++	+++
				Mortality	Yes	Yes	No	Yes
rMA15 GD03-S	rMA15 bearing the GD03 spike	24	19	Weight loss	++	+++	N/A	N/A
				Virus replication in lung	++	+++	N/A	N/A
				Mortality	No	Yes	N/A	N/A

N/A: Not accessed; AA Δs: amino acid changes

13 SARS Coronavirus Pathogenesis and Therapeutic Treatment Design

Fig. 13.2 SARS-CoV reverse genetics, mouse adapting mutations and the construction of SARS-CoV bearing heterologous spike proteins. The SARS-CoV infectious cDNA clone for the epidemic strain, SARS Urbani, developed in the Baric laboratory divides the 29,727 bp virus genome over six cDNA plasmid clones. To construct a SARS-CoV bearing a heterologous spike (S) protein, synthetic biology can be harnessed and the heterologous S gene can be inserted into the infectious clone using standard molecular biology techniques. When the recombinant virus is constructed, only the S gene is heterologous, resulting in a SARS-CoV bearing a heterologous S protein. The locations of the mouse-adapting mutations reported by Roberts et al. (2006) are marked within the schematic of the SARS-CoV genome (nsp5, nsp5, nsp9, nsp13, Spike RBD, and M)

protection (Glass et al. 2004). In support of Glass et al., Hogan et al. demonstrated that STAT1, a key modulator of INF α/β, λ, and γ signaling, was required for the resolution of SARS infection, once again implicating the importance of the innate response in the clearance of SARS-CoV (Hogan et al. 2004). The induction of INF in mice seems to be dependent on mouse strain, where BALB/c mice induce type I INF following SARS-CoV infection as measured by microarray (Rockx et al. 2009) and ELISA while the induction of INF is undetectable in C57BL/6 mice (Sheahan et al. 2008) (Roberts et al. 2005a). We have recently developed a C57BL/6 mouse model of acute SARS-CoV pathogenesis with significant morbidity (12–15% loss in body weight) and complete recovery due to the activation of the innate response and the recruitment of inflammatory leukocytes to the lung (see Sect. 13.2.5) (Sheahan et al. 2008). These immunologic and pathologic discrepancies between mouse strains are unfortunate caveats of animal models of viral disease. Moreover, no animal model of SARS-CoV pathogenesis to date has fully recapitulated both the innate and adaptive immunopathological aspects SARS-CoV disease. Nevertheless, mouse models of SARS-CoV pathogenesis faithfully recapitulate many aspects of acute human disease like virus replication, the induction of inflammatory cytokines, migration of immune cells into pulmonary tissues, virus- and immune-cell-mediated lung pathology, and weight loss. Also, unlike human and NHP in vivo models, transgenic "knock out" mice provide the opportunity to evaluate the role of single genes in viral pathogenesis. Recently, a novel panel of genetically dissimilar recombinant inbred mice has been developed called "the collaborative cross," which will allow for the elucidation of genetic pathways involved in multigenic complex traits (Churchill et al. 2004). The genetic diversity within the collaborative cross is similar to that present in the human population and, using statistical and genomic analysis, genes responsible for observed phenotypes are elucidated. Rather than evaluate the role of a single gene in SARS-CoV pathogenesis in "knock out" mice, the use of the collaborative cross for SARS-CoV pathogenesis may uncover roles for multiple host genes involved in the progression or prevention of disease.

13.2.5 In Vitro Models of SARS-CoV Pahtogenesis

Perhaps the most simplified models within which to study SARS-CoV pathogenesis are in vitro models. Though in vitro models are less complicated than in vivo models, the resultant data and relevance to SARS-CoV pathogenesis is hotly debated. There is a seeming incongruity between in vivo human, primate and mouse data where the induction of INF is observed while the in vitro infection of various primary or immortalized interferon-competent cell types fail to induce or produce INF. These in vitro experiments are further complicated by the notion that several viral genes have been implicated as active INF antagonists (Frieman et al. 2007; Kopecky-Bromberg et al. 2007; Zust et al. 2007). The in vitro data regarding SARS-CoV innate immune activation is reviewed below.

The infection of interferon-competent primary human airway epithelial cells (HAE) with SARS-CoV does not result in the induction or secretion of INF but does result in the secretion of inflammatory chemokines IL-6, MCP-1 and IP-10 (Sims unpublished data) (Frieman et al. 2007). Pathological evaluation of lung tissue from lethal SARS-CoV and in vitro data suggests that airway epithelial cells are the primary target for SARS-CoV infection yet do not induce INF in vitro (Sims et al. 2005; Tse et al. 2004; Ye et al. 2007). In NHP studies by Haagmans and colleagues, it was demonstrated that epithelial cells adjacent to infected epithelial cells stained positive for INFβ, suggesting a possible bystander activation effect. It may be that viral INF antagonists suppress the antiviral-sensing network in the infected cell but eventually neighboring cells are activated by circulating interferons released from nonpermissive cells or through the sensing of viral proteins or genomic RNA released into the extracellular milieu as a result of viral-induced cell lysis.

Several studies focusing on infection of professional antigen-presenting cells like dendritic cells and macrophages have also produced controversial data. Most studies utilize human PBMC-derived macrophages or DCs where CD14+ cells are isolated and differentiated in the presence of cytokines (Mø = GM-CSF, DC = IL-4, GM-CSF) in vitro, after which the cell populations resemble "macrophages" and "dendritic cells" by cell surface staining profiles (Cheung et al. 2005; Frieman et al. 2007; Law et al. 2005). When SARS-CoV is added to these Mø and DC populations, a productive infection does not ensue; INF is not induced but several other inflammatory cytokines are induced and secreted (MIP1-α, IP-10, MCP-1) (Cheung et al. 2005; Frieman et al. 2007; Law et al. 2005). A possible explanation for the disparity between the in vitro and in vivo data regarding INF induction in SARS-CoV infection is presented by Cervantes-Barragan and colleagues where they show key differences in conventional (cDC) and plasmacytoid dendritic cell (pDC) populations in response to SARS-CoV infection (Cervantes-Barragan et al. 2007). pDCs differ from cDCs in their surface characteristics (cDC = CD11c+, B220−, pDC = B220+, CD11c-low, PDCA-1+) and function where pDCs are the major source of INFα in both humans and mice (Asselin-Paturel et al. 2001; Cella et al. 1999; Cervantes-Barragan et al. 2007; Siegal et al. 1999). Unlike most investigators who artificially differentiate Mø and DC from CD14+ precursor cells isolated from PBMCs, Cervantes-Barragan and colleagues isolated cDC and pDC populations directly from human blood which were subsequently incubated with SARS-CoV and demonstrated that, unlike cDC, pDC induced INFβ transcription and produced large amounts of INFα protein in the cell media (Cervantes-Barragan et al. 2007). In the assessment of both pDC and cDC populations side by side, Cervantes-Barragan provide a possible explanation as to why previous studies of SARS-CoV "infection" of dendritic cells failed to induce INF, especially since the differentiation protocol used by Law, Tseng, and Cheung results in the differentiation of a more cDC-like cell (Cervantes-Barragan et al. 2007; Cheung et al. 2005; Law et al. 2005; Tseng et al. 2005). Cervantes-Barragan also utilized in vitro differentiated mouse cDC and pDC that were used in mouse hepatitis virus (MHV) experiments. Interestingly, the cytokine used by Cervantes-Barragan to generate cDC (GM-CSF only) from CD14+ cells was used by other investigators to create Mø, though the generation of the pDC using Flt3-L was not employed by either Law, Tseng or Cheung (Cervantes-Barragan et al. 2007; Cheung et al. 2005; Law et al. 2005; Tseng et al. 2005). Though the work of Cervantes-Barragan helps clarify some of the discrepancies seen in studies of the innate immune response to SARS-CoV, the body of work is deficient in demonstrating the generation of infectious virus through infection of cDCs or in suggesting a mechanism of SARS-CoV binding and entry, since ACE2 expression in cDCs was not addressed (Cervantes-Barragan et al. 2007; Cheung et al. 2005; Law et al. 2005; Tseng et al. 2005). Of note, studies have demonstrated that the lectins, DC-SIGN/L-SIGN, are coreceptors for SARS-CoV docking and entry and these receptors are often found on DCs and other APCs (Jeffers et al. 2004). Nevertheless, it will be interesting to see whether future mouse/NHP studies definitively demonstrate a role for pDCs in SARS-CoV pathogenesis.

13.2.6 SARS-CoV and MyD88

Toll-like receptors (TLRs) and Nod-like receptors (NLRs) are examples of host cell proteins that recognize pathogen-associated molecular patterns (PAMPs) (O'Neill and Bowie 2007; Uehara et al. 2007). MyD88 is an important adaptor protein required for the perpetuation of almost all TLR proinflammatory signals as well as interleukin-1 and -18 receptor (IL-1R1, IL-18R1) signaling events (O'Neill and Bowie 2007). Recent data has implicated MyD88 in both the progression and prevention of viral disease. Infection of MyD88-deficient mice with respiratory syncytial virus (RSV) or vesicular stomatitis virus (VSV) results in an exacerbation of disease, while infection with reovirus results in a similar clinical course to that seen in WT mice (Johansson et al. 2007; Phipps et al. 2007; Rudd et al. 2007; Zhou et al. 2007b). We have recently developed a mouse model for SARS-CoV pathogenesis that recapitulates aspects of the acute human infection where wild-type C57BL/6 mice infected with 10^5 pfu of rMA15 (recombinant mouse adapted SARS-CoV) experience a significant but transient weight loss (12–15% by 3–4 dpi), high titer virus replication ($>10^8$ pfu/g 1 and 2 dpi), inflammation in the lung with the induction of proinflammatory chemokines and cytokines with a marked recruitment of inflammatory monocytes to the infected lung, and virus clearance and recovery from disease by 7 dpi (Fig. 13.3) (Sheahan et al. 2008). Interestingly, infection of age- and sex-matched MyD88-deficient mice results in a failure to control virus replication with significantly higher lung titers over time, a delay in the induction of inflammatory gene transcription, a delay in inflammatory leukocyte recruitment, and 90% mortality by 6 dpi (Sheahan et al. 2008). The receptor providing the protective signal through MyD88 remains to be elucidated though we have ruled out both IL-1R1 and IL-18R1. These data suggest that the MyD88-dependent induction of innate proinflammatory chemokines and cytokines and the subsequent recruitment of inflammatory leukocytes are required for protection from

	WT C57Bl/6	MyD88-/-
Peak Titer	2dpi 10^9 pfu/g	2dpi 10^9 pfu/g
Virus clearance	by 7dpi	Persistant infection > 10^7 pfu/g until death
Weight Loss	12-15% by 3-4dpi	12-15% by 3-4dpi
Pulmonary proinflammatory gene transcription	Massive upregulation of IL-1, IL-6, TNF, CCL2, CCL3, and CCL5 by 2dpi	Minimal upregulation of IL-1, IL-6, TNF, CCL2, CCL3, and CCL5 by 4dpi
Recruitment of inflammtory leukocytes	Moncytes and Neutrophils by 2dpi	Moncytes and Neutrophils by 4dpi
Lung Pathology	Mild bronchiolitis, inflammatory cell recruitment (2dpi)	Severe denuding bronchioltis, delayed inflammatory cell recruitment (4dpi)
Mortality	No	Yes

Fig. 13.3 Phenotypic differences between mouse-adapted SARS-CoV (rMA15)-infected WT C57BL/6 or mice deficient in the gene *MyD88*. MyD88 is an important adapter protein for almost all toll-like receptors, IL-1R1, and IL-18R1 proinflammatory signaling events. Mice deficient in MyD88 are far more susceptible to rMA15 infection

SARS-CoV-induced mortality, and future studies may elucidate the precise interaction between the virus and host responsible for the MyD88-dependent protective signal. Furthermore, future genetic and epidemiological studies of SARS-CoV-infected persons may reveal a role for MyD88- and MyD88-related gene polymorphisms in SARS-CoV disease.

13.2.7 SARS-CoV and the Renin–Angiotensin System

The cellular receptor for SARS-CoV infection, angiotensin I converting enzyme 2 (ACE2), serves as a prime example of a cellular protein strictly required for the virus to gain entry into the host cell while also serving an important function in host physiology and perhaps viral pathogenesis and disease. Within a year of discovering SARS-CoV, ACE2 was identified as the chief virus receptor utilized to gain entry into the host cell, though other attachment factors have also been proposed (Jeffers et al. 2004; Li et al. 2003). A second human coronavirus, NL63, also uses ACE2 as a receptor for docking and entry (Pyrc et al. 2006). Isolation and in vitro expression of ACE2 molecules from various species such as mouse, civet and human have also helped elucidate important facets of epidemic and zoonotic SARS-CoV S and ACE2 interactions, virus host range expansion and the evolution of the epidemic strain (Li et al. 2006). ACE2 is expressed within lung epithelia, type I/II pneumocytes (the primary cellular targets of SARS-CoV) as well as within the intestinal epithelium, vascular endothelium, heart, kidney, and testis (Donoghue et al. 2000; Hamming et al. 2004). ACE2 and angiotensin I converting enzyme (ACE) are key regulators of the renin–angiotensin system (RAS), which helps control cardiovascular function by maintaining the body's blood pressure and electrolyte balance (Fig. 13.4) (Kuba et al. 2006b; Nicholls et al. 1998). ACE and ACE2 are metalloproteases with differing vasoactive peptide substrate specificities and as a result have disparate and antagonistic roles in maintaining physiologic homeostasis (Kuba et al. 2006b). ACE cleaves the peptide ANG I into ANG II, which has vasoconstrictive effects inducing hypertension while also inducing cell proliferation and fibrosis (Kuba et al. 2006b; Turner and Hooper 2002). In contrast, ACE2 processes ANG I into ANG 1-9 and further processes ANG II into the peptide ANG 1-7 which acts as a vasodilator while also being antiproliferative and apoptotic (Donoghue et al. 2000; Kuba et al. 2006b; Tipnis et al. 2000; Vickers et al. 2002). Current in vitro data suggests that ANG II ligation and signaling through angiotensin receptor 1a (AT1aR) can result in the production of proimflammatory cytokines (TNFα, IL-1β, IL-6, MCP-1, etc.), fibrosis, and cell proliferation (McAllister-Lucas et al. 2007). In vivo models of liver fibrosis or ALI support the above in vitro data, suggesting that ANG II exacerbates pathology and that this pathology is ameliorated by ACE2-related signals. In an acid aspiration model of ALI/ARDS, Imai et al. demonstrated that ACE2 protected mice from injury while ACE, ANG II and AT1aR promoted disease pathology (Imai et al. 2005). Similarly, Herath et al. demonstrated that ACE2 and ANG 1-7 counteracted the detrimental

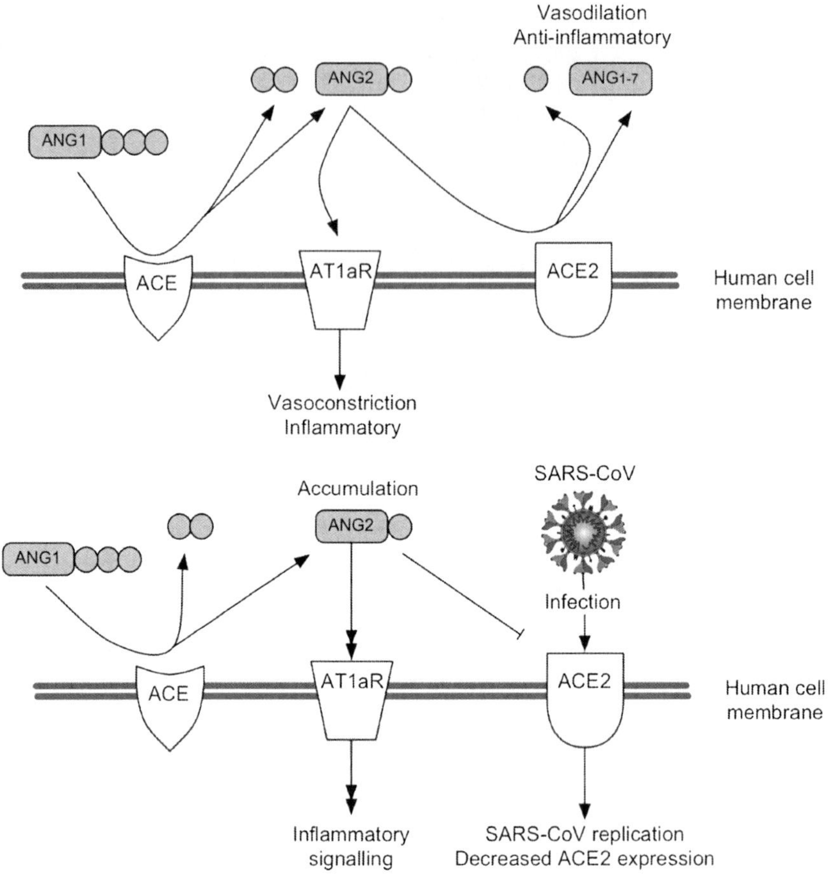

Fig. 13.4 The SARS-CoV receptor, angiotensin I converting enzyme 2 (ACE2), in virus entry and pathogenesis. (**a**) ACE2 and angiotensin I converting enzyme (ACE) are key regulators of the renin–angiotensin system (RAS), which helps control cardiovascular function by maintaining the body's blood pressure and electrolyte balance. Current in vitro data suggests that ANG II ligation and signaling through angiotensin receptor 1a (AT1aR) can result in the production of proinflammatory cytokines (TNFα, IL-1β, IL-6, MCP-1, etc.), fibrosis, and cell proliferation. (**b**) Infection of the lung by SARS-CoV disrupts RAS homeostasis. Current data suggests that SARS-CoV infection or SARS S protein decreases levels of ACE2 within the lung, thereby removing a key regulator and processor of the proinflammatory ANG II peptide whose excess contributes to more severe disease

effects of ANG II in liver disease in rats (Herath et al. 2007). These data suggest a duality of RAS contributing to both homeostasis and immunopathology.

Perhaps the most interesting facet of the SARS-CoV and ACE2 relationship resides in the possible effect of virus infection on the local pulmonary disruption of RAS homeostasis (Fig. 13.4). Within a mouse model of SARS-CoV replication, Kuba et al. demonstrated that SARS-CoV infection diminished levels of ACE2 within the lung (Kuba et al. 2005). Recombinant SARS-CoV S protein delivered intraperitoneally similarly reduced levels of ACE2 within the lung. Furthermore, within an acid aspiration model of acute lung injury (ALI), the administration of SARS S recombinant protein exacerbated ALI as measured by changes in lung elastance and the accumulation of edema within a 3 h period post acid injury (Kuba et al. 2005). It was proposed that SARS-CoV or SARS S decreased levels of ACE2 within the lung, thereby removing a key regulator and processor of the proinflammatory ANG II peptide whose excess contributed to more severe disease through proinflammatory signaling through AT1aR (Kuba et al. 2005). Recent data by Haga et al. suggest a mechanism for the downregulation of ACE2 in the lung and resultant increase in lung tissue damage. They found that the cleavage of ACE2 ectodomain on the cell surface was mediated by SARS-CoV S and TNF-α converting enzyme while the cytoplasmic tail of ACE2 simultaneously triggered the production of the tissue-destroying cytokine, TNF-α (Haga et al. 2008). Taken together, these data provide an interesting insight into possible RAS involvement in SARS-CoV pathogenesis and the progression of ALI to ARDS seen in more severe cases of SARS-CoV. The Kuba et al. model of acid aspiration-induced ALI is very acute (injury assessed within a 3 h window) compared with virus-induced lung injury where phenotypes evolve over many hours to days. Further, the acid aspiration model exists outside the context of virus infection and the induction of the innate and adaptive immune response, which were both found to contribute considerably to SARS-CoV pathogenesis in humans. Recently, an hACE2 transgenic mouse was created where hACE2 expression is targeted to epithelial cells via the K18 promoter (McCray et al. 2007). Although the K18 promoter targeted hACE2 expression to the lung epithelium, these mice also expressed a large amount of ACE2 in their central nervous system (CNS) epithelia. As such, infection of these mice with the epidemic strain, SARS Urbani, resulted in 100% mortality, most likely due to replication and infection within the CNS. Since CNS manifestations were not the chief pathological observation in human SARS-CoV infection, these data suggest that hACE2 transgenic mice under the control of nonlung cell-specific promoters may have limitations in determining pathways of SARS-CoV pathogenesis within the host. Although the K18 hACE2 mice most likely succumbed to infection of the CNS, the increased amounts of pulmonary ACE2 was not protective of lung pathology as predicted by the model presented by Kuba et al. Since K18 hACE2 transgenic mice are extremely susceptible to SARS-CoV infection, they may be useful as a highly stringent model to assess vaccine efficacy and challenge in the future. Given the interesting but conflicting data regarding RAS and SARS-CoV pathogenesis, future evaluation within current animal models of SARS-CoV pathogenesis may help resolve this discrepancy. It will be interesting to see if SARS-CoV infection of ACE- or AT1aR-deficient mice modulate the development of severe SARS-CoV disease. The use of commercially available drugs that block AT1aR (Telmisartan) or ACE (A0773) in the context of SARS-CoV infection may also provide interesting information on the involvement of the RAS system in SARS-CoV pathogenesis. Lastly, the uncoupling of ACE2 physiological function and SARS-CoV receptor function through the generation of catalytically inactive ACE2 "knock in" mice would allow for SARS-CoV infection, but

ACE2 cleavage of ANGII could not occur. One would predict that the infection of these catalytically inactive ACE2 mice would experience more severe disease due to the absence of the protective ACE2 metabolism of the proinflammatory ANGII.

13.3 SARS-CoV Therapeutic Design

It is clear that vaccination and passive immunization technologies are among the most important public health interventions in the past 200 years contributing to the complete eradication of smallpox (Marasco and Sui 2007; Plotkin 1999). Though vaccination campaigns have eradicated polio and measles in developed nations, these diseases and other vaccine-preventable diseases continue to plague developing nations (WHO 2008a, 2008b, 2008c). Rapidly and newly emerging infectious diseases like SARS-CoV provide unpredictable scenarios for the field of vaccinology, where diseases never before seen in human populations arise and spread rapidly while reagents necessary for vaccine development do not yet exist. Emerging viruses arising from zoonotic pools are especially problematic, as vaccines and therapeutics targeted against previously evolved strains might not function against strains associated with contemporary outbreaks of disease. As seen with the SARS-CoV epidemic, isolation of the virus allowed for the rapid generation of "killed," DNA and viral vectored vaccines within a year of the start of the epidemic (Sui et al. 2004; Tang et al. 2004). Given the sequence diversity in bat SARS-CoV reservoirs, it is likely that these killed vaccines will fail against newly emerged strains that arise in the future, especially since antisera directed against bat S glycoproteins do not neutralize human epidemic strains (Becker et al. 2008). Like vaccination, the practice of passive immunization to prevent infection or curtail established disease was first shown by Robert Koch in the late 1800s, when he demonstrated that sheep antisera against diptheria toxin could protect against death in humans (Marasco and Sui 2007). More recently, technologies like phage display and memory B cell immortalization have been developed to produce sufficient quantities of human monoclonal antibodies (hu-mABs) directed against specific viral antigens (Marasco and Sui 2007). These technologies allowed for the rapid development of neutralizing hu-mABs directed against SARS-CoV within a year of the beginning of the epidemic (Sui et al. 2004). We will discuss the problems associated with SARS-CoV vaccination and passive immunization therapies, which fall into three categories that include (a) SARS-CoV antigenic variation and therapy efficacy, (b) the complications of immunosenescence and SARS-CoV vaccine efficacy, and (c) SARS-CoV vaccine immunopotentiation of lung pathology.

13.3.1 SARS-CoV Antigenic Variation and Therapy Efficacy

Within humans infected by the SARS-CoV, multiple antigens were targeted by the adaptive immune response. Although T cell responses directed against nonstructural replicase proteins have been measured in convalescent patients, the majority of both T and B cell responses were directed against structural proteins (Li et al. 2008; Qiu et al. 2005; Yang et al. 2007). In a study exploring T cell responses in SARS-CoV patients by Li et al., the most frequently targeted peptides resided within the S and ORF 3a proteins, with lesser responses to E, M, and N proteins (Li et al. 2008). In studies evaluating the humoral response to SARS antigens in patient samples, antibody responses to S, ORF 3a, N and ORF9b were measured but only antibodies targeting the S protein were capable of neutralization (Qiu et al. 2005). Most studies demonstrate that only antibodies directed against the S glycoprotein are capable of neutralizing SARS-CoV, although conflicting data regarding anti-ORF-3a antibody neutralization has also been reported (Akerstrom et al. 2006; Cameron et al. 2007; Qiu et al. 2005; Yount et al. 2005).

The body of work related to SARS-CoV vaccine development is astounding. Unfortunately, many studies only evaluate the immune response to antigen and fail to evaluate the vaccine efficacy through SARS-CoV virus challenge (Bai et al. 2008; Jin et al. 2005; Liu et al. 2005; Zhang et al. 2005). Inactivated whole virus and vectored SARS-CoV vaccine trials in a number of different animals models have demonstrated that the SARS-CoV spike glycoprotein (S) is the critical component of protective immunity and the passive transfer of SARS-CoV S-specific sera is sufficient to provide protection from infection and disease caused by a homologous SARS-CoV strain (Buchholz et al. 2004; Deming et al. 2006; He et al. 2006; Kapadia et al. 2005; Qin et al. 2006; Spruth et al. 2006; Subbarao et al. 2004; Wang et al. 2005; Yang et al. 2004; Zhou et al. 2005, 2007a). However, current animal models universally display a very acute SARS-CoV-like disease and most vaccines have only been evaluated in the context of virus replication without severe acute lung injury (Chu et al. 2008; Deming et al. 2006; Haagmans et al. 2004; Haagmans and Osterhaus 2006; Roberts et al. 2006, 2007; Roberts and Subbarao 2006; Subbarao and Roberts 2006). As such, these models may underrepresent the importance of the cell-mediated and humoral responses in controlling more prolonged infection and pathogenesis as seen in human cases of SARS-CoV. In fact, the development of a SARS-CoV animal model that recapitulates both acute and prolonged infection with the development of adaptive immunity would greatly benefit the study of SARS-CoV pathogenesis and SARS-CoV vaccine development. Nevertheless, several replication, mouse-adapted lethal, and age-related models of acute SARS-CoV pathogenesis now exist and are currently the most effective systems within which to assess vaccine and passive immunization efficacy (Deming et al. 2006; Roberts et al. 2007, 2005a, 2005b; Rockx et al. 2007).

13.3.2 Animal Models to Assess Passive Immunization Therapy Efficacy Against Divergent SARS-CoV Antigens

Effective human monoclonal antibodies (hu-mAbs) can be utilized for the prophylactic or acute treatment of viral infections (Marasco and Sui 2007). Since the generation of effective vaccines and vaccine-induced immunity can be time-consuming, the production of broadly neutralizing hu-mAbs targeting emerging viral diseases may be valuable for use in healthcare workers and vulnerable populations in emerging viral outbreak situations in immunologically naïve populations. The past emergence of SARS-CoV from zoonotic pools and the continued circulation of SARS-like viruses within bat populations provides the perfect situation for the development of hu-mAb therapies to protect against future emergence. Due to the unknown antigenic identity of future emergent SARS-CoV, we are presented with a difficult problem: which cross-neutralizing epitopes should be targeted by passive immunization therapies in order to effectively treat future emergence of SARS-CoV? As with vaccination, the most successful hu-mABs for passive immunization against SARS-CoV should broadly neutralize all current and future SARS-CoV strains. One of the first hu-mABs developed against SARS-CoV, 80R, was effective in neutralizing pseudovirus-bearing epidemic (Tor2) and civet (SZ3) S proteins but was not as effective against pseudovirus bearing the civet-like GD03 S in vitro (Sui et al. 2005). Using hu-mAB m396, Zhu et al. reported complete neutralization of SARS Urbani and SARS-CoV bearing GD03 S (icGD03-S) but was less effective at neutralizing the SARS-CoV bearing the SZ16-K479N S (icSZ16-S K479N, 4 log reduction in virus lung titer as compared to control hu-mAB) in passive transfer experiments in young BALB/c mice (Zhu et al. 2007). In contrast to the m396 antibody, Zhu et al. also reported complete protection from virus replication (day 2 post infection) against SARS Urbani, icGD03-S, and icSZ16-S K479N using similar doses of hu-mAB S230.15 (Zhu et al. 2007). Rockx et al. also demonstrated the cross-reactivity and potent neutralizing ability of hu-mAB S230.15, S227.14, and S109.8 in passive transfer studies in mice (Rockx et al. 2008). S230.15, S227.14, and S109.8 effectively neutralized SARS Urbani and recombinant SARS-CoV bearing the early epidemic phase GZ02 S, though all were slightly less efficacious against recombinant SARS-CoV bearing the civet HC/SZ/61/03 S glycoprotein (2 log reduction in virus lung titers on day 2 post infection as compared to control Ab) (Rockx et al. 2008). Though S230.15 and S227.14 did not protect against virus replication in HC/SZ/61/03-challenged mice, both antibodies protected Urbani-, GZ02-, and HC/SZ/61/03-nfected mice from clinical signs of disease and death (Rockx et al. 2008). These data highlight the importance of using more than one strain of SARS-CoV when evaluating SARS-CoV therapies, since the hu-mABs discussed above provided varying degrees of protection from replication depending on the SARS-CoV S variant that was employed in the in vitro or in vivo assay. Also, these data suggest that the complete abrogation of replication may not be necessary to protect from clinical signs of SARS-CoV disease. Importantly, escape mutants generated from hu-mAB S227.14, S230.15, and S109.8 were significantly attenuated in mice.

13.3.3 Animal Models to Assess Vaccine Immunization Therapy Efficacy Against Divergent SARS-CoV Antigens

Since the epidemic strain may no longer exist in nature, vaccination with epidemic strain antigens followed by challenge with the epidemic strain may not be the most biologically and medically relevant design. Due to the complications of designing a vaccine against future emergence of SARS-CoV whose antigenic identity is unknown, several difficult questions arise in the development of effective SARS-CoV therapies: Which SARS-CoV antigen or pool of antigens will provide the greatest degree of cross-protection if the vaccination is to prevent disease from future emergence of SARS-CoV? Which vaccine formulation will be effective in the elderly? Lastly, which SARS-CoV strain(s) should be employed as challenge virus to assess vaccine efficacy? We can begin to answer these questions within current animal models of SARS-CoV pathogenesis.

Several animal models (see Sect. 13.2.1) and common vaccine formulations have been utilized in SARS-CoV vaccine development and these data are summarized in Table 13.2. Major approaches to SARS-CoV vaccine platform development include whole killed, recombinant viral vector, DNA, live-attenuated SARS-CoV, and recombinant protein subunit vaccines (Bisht et al. 2004, 2005; Chen et al. 2005; Czub et al. 2005; Darnell et al. 2007; Deming et al. 2006; Du et al. 2007, 2008a, 2008b; Faber et al. 2005; Gai et al. 2008; He et al. 2004; Jin et al. 2005; Kapadia et al. 2005, 2008; Kobinger et al. 2007; Lamirande et al. 2008; Liniger et al. 2008; Martin et al. 2008; Qin et al. 2006; Qu et al. 2005; See et al. 2006, 2008; Spruth et al. 2006; Tsunetsugu-Yokota et al. 2007; Wang et al. 2005; Weingartl et al. 2004; Yang et al. 2004; Zhang et al. 2005; Zhou et al. 2005; Zhu et al. 2004). Ideally, animal models used to assess protective vaccine efficacy would display virus replication, pathology, morbidity, and mortality. Though all of the models utilized to assess vaccine efficacy demonstrate virus replication and lung manifestations of disease, very few demonstrate virus-induced morbidity and severe lung pathology, and none demonstrate mortality (Table 13.2). In turn, the more stringent animal models (mouse-adapted SARS-CoV, hACE2 transgenic mice, etc.) that demonstrate morbidity and mortality that are currently available need to be employed in order to provide a thorough evaluation of vaccine protective efficacy. Moreover, very few of the above vaccine studies assessed protective vaccine efficacy with a SARS-CoV virus challenge, without which the utility and success of the vaccine remains unknown. Furthermore, since future SARS-CoV emergence will most likely differ in antigenic identity as compared to epidemic strain (SARS Urbani), vaccine challenge strains should ideally be antigenically distinct from the vaccine antigen(s), thus allowing for the important assessment of vaccine cross-protection. To our knowledge, only two SARS-CoV vaccine studies to date have assessed protective efficacy using a heterologous challenge virus (Deming et al. 2006; Lamirande et al. 2008). Successful vaccination in aged populations is a necessary goal for SARS-CoV vaccine platforms and, unfortunately, only one study to date has focused on developing effective vaccines in this most vulnerable population (Deming et al. 2006) |This conundrum is discussed below.

13.3.4 *SARS-CoV Vaccine Efficacy in Immunosenescent Populations*

As mentioned above, the immunosenescence that occurs with aging can hamper both the innate and adaptive immune responses whose collaboration is necessary

Table 13.2 Examples of common approaches to SARS-CoV vaccine development

Platform	Vaccine formulation(s)	Vaccine antigen(s)	Animal model	Metrics of vaccine efficacy							
				T cell response measured	Antibody response measured	Neutralizing antibody measured	Homologous or heterologous challenge	Virus replication measured	Morbidity	Mortality	References
Killed SARS-CoV	FI, UV, or β-propiolactone ± alum	Whole virus	Mouse	Yes	Yes	Yes	Homologous (2/7 studies)	Yes	No	No	Gai et al. (2008); Qu et al. (2005); See et al. (2006); Spruth et al. (2006); Tsunetsugu-Yokota et al. (2007); Xiong et al. (2004); Zhang et al. (2005)
	FI or β-propiolactone ± alum	Whole virus	Ferret	No	Yes	Yes	Homologous (2/2)	Yes	Yes	No	Darnell et al. (2007); See et al. (2008)
	β-propiolactone ± alum	Whole virus	Non-human primate	No	Yes	Yes	Homologous (2/2)	Yes	Yes	No	Qin et al. (2006); Zhou et al. (2005)
Recombinant viral vector	MVA, VRP, MV, RV, VSV, AAV	S or N	Mouse	Yes	Yes	Yes	Homologous (2/8) and heterologous (1/8, Deming et al.)	Yes	Yes (Deming et al.)	No	Bisht et al. (2004); Chen et al. (2005); Deming et al. (2006); Du et al., (2008); Faber et al. (2005); Kapadia et al. (2005); Kapadia et al. (2008); Liniger et al. (2008)
	Adenovirus, MVA	S or N	Ferret	No	Yes	Yes	Homologous (3/3)	Yes	Yes	No	Czub et al. (2005); Kobinger et al. (2007); Weingartl et al. (2004)

Adenovirus, MVA	S or N	Non-human primate	No	Yes	Yes	Homologous (1/2)	Yes	No	No	Chen et al. (2005); Kobinger et al. (2007)
DNA	DNA plasmid	S, N, M, E, +/− adjuvant	Mouse	Yes	Yes	Yes	Homologous (1/4)	Yes	No	No	Jin et al. (2005); Wang et al. (2005); Yang et al. (2004); Zhu et al. (2004)
	DNA plasmid	S	Human	Yes	Yes	Yes	No	No	No	No	Martin et al. (2008)
Live attenuated	SARS-CoV deleted for E	Whole virus	Hamster	No	Yes	Yes	Heterologous	Yes	Yes	No	Lamirande et al. (2008)
Subunit	SARS S RBD	Recombinant protein	Mouse	No	Yes	Yes	Homologous (1/2, Du et al.)	Yes (Du et al.)	No	No	Du et al. (2007); He et al. (2004)
	SARS S 14-762aa ± saponin or a Ribi adjuvant	Recombinant protein	Mouse	No	Yes	Yes	Homologous	Yes	No	No	Bisht et al. (2005)

FI: Formalin inactivated; UV: Ultraviolet radiation; MVA: Modified vaccinia virus Ankara; VRP: Venezuelan equine encephalitis virus replicon particle; MV: Modified measles virus; RV: Modified rhabdovirus; VSV: Modified vesicular stomatitis virus; AAV: adeno-associated virus for efficient vaccination. The SARS-CoV epidemic was particularly harsh on aged populations where mortality ranged between 25 and 55% in people over the age of 65 (Booth et al. 2003; Leung et al. 2004; Liang et al. 2004; Peiris et al. 2003a). If a SARS-CoV-like virus were to reemerge in the future, it would be imperative that current vaccination strategies were successful in the most vulnerable populations. Unfortunately, the successful vaccination of elderly populations is a difficult and unpredictable task due to immunosenescence (Bernstein et al. 1999, 1998; Eaton et al. 2004; Effros 2007; Goodwin et al. 2006; Goronzy et al. 2001; Gruver et al. 2007; Haynes and Swain 2006; Pawelec and Larbi 2008; Vallejo 2005; Vasto et al. 2006). Much of the research related to vaccination of the immunosenescent has been performed with influenza. Current models predict that influenza vaccine efficacy in elderly populations ranges from 17 to 53% while the vaccine in young adults is 70–90% effective and the discrepancy seems to be a result of senescent immune system malfunction on multiple levels (Goodwin et al. 2006). Defects in antigen presentation, T cell activation, and cytokine secretion affect the generation of effective adaptive immune system helper (T helper or Th) cells and effector (B cells and cytotoxic T cells) cells resulting in diminished vaccine efficacy (Eaton et al. 2004; Effros 2007; Fujihashi et al. 2000; Goodwin et al. 2006; Goronzy et al. 2001; Haynes and Swain 2006; McElhaney et al. 2005; Vallejo 2005; Wang et al. 1995). Current research suggests that some defects of the senescent immune system can be overcome through administration of cytokines (IL-2) or adjuvants (MF59, CpG DNA) during vaccination that effectively activate APCs/Th cells, thereby increasing the probability of generating appropriate effector cells required for successful vaccination (Haynes et al. 2004, 1999; Higgins et al. 1996; Pulendran and Ahmed 2006; Thompson et al. 2006). Since influenza, West Nile virus, and SARS-CoV infection all produce disproportionately more disease in the elderly, the development of successful vaccine strategies in the elderly has a broad public health application (Anonymous 1995; Leung et al. 2004; Murray et al. 2006).

As in SARS-CoV-infection of aged humans, the infection of aged mice with SARS Urbani resulted in more severe disease as compared to similar infection of young adult mice. In senescent mice, both virus replication and lung pathology was enhanced but the virus was eventually cleared, suggesting that components of the aged immune system were less effective at controlling virus replication. Though the senescent mouse model does not fully recapitulate SARS-CoV acute and extended cell-mediated pathogenesis seen in humans, it serves as a useful model to study the effects of immunosenescence on vaccine efficacy. In 2006, Deming et al. demonstrated that a Venezuelan equine encephalitis virus replicon particle expressing SARS Urbani S (VRP-S) vaccine provided complete protection from replication of a SARS-CoV bearing a zoonotic heterologous GD03 S, but protection was variable in senescent mice (Deming et al. 2006). Due to the lack of significant morbidity and mortality in the SARS-CoV replication models, previous vaccine studies were unable to assess protection from disease or death and could only speculate that diminishing virus replication would diminish disease. Nevertheless, heterogeneity between antigen and challenge virus provides a more stringent, thorough, biologically and medically relevant model within which to assess vaccine efficacy. Therefore, employing an antigenically diverse panel of SARS-CoV antigens for vaccination coupled with the use of a similarly diverse lethal challenge virus panel may represent the most pertinent and relevant strategy for assessing vaccine efficacy (Fig. 13.3). Moreover, the robustness of the newly developed MA15 lethal BALB/c model would allow for the assessment of vaccines to induce protection from not only replication but also disease and mortality. Using VSV vectors expressing Urbani S (rVSV-S), Vogel et al. obtained similar results in aged mice where neutralization titers in vaccinated aged mice were low and did not provide protection from replication upon homologous SARS Urbani challenge (Vogel et al. 2007). Similar to the situation observed in humans, these data suggest that vaccination of young mice induces a robust and cross-protective IgG response while the IgG response in aged animals is depressed in both magnitude and cross-reactivity. As compared with young animals, our data indicate that aged mice have ~10–20-fold reduced neutralization titers against homologous viruses and 100–400-fold reduced titers against closely related heterologous viruses (Deming et al. 2006). The underlying mechanisms of vaccine failure in the elderly should evolve into the major focus for future SARS-CoV vaccine research.

13.3.5 SARS-CoV Vaccine Immunopotentiation

Effective vaccination induces specific protective immunity that confers protection against future disease. Unfortunately, vaccination can sometimes exacerbate disease upon natural infection with the pathogen the vaccine was designed to protect against. This phenomenon of vaccine-induced immune pathology is often called "immunopotentiation" of disease (Werle et al. 1999). Both measles (MV) and RSV are paramyxovirus respiratory pathogens that cause significant morbidity and mortality in infants that might be prevented through the development of effective vaccines (Polack et al. 1999; Varga et al. 2001). In the 1960s, two infamous examples of immunopotentiation of disease surfaced with formalin-inactivated MV (FI-MV) and RSV (FI-RSV) vaccines (Polack et al. 1999; Varga et al. 2001). Infants with no prior exposure to RSV who were vaccinated with the FI-RSV vaccine developed virus-specific antibody but were not protected from subsequent natural RSV infection (Durbin and Durbin 2004). In fact, FI-RSV-vaccinated children infected with RSV suffered from enhanced RSV disease requiring hospitalization, and a few children died from infection (Durbin and Durbin 2004). The pathological hallmark of FI-RSV immunopotentiation was lung and peripheral eosinophilia, which is rarely seen in the natural course of RSV infection (Durbin and Durbin 2004; Varga et al. 2001). Similar to FI-RSV, protective immunity waned in infants shortly after vaccination with FI-MV and subsequent natural measles infection resulted in more severe disease with eosinophilia uncharacteristic of natural measles infection (Polack et al. 1999). Due to these severe vaccine-associated disease complications, both vaccines were withdrawn and research efforts focused on the elucidation of the underlying molecular mechanisms of these vaccine-induced pathologies. Within mouse and nonhuman primate animal models, both FI-RSV and FI-MV were found to induce an atypical Th2 adaptive immune response not seen in natural infection (De Swart et al. 2002; Durbin and Durbin 2004; Polack et al. 1999; Varga et al. 2001). The effects of the Th2 responses generated by FI-RSV and FI-MV differ. FI-RSV vaccination induces a Th2 allergic immune response with T cells secreting cytokines (IL-13, IL-5) that upregulate the production of the potent eosinophil chemotactic molecule eotaxin (De Swart et al. 2002; Durbin and Durbin 2004; Varga et al. 2001). Natural infection of vaccinated individuals is thought to have been exacerbated by this atypical allergic immune response in the lung (De Swart et al. 2002; Durbin and Durbin 2004; Varga et al. 2001). Interestingly, similar results are achieved in macaques using formalin-inactivated human metapneumovirus vaccines (FI-hMPV) followed by hMPV challenge (de Swart et al. 2007). With FI-MV vaccination, non-human primate models suggest that the associated allergic Th2 response generated after MV challenge recruits eosinophils to sites of virus replication with disease pathology in part mediated by immune complex deposition (Polack et al. 1999).

Vaccine-induced immunopotentiation has also been observed in coronavirus with vaccinia virus vectored feline infectious peritonitis virus (FIPV) vaccines. Vennema et al. observed that vaccination with recombinant vaccinia virus expressing FIPV S (vFS) induced short-lived immunity in kittens (Vennema et al. 1990). When challenged, vFS-immunized animals suffered from much more severe disease than those receiving a control vaccine. vFS vaccine-induced immunopotentiation was suspected to be a result of antibody-dependent enhancement (ADE) of virus infection where subneutralizing antibody coating FIPV virions allowed for the entry and productive infection of cells (e.g., macrophages) not normally targeted during natural infection.

Vaccine-induced immunopotentiation of disease has also been shown with vectored vaccines expressing SARS-CoV N protein. Deming et al. demonstrated that mice vaccinated with VRP-based vaccines expressing the SARS-CoV N gene were not protected from infection and developed enhanced lung immunopathology upon challenge with eosinophilia not seen in control mice, and these data have recently been confirmed by Yasui et al. (Deming et al. 2006; Yasui et al. 2008). These data suggest that the N protein not only fails to provide protection from disease in these acute replication models, it also promotes enhanced disease pathogenesis in the lung. The evaluation of several formalin-inactivated SARS-CoV vaccines suggest these vaccines primarily induce a Th2 response while natural SARS-CoV infection of humans induces primarily a Th1 response (Spruth et al. 2006; Tsunetsugu-Yokota et al. 2007; Wong et al. 2004). Given the data from both FI-RSV and FI-MV vaccination where the alteration of the natural Th response induced more severe disease upon challenge, caution should be used in developing vectored vaccines containing SARS-CoV N or formalin-inactivated SARS-CoV vaccines which may promote rather than prevent disease. In support of this viewpoint, we have shown that killed SARS vaccines in the presence or absence of alum induce extensive lung pathology, especially in the lungs of aged animals inoculated with a heterologous lethal challenge viruses (Deming et al. 2009, in preparation).

13.4 Conclusion

Due to the ever-increasing human population, increasing wild-life habitat destruction for human inhabitation, the demand for exotic animals for food, and the inability of humans to control or successfully track zoonotic diseases in wild animal populations, the emergence of novel viral pathogens from zoonotic pools will continue to threaten human global public health. The development of antiviral therapies against viral pathogens that might emerge in the future is a difficult multifaceted problem, but it is critical for improving global health. SARS-CoV was the first significant emerging virus of the twenty-first century. The availability of reverse genetics, time-ordered sequence variation of animal and human strains, robust availability of biochemical reagents, and age-related animal models provide a unique opportunity to study many basic aspects of novel virus emergence and antigenic diversity, pathogenesis, antiviral therapy development, and vaccine immunopotentiation of disease. As SARS-CoV vaccines must provide broad protection against the larger zoonotic pool, successful vaccine strategies may provide a template for developing broadly reactive vaccines against other emerging viruses, like filoviruses, Nipah virus, NL63, HKU1, and avian influenza viruses. Importantly, SARS-CoV pathogenesis is exacerbated in the immunosenescent, a population that suffers a disproportionate disease burden from other emerging viruses. Through the use of aged models of SARS-CoV pathogenesis and vaccine efficacy, the immunological deficiencies of the aged immune system and/or the variables required for successful vaccination may be elucidated. These data may be applied to improve vaccines for other viral pathogens that cause a disproportionate disease burden in vulnerable populations like the elderly (West Nile virus, influenza, norovirus, SARS-CoV, RSV, etc.). In the past, the use of whole killed vaccines for vaccination has been successful in preventing disease but has also contributed to immunopotentiation of disease, and the mechanisms for this exacerbation of disease are not completely understood. Uncovering the mechanisms of SARS-CoV nucleocapsid-induced immunopotentiation may reveal common host pathways with other vaccine formulations (e.g., FI-RSV, FI-MV) that mediate vaccine-related pathologies. Alternatively, the unique genetic differences between coronaviruses and paramyxoviruses may reveal entirely new pathways for virus–host interactions that potentiate vaccine-induced immune pathology. Thus, current models of SARS-CoV pathogenesis can be employed to study the many difficult problems associated with the development of effective therapies for emerging pathogens, and future studies may provide the solutions that will prepare us for future SARS-CoV emergence or the emergence of yet unknown viral pathogens.

References

Achtman M, Zurth K, Morelli G, Torrea G, Guiyoule A, Carniel E (1999) Yersinia pestis, the cause of plague, is a recently emerged clone of Yersinia pseudotuberculosis. Proc Natl Acad Sci USA 96:14043–14048

Akerstrom S, Tan YJ, Mirazimi A (2006) Amino acids 15–28 in the ectodomain of SARS coronavirus 3a protein induces neutralizing antibodies. FEBS Lett 580:3799–3803

Anonymous (1995) From the centers for disease control and prevention. Pneumonia and influenza death rates–United States, 1979–1994. JAMA 274:532

Antonio GE, Wong KT, Hui DS, Wu A, Lee N, Yuen EH, Leung CB, Rainer TH, Cameron P, Chung SS, Sung JJ, Ahuja AT (2003) Thin-section CT in patients with severe acute respiratory syndrome following hospital discharge: preliminary experience. Radiology 228:810–815

Asselin-Paturel C, Boonstra A, Dalod M, Durand I, Yessaad N, Dezutter-Dambuyant C, Vicari A, O'Garra A, Biron C, Briere F, Trinchieri G (2001) Mouse type I IFN-producing cells are immature APCs with plasmacytoid morphology. Nat Immunol 2:1144–1150

Avendano M, Derkach P, Swan S (2003) Clinical course and management of SARS in health care workers in Toronto: a case series. CMAJ 168:1649–1660

Bai B, Lu X, Meng J, Hu Q, Mao P, Lu B, Chen Z, Yuan Z, Wang H (2008) Vaccination of mice with recombinant baculovirus expressing spike or nucleocapsid protein of SARS-like coronavirus generates humoral and cellular immune responses. Mol Immunol 45:868–875

Bauer TT, Ewig S, Rodloff AC, Muller EE (2006a) Acute respiratory distress syndrome and pneumonia: a comprehensive review of clinical data. Clin Infect Dis 43:748–756

Bauer TT, Ewig S, Rodloff AC, Muller EE (2006b) ARDS and pneumonia: a comprehensive review of clinical data. Clin Infect Dis 43:748–756

Becker MM, Graham RL, Donaldson EF, Rockx B, Sims AC, Sheahan T, Pickles RJ, Corti D, Johnston RE, Baric RS, Denison MR (2008) Synthetic recombinant bat SARS-like coronavirus is infectious in cultured cells and in mice. Proc Natl Acad Sci USA 105: 19944–19949

Bernstein ED, Gardner EM, Abrutyn E, Gross P, Murasko DM (1998) Cytokine production after influenza vaccination in a healthy elderly population. Vaccine 16:1722–1731

Bernstein E, Kaye D, Abrutyn E, Gross P, Dorfman M, Murasko DM (1999) Immune response to influenza vaccination in a large healthy elderly population. Vaccine 17:82–94

Bisht H, Roberts A, Vogel L, Bukreyev A, Collins PL, Murphy BR, Subbarao K, Moss B (2004) Severe acute respiratory syndrome coronavirus spike protein expressed by attenuated vaccinia virus protectively immunizes mice. Proc Natl Acad Sci USA 101:6641–6646

Bisht H, Roberts A, Vogel L, Subbarao K, Moss B (2005) Neutralizing antibody and protective immunity to SARS coronavirus infection of mice induced by a soluble recombinant polypeptide containing an N-terminal segment of the spike glycoprotein. Virology 334:160–165

Booth CM, Matukas LM, Tomlinson GA, Rachlis AR, Rose DB, Dwosh HA, Walmsley SL, Mazzulli T, Avendano M, Derkach P, Ephtimios IE, Kitai I, Mederski BD, Shadowitz SB, Gold WL, Hawryluck LA, Rea E, Chenkin JS, Cescon DW, Poutanen SM, Detsky AS (2003) Clinical features and short-term outcomes of 144 patients with SARS in the greater Toronto area. JAMA 289:2801–2809

Buchholz UJ, Bukreyev A, Yang L, Lamirande EW, Murphy BR, Subbarao K, Collins PL (2004) Contributions of the structural proteins of severe acute respiratory syndrome coronavirus to protective immunity. Proc Natl Acad Sci USA 101:9804–9809

Cameron MJ, Ran L, Xu L, Danesh A, Bermejo-Martin JF, Cameron CM, Muller MP, Gold WL, Richardson SE, Poutanen SM, Willey BM, DeVries ME, Fang Y, Seneviratne C, Bosinger SE, Persad D, Wilkinson P, Greller LD, Somogyi R, Humar A, Keshavjee S, Louie M, Loeb MB, Brunton J, McGeer AJ, Kelvin DJ (2007) Interferon-mediated immunopathological events are associated with atypical innate and adaptive immune responses in patients with severe acute respiratory syndrome. J Virol 81:8692–8706

Cella M, Jarrossay D, Facchetti F, Alebardi O, Nakajima H, Lanzavecchia A, Colonna M (1999) Plasmacytoid monocytes migrate to inflamed lymph nodes and produce large amounts of type I interferon. Nat Med 5:919–923

Cervantes-Barragan L, Zust R, Weber F, Spiegel M, Lang KS, Akira S, Thiel V, Ludewig B (2007) Control of coronavirus infection through plasmacytoid dendritic-cell-derived type I interferon. Blood 109:1131–1137

Chen Z, Zhang L, Qin C, Ba L, Yi CE, Zhang F, Wei Q, He T, Yu W, Yu J, Gao H, Tu X, Gettie A, Farzan M, Yuen KY, Ho DD (2005) Recombinant modified vaccinia virus Ankara expressing the spike glycoprotein of severe acute respiratory syndrome coronavirus induces protective neutralizing antibodies primarily targeting the receptor binding region. J Virol 79:2678–2688

Chen Y, Chan VS, Zheng B, Chan KY, Xu X, To LY, Huang FP, Khoo US, Lin CL (2007) A novel subset of putative stem/progenitor CD34 + Oct-4+ cells is the major target for SARS coronavirus in human lung. J Exp Med 204:2529–2536

Cheung CY, Poon LL, Ng IH, Luk W, Sia SF, Wu MH, Chan KH, Yuen KY, Gordon S, Guan Y, Peiris JS (2005) Cytokine responses in severe acute respiratory syndrome coronavirus-infected macrophages in vitro: possible relevance to pathogenesis. J Virol 79:7819–7826

Chinese SMEC (2004) Molecular evolution of the SARS coronavirus during the course of the SARS epidemic in China. Science 303:1666–1669

Christian MD, Poutanen SM, Loutfy MR, Muller MP, Low DE (2004) Severe acute respiratory syndrome. Clin Infect Dis 38:1420–1427

Chu YK, Ali GD, Jia F, Li Q, Kelvin D, Couch RC, Harrod KS, Hutt JA, Cameron C, Weiss SR, Jonsson CB (2008) The SARS-CoV ferret model in an infection-challenge study. Virology 374 (1):151–163

Churchill GA, Airey DC, Allayee H, Angel JM, Attie AD, Beatty J, Beavis WD, Belknap JK, Bennett B, Berrettini W, Bleich A, Bogue M, Broman KW, Buck KJ, Buckler E, Burmeister M, Chesler EJ, Cheverud JM, Clapcote S, Cook MN, Cox RD, Crabbe JC, Crusio WE, Darvasi A, Deschepper CF, Doerge RW, Farber CR, Forejt J, Gaile D, Garlow SJ, Geiger H, Gershenfeld H, Gordon T, Gu J, Gu W, de Haan G, Hayes NL, Heller C, Himmelbauer H, Hitzemann R, Hunter K, Hsu HC, Iraqi FA, Ivandic B, Jacob HJ, Jansen RC, Jepsen KJ, Johnson DK, Johnson TE, Kempermann G, Kendziorski C, Kotb M, Kooy RF, Llamas B, Lammert F, Lassalle JM, Lowenstein PR, Lu L, Lusis A, Manly KF, Marcucio R, Matthews D, Medrano JF, Miller DR, Mittleman G, Mock BA, Mogil JS, Montagutelli X, Morahan G, Morris DG, Mott R, Nadeau JH, Nagase H, Nowakowski RS, O'Hara BF, Osadchuk AV, Page GP, Paigen B, Paigen K, Palmer AA, Pan HJ, Peltonen-Palotie L, Peirce J, Pomp D, Pravenec M, Prows DR, Qi Z, Reeves RH, Roder J, Rosen GD, Schadt EE, Schalkwyk LC, Seltzer Z, Shimomura K, Shou S, Sillanpaa MJ, Siracusa LD, Snoeck HW, Spearow JL, Svenson K et al (2004) The Collaborative Cross, a community resource for the genetic analysis of complex traits. Nat Genet 36:1133–1137

Czub M, Weingartl H, Czub S, He R, Cao J (2005) Evaluation of modified vaccinia virus Ankara based recombinant SARS vaccine in ferrets. Vaccine 23:2273–2279

Dahlem P, van Aalderen WM, Bos AP (2007) Pediatric acute lung injury. Paediatr Respir Rev 8:348–362

Darnell ME, Plant EP, Watanabe H, Byrum R, St Claire M, Ward JM, Taylor DR (2007) Severe acute respiratory syndrome coronavirus infection in vaccinated ferrets. J Infect Dis 196:1329–1338

de Lang A, Baas T, Teal T, Leijten LM, Rain B, Osterhaus AD, Haagmans BL, Katze MG (2007) Functional genomics highlights differential induction of antiviral pathways in the lungs of SARS-CoV-infected macaques. PLoS Pathog 3:e112

De Swart RL, Kuiken T, Timmerman HH, van Amerongen G, Van Den Hoogen BG, Vos HW, Neijens HJ, Andeweg AC, Osterhaus AD (2002) Immunization of macaques with formalin-inactivated respiratory syncytial virus (RSV) induces interleukin-13-associated hypersensitivity to subsequent RSV infection. J Virol 76:11561–11569 de Swart RL, van den Hoogen BG, Kuiken T, Herfst S, van Amerongen G, Yuksel S, Sprong L, Osterhaus AD (2007) Immunization of macaques with formalin-inactivated human metapneumovirus induces hypersensitivity to hMPV infection. Vaccine 25:8518–8528

Deming D, Sheahan T, Heise M, Yount B, Davis N, Sims A, Suthar M, Harkema J, Whitmore A, Pickles R, West A, Donaldson E, Curtis K, Johnston R, Baric R (2006) Vaccine efficacy in senescent mice challenged with recombinant SARS-CoV bearing epidemic and zoonotic spike variants. PLoS Med 3:e525

Donoghue M, Hsieh F, Baronas E, Godbout K, Gosselin M, Stagliano N, Donovan M, Woolf B, Robison K, Jeyaseelan R, Breitbart RE, Acton S (2000) A novel angiotensin-converting enzyme-related carboxypeptidase (ACE2) converts angiotensin I to angiotensin 1–9. Circ Res 87:E1–E9

Du L, Zhao G, He Y, Guo Y, Zheng BJ, Jiang S, Zhou Y (2007) Receptor-binding domain of SARS-CoV spike protein induces long-term protective immunity in an animal model. Vaccine 25:2832–2838

Du L, Zhao G, Lin Y, Chan C, He Y, Jiang S, Wu C, Jin DY, Yuen KY, Zhou Y, Zheng BJ (2008a) Priming with rAAV encoding RBD of SARS-CoV S protein and boosting with RBD-specific peptides for T cell epitopes elevated humoral and cellular immune responses against SARS-CoV infection. Vaccine 26:1644–1651

Du L, Zhao G, Lin Y, Sui H, Chan C, Ma S, He Y, Jiang S, Wu C, Yuen KY, Jin DY, Zhou Y, Zheng BJ (2008b) Intranasal vaccination of recombinant adeno-associated virus encoding receptor-binding domain of severe acute respiratory syndrome coronavirus (SARS-CoV) spike protein induces strong mucosal immune responses and provides long-term protection against SARS-CoV infection. J Immunol 180:948–956

Durbin JE, Durbin RK (2004) Respiratory syncytial virus-induced immunoprotection and immunopathology. Viral Immunol 17:370–380

Eaton SM, Burns EM, Kusser K, Randall TD, Haynes L (2004) Age-related defects in CD4 T cell cognate helper function lead to reductions in humoral responses. J Exp Med 200:1613–1622

Effros RB (2007) Role of T lymphocyte replicative senescence in vaccine efficacy. Vaccine 25:599–604

Faber M, Lamirande EW, Roberts A, Rice AB, Koprowski H, Dietzschold B, Schnell MJ (2005) A single immunization with a rhabdovirus-based vector expressing severe acute respiratory syndrome coronavirus (SARS-CoV) S protein results in the production of high levels of SARS-CoV-neutralizing antibodies. J Gen Virol 86:1435–1440

Frieman M, Yount B, Heise M, Kopecky-Bromberg SA, Palese P, Baric RS (2007) Severe acute respiratory syndrome coronavirus ORF6 antagonizes STAT1 function by sequestering nuclear import factors on the rough endoplasmic reticulum/Golgi membrane. J Virol 81:9812–9824

Fujihashi K, Koga T, McGhee JR (2000) Mucosal vaccination and immune responses in the elderly. Vaccine 18:1675–1680

Gai W, Zou W, Lei L, Luo J, Tu H, Zhang Y, Wang K, Tien P, Yan H (2008) Effects of different immunization protocols and adjuvant on antibody responses to inactivated SARS-CoV vaccine. Viral Immunol 21:27–37

Glass WG, Subbarao K, Murphy B, Murphy PM (2004) Mechanisms of host defense following severe acute respiratory syndrome-coronavirus (SARS-CoV) pulmonary infection of mice. J Immunol 173:4030–4039

Gonzalez JP, Pourrut X, Leroy E (2007) Ebolavirus and other filoviruses. Curr Top Microbiol Immunol 315:363–387

Goodwin K, Viboud C, Simonsen L (2006) Antibody response to influenza vaccination in the elderly: a quantitative review. Vaccine 24:1159–1169

Goronzy JJ, Fulbright JW, Crowson CS, Poland GA, O'Fallon WM, Weyand CM (2001) Value of immunological markers in predicting responsiveness to influenza vaccination in elderly individuals. J Virol 75:12182–12187

Gruver AL, Hudson LL, Sempowski GD (2007) Immunosenescence of ageing. J Pathol 211:144–156

Gu J, Gong E, Zhang B, Zheng J, Gao Z, Zhong Y, Zou W, Zhan J, Wang S, Xie Z, Zhuang H, Wu B, Zhong H, Shao H, Fang W, Gao D, Pei F, Li X, He Z, Xu D, Shi X, Anderson VM, Leong AS (2005) Multiple organ infection and the pathogenesis of SARS. J Exp Med 202:415–424

Guan Y, Zheng BJ, He YQ, Liu XL, Zhuang ZX, Cheung CL, Luo SW, Li PH, Zhang LJ, Guan YJ, Butt KM, Wong KL, Chan KW, Lim W, Shortridge KF, Yuen KY, Peiris JS, Poon LL (2003) Isolation and characterization of viruses related to the SARS coronavirus from animals in southern China. Science 302:276–278

Guidot DM, Folkesson HG, Jain L, Sznajder JI, Pittet JF, Matthay MA (2006) Integrating acute lung injury and regulation of alveolar fluid clearance. Am J Physiol Lung Cell Mol Physiol 291:L301–L306

Haagmans BL, Osterhaus AD (2006) Nonhuman primate models for SARS. PLoS Med 3:e194

Haagmans BL, Kuiken T, Martina BE, Fouchier RA, Rimmelzwaan GF, van Amerongen G, van Riel D, de Jong T, Itamura S, Chan KH, Tashiro M, Osterhaus AD (2004) Pegylated interferon-alpha protects type 1 pneumocytes against SARS coronavirus infection in macaques. Nat Med 10:290–293

Haga S, Yamamoto N, Nakai-Murakami C, Osawa Y, Tokunaga K, Sata T, Yamamoto N, Sasazuki T, Ishizaka Y (2008) Modulation of TNF-alpha-converting enzyme by the spike protein of SARS-CoV and ACE2 induces TNF-alpha production and facilitates viral entry. Proc Natl Acad Sci USA 105:7809–7814

Hamming I, Timens W, Bulthuis ML, Lely AT, Navis GJ, van Goor H (2004) Tissue distribution of ACE2 protein, the functional receptor for SARS coronavirus. A first step in understanding SARS pathogenesis. J Pathol 203:631–637

Haynes L, Swain SL (2006) Why aging T cells fail: implications for vaccination. Immunity 24:663–666

Haynes L, Linton PJ, Eaton SM, Tonkonogy SL, Swain SL (1999) Interleukin 2, but not other common gamma chain-binding cytokines, can reverse the defect in generation of CD4 effector T cells from naive T cells of aged mice. J Exp Med 190:1013–1024

Haynes L, Eaton SM, Burns EM, Rincon M, Swain SL (2004) Inflammatory cytokines overcome age-related defects in CD4 T cell responses in vivo. J Immunol 172:5194–5199

He Y, Zhou Y, Liu S, Kou Z, Li W, Farzan M, Jiang S (2004) Receptor-binding domain of SARS-CoV spike protein induces highly potent neutralizing antibodies: implication for developing subunit vaccine. Biochem Biophys Res Commun 324:773–781

He Y, Li J, Heck S, Lustigman S, Jiang S (2006) Antigenic and immunogenic characterization of recombinant baculovirus-expressed severe acute respiratory syndrome coronavirus spike protein: implication for vaccine design. J Virol 80:5757–5767

Herath CB, Warner FJ, Lubel JS, Dean RG, Jia Z, Lew RA, Smith AI, Burrell LM, Angus PW (2007) Upregulation of hepatic angiotensin-converting enzyme 2 (ACE2) and angiotensin-(1–7) levels in experimental biliary fibrosis. J Hepatol 47:387–395

Higgins DA, Carlson JR, Van Nest G (1996) MF59 adjuvant enhances the immunogenicity of influenza vaccine in both young and old mice. Vaccine 14:478–484

Hogan RJ, Gao G, Rowe T, Bell P, Flieder D, Paragas J, Kobinger GP, Wivel NA, Crystal RG, Boyer J, Feldmann H, Voss TG, Wilson JM (2004) Resolution of primary severe acute respiratory syndrome-associated coronavirus infection requires Stat1. J Virol 78:11416–11421

Hwang DM, Chamberlain DW, Poutanen SM, Low DE, Asa SL, Butany J (2005) Pulmonary pathology of severe acute respiratory syndrome in Toronto. Mod Pathol 18:1–10

Imai Y, Kuba K, Rao S, Huan Y, Guo F, Guan B, Yang P, Sarao R, Wada T, Leong-Poi H, Crackower MA, Fukamizu A, Hui CC, Hein L, Uhlig S, Slutsky AS, Jiang C, Penninger JM (2005) Angiotensin-converting enzyme 2 protects from severe acute lung failure. Nature 436:112–116

Jeffers SA, Tusell SM, Gillim-Ross L, Hemmila EM, Achenbach JE, Babcock GJ, Thomas WD Jr, Thackray LB, Young MD, Mason RJ, Ambrosino DM, Wentworth DE, Demartini JC, Holmes KV (2004) CD209L (L-SIGN) is a receptor for severe acute respiratory syndrome coronavirus. Proc Natl Acad Sci USA 101:15748–15753

Jin H, Xiao C, Chen Z, Kang Y, Ma Y, Zhu K, Xie Q, Tu Y, Yu Y, Wang B (2005) Induction of Th1 type response by DNA vaccinations with N, M, and E genes against SARS-CoV in mice. Biochem Biophys Res Commun 328:979–986

Johansson C, Wetzel JD, He J, Mikacenic C, Dermody TS, Kelsall BL (2007) Type I interferons produced by hematopoietic cells protect mice against lethal infection by mammalian reovirus. J Exp Med 204:1349–1358

Jung K, Alekseev KP, Zhang X, Cheon DS, Vlasova AN, Saif LJ (2007) Altered pathogenesis of porcine respiratory coronavirus in pigs due to immunosuppressive effects of dexamethasone: implications for corticosteroid use in treatment of severe acute respiratory syndrome coronavirus. J Virol 81:13681–13693

Kapadia SU, Rose JK, Lamirande E, Vogel L, Subbarao K, Roberts A (2005) Long-term protection from SARS coronavirus infection conferred by a single immunization with an attenuated VSV-based vaccine. Virology 340:174–182

Kapadia SU, Simon ID, Rose JK (2008) SARS vaccine based on a replication-defective recombinant vesicular stomatitis virus is more potent than one based on a replication-competent vector. Virology 376:165–172

Kaur P, Ponniah M, Murhekar MV, Ramachandran V, Ramachandran R, Raju HK, Perumal V, Mishra AC, Gupte MD (2008) Chikungunya outbreak, South India, 2006. Emerg Infect Dis 14:1623–1625

Kobinger GP, Figueredo JM, Rowe T, Zhi Y, Gao G, Sanmiguel JC, Bell P, Wivel NA, Zitzow LA, Flieder DB, Hogan RJ, Wilson JM (2007) Adenovirus-based vaccine prevents pneumonia in ferrets challenged with the SARS coronavirus and stimulates robust immune responses in macaques. Vaccine 25:5220–5231

Kopecky-Bromberg SA, Martinez-Sobrido L, Frieman M, Baric RA, Palese P (2007) Severe acute respiratory syndrome coronavirus open reading frame (ORF) 3b, ORF 6, and nucleocapsid proteins function as interferon antagonists. J Virol 81:548–557

Ksiazek TG, Erdman D, Goldsmith CS, Zaki SR, Peret T, Emery S, Tong S, Urbani C, Comer JA, Lim W, Rollin PE, Dowell SF, Ling AE, Humphrey CD, Shieh WJ, Guarner J, Paddock CD, Rota P, Fields B, DeRisi J, Yang JY, Cox N, Hughes JM, LeDuc JW, Bellini WJ, Anderson LJ (2003) A novel coronavirus associated with severe acute respiratory syndrome. N Engl J Med 348:1953–1966

Kuba K, Imai Y, Rao S, Gao H, Guo F, Guan B, Huan Y, Yang P, Zhang Y, Deng W, Bao L, Zhang B, Liu G, Wang Z, Chappell M, Liu Y, Zheng D, Leibbrandt A, Wada T, Slutsky AS, Liu D, Qin C, Jiang C, Penninger JM (2005) A crucial role of angiotensin converting enzyme 2 (ACE2) in SARS coronavirus-induced lung injury. Nat Med 11:875–879

Kuba K, Imai Y, Penninger JM (2006a) Angiotensin-converting enzyme 2 in lung diseases. Curr Opin Pharmacol 6(3):271–276

Kuba K, Imai Y, Rao S, Jiang C, Penninger JM (2006b) Lessons from SARS: control of acute lung failure by the SARS receptor ACE2. J Mol Med 84:814–820

Lamirande EW, DeDiego ML, Roberts A, Jackson JP, Alvarez E, Sheahan T, Shieh WJ, Zaki SR, Baric R, Enjuanes L, Subbarao K (2008) A live attenuated severe acute respiratory syndrome coronavirus is immunogenic and efficacious in golden Syrian hamsters. J Virol 82:7721–7724

Lau SK, Woo PC, Li KS, Huang Y, Tsoi HW, Wong BH, Wong SS, Leung SY, Chan KH, Yuen KY (2005) Severe acute respiratory syndrome coronavirus-like virus in Chinese horseshoe bats. Proc Natl Acad Sci USA 102:14040–14045

Law HK, Cheung CY, Ng HY, Sia SF, Chan YO, Luk W, Nicholls JM, Peiris JS, Lau YL (2005) Chemokine up-regulation in SARS-coronavirus-infected, monocyte-derived human dendritic cells. Blood 106:2366–2374

Lee N, Chan PK, Ip M, Wong E, Ho J, Ho C, Cockram CS, Hui DS (2006) Anti-SARS-CoV IgG response in relation to disease severity of severe acute respiratory syndrome. J Clin Virol 35:179–184

Leroy EM, Kumulungui B, Pourrut X, Rouquet P, Hassanin A, Yaba P, Delicat A, Paweska JT, Gonzalez JP, Swanepoel R (2005) Fruit bats as reservoirs of Ebola virus. Nature 438:575–576

Leung GM, Hedley AJ, Ho LM, Chau P, Wong IO, Thach TQ, Ghani AC, Donnelly CA, Fraser C, Riley S, Ferguson NM, Anderson RM, Tsang T, Leung PY, Wong V, Chan JC, Tsui E, Lo SV, Lam TH (2004) The epidemiology of severe acute respiratory syndrome in the 2003 Hong Kong epidemic: an analysis of all 1755 patients. Ann Intern Med 141:662–673

Li W, Moore MJ, Vasilieva N, Sui J, Wong SK, Berne MA, Somasundaran M, Sullivan JL, Luzuriaga K, Greenough TC, Choe H, Farzan M (2003) Angiotensin-converting enzyme 2 is a functional receptor for the SARS coronavirus. Nature 426:450–454

Li F, Li W, Farzan M, Harrison SC (2005a) Structure of SARS coronavirus spike receptor-binding domain complexed with receptor. Science 309:1864–1868

Li W, Zhang C, Sui J, Kuhn JH, Moore MJ, Luo S, Wong SK, Huang IC, Xu K, Vasilieva N, Murakami A, He Y, Marasco WA, Guan Y, Choe H, Farzan M (2005b) Receptor and viral determinants of SARS-coronavirus adaptation to human ACE2. EMBO J 24:1634–1643

Li W, Wong SK, Li F, Kuhn JH, Huang IC, Choe H, Farzan M (2006) Animal origins of the severe acute respiratory syndrome coronavirus: insight from ACE2-S-protein interactions. J Virol 80:4211–4219

Li Y, Carroll DS, Gardner SN, Walsh MC, Vitalis EA, Damon IK (2007) On the origin of smallpox: correlating variola phylogenics with historical smallpox records. Proc Natl Acad Sci USA 104:15787–15792

Li CK, Wu H, Yan H, Ma S, Wang L, Zhang M, Tang X, Temperton NJ, Weiss RA, Brenchley JM, Douek DC, Mongkolsapaya J, Tran BH, Lin CL, Screaton GR, Hou JL, McMichael AJ, Xu XN (2008) T cell responses to whole SARS coronavirus in humans. J Immunol 181:5490–5500

Liang W, Zhu Z, Guo J, Liu Z, Zhou W, Chin DP, Schuchat A (2004) Severe acute respiratory syndrome, Beijing, 2003. Emerg Infect Dis 10:25–31

Liniger M, Zuniga A, Tamin A, Azzouz-Morin TN, Knuchel M, Marty RR, Wiegand M, Weibel S, Kelvin D, Rota PA, Naim HY (2008) Induction of neutralising antibodies and cellular immune responses against SARS coronavirus by recombinant measles viruses. Vaccine 26:2164–2174

Liu RY, Wu LZ, Huang BJ, Huang JL, Zhang YL, Ke ML, Wang JM, Tan WP, Zhang RH, Chen HK, Zeng YX, Huang W (2005) Adenoviral expression of a truncated S1 subunit of SARS-CoV spike protein results in specific humoral immune responses against SARS-CoV in rats. Virus Res 112:24–31

Marasco WA, Sui J (2007) The growth and potential of human antiviral monoclonal antibody therapeutics. Nat Biotechnol 25:1421–1434

Martin JE, Louder MK, Holman LA, Gordon IJ, Enama ME, Larkin BD, Andrews CA, Vogel L, Koup RA, Roederer M, Bailer RT, Gomez PL, Nason M, Mascola JR, Nabel GJ, Graham BS (2008) A SARS DNA vaccine induces neutralizing antibody and cellular immune responses in healthy adults in a Phase I clinical trial. Vaccine 26:6338–6343

McAllister-Lucas LM, Ruland J, Siu K, Jin X, Gu S, Kim DS, Kuffa P, Kohrt D, Mak TW, Nunez G, Lucas PC (2007) CARMA3/Bcl10/MALT1-dependent NF-kappaB activation mediates angiotensin II-responsive inflammatory signaling in nonimmune cells. Proc Natl Acad Sci USA 104:139–144

McCray PB Jr, Pewe L, Wohlford-Lenane C, Hickey M, Manzel L, Shi L, Netland J, Jia HP, Halabi C, Sigmund CD, Meyerholz DK, Kirby P, Look DC, Perlman S (2007) Lethal infection of K18-hACE2 mice infected with severe acute respiratory syndrome coronavirus. J Virol 81:813–821

McElhaney JE, Hooton JW, Hooton N, Bleackley RC (2005) Comparison of single versus booster dose of influenza vaccination on humoral and cellular immune responses in older adults. Vaccine 23:3294–3300

Murray K, Baraniuk S, Resnick M, Arafat R, Kilborn C, Cain K, Shallenberger R, York TL, Martinez D, Hellums JS, Hellums D, Malkoff M, Elgawley N, McNeely W, Khuwaja SA, Tesh RB (2006) Risk factors for encephalitis and death from West Nile virus infection. Epidemiol Infect 134:1325–1332

Nguyen DC, Uyeki TM, Jadhao S, Maines T, Shaw M, Matsuoka Y, Smith C, Rowe T, Lu X, Hall H, Xu X, Balish A, Klimov A, Tumpey TM, Swayne DE, Huynh LP, Nghiem HK,

Nguyen HH, Hoang LT, Cox NJ, Katz JM (2005) Isolation and characterization of avian influenza viruses, including highly pathogenic H5N1, from poultry in live bird markets in Hanoi, Vietnam, in 2001. J Virol 79:4201–4212

Nicholls MG, Richards AM, Agarwal M (1998) The importance of the renin-angiotensin system in cardiovascular disease. J Hum Hypertens 12:295–299

Nicholls JM, Butany J, Poon LL, Chan KH, Beh SL, Poutanen S, Peiris JS, Wong M (2006) Time course and cellular localization of SARS-CoV nucleoprotein and RNA in lungs from fatal cases of SARS. PLoS Med 3:e27

O'Neill LA, Bowie AG (2007) The family of five: TIR-domain-containing adaptors in Toll-like receptor signalling. Nat Rev Immunol 7:353–364

Parrish CR, Kawaoka Y (2005) The origins of new pandemic viruses: the acquisition of new host ranges by canine parvovirus and influenza A viruses. Annu Rev Microbiol 59: 553–586

Pawelec G, Larbi A (2008) Immunity and ageing in man: Annual review 2006/2007. Exp Gerontol 43:34–38

Peiris JS, Chu CM, Cheng VC, Chan KS, Hung IF, Poon LL, Law KI, Tang BS, Hon TY, Chan CS, Chan KH, Ng JS, Zheng BJ, Ng WL, Lai RW, Guan Y, Yuen KY (2003a) Clinical progression and viral load in a community outbreak of coronavirus-associated SARS pneumonia: a prospective study. Lancet 361:1767–1772

Peiris JS, Yuen KY, Osterhaus AD, Stohr K (2003b) The severe acute respiratory syndrome. N Engl J Med 349:2431–2441

Phipps S, Lam CE, Mahalingam S, Newhouse M, Ramirez R, Rosenberg HF, Foster PS, Matthaei KI (2007) Eosinophils contribute to innate antiviral immunity and promote clearance of respiratory syncytial virus. Blood 110:1578–1586

Plotkin SA (1999) Vaccination against the major infectious diseases. C R Acad Sci III 322: 943–951

Polack FP, Auwaerter PG, Lee SH, Nousari HC, Valsamakis A, Leiferman KM, Diwan A, Adams RJ, Griffin DE (1999) Production of atypical measles in rhesus macaques: evidence for disease mediated by immune complex formation and eosinophils in the presence of fusion-inhibiting antibody. Nat Med 5:629–634

Pulendran B, Ahmed R (2006) Translating innate immunity into immunological memory: implications for vaccine development. Cell 124:849–863

Pyrc K, Berkhout B, van der Hoek L (2006) The novel human coronaviruses NL63 and HKU1. J Virol 81(7):3051–3057

Qin E, Shi H, Tang L, Wang C, Chang G, Ding Z, Zhao K, Wang J, Chen Z, Yu M, Si B, Liu J, Wu D, Cheng X, Yang B, Peng W, Meng Q, Liu B, Han W, Yin X, Duan H, Zhan D, Tian L, Li S, Wu J, Tan G, Li Y, Li Y, Liu Y, Liu H, Lv F, Zhang Y, Kong X, Fan B, Jiang T, Xu S, Wang X, Li C, Wu X, Deng Y, Zhao M, Zhu Q (2006) Immunogenicity and protective efficacy in monkeys of purified inactivated Vero-cell SARS vaccine. Vaccine 24:1028–1034

Qiu M, Shi Y, Guo Z, Chen Z, He R, Chen R, Zhou D, Dai E, Wang X, Si B, Song Y, Li J, Yang L, Wang J, Wang H, Pang X, Zhai J, Du Z, Liu Y, Zhang Y, Li L, Wang J, Sun B, Yang R (2005) Antibody responses to individual proteins of SARS coronavirus and their neutralization activities. Microbes Infect 7:882–889

Qu D, Zheng B, Yao X, Guan Y, Yuan ZH, Zhong NS, Lu LW, Xie JP, Wen YM (2005) Intranasal immunization with inactivated SARS-CoV (SARS-associated coronavirus) induced local and serum antibodies in mice. Vaccine 23:924–931

Reghunathan R, Jayapal M, Hsu LY, Chng HH, Tai D, Leung BP, Melendez AJ (2005) Expression profile of immune response genes in patients with severe acute respiratory syndrome. BMC Immunol 6:2

Roberts A, Subbarao K (2006) Animal models for SARS. Adv Exp Med Biol 581:463–471

Roberts A, Paddock C, Vogel L, Butler E, Zaki S, Subbarao K (2005a) Aged BALB/c mice as a model for increased severity of severe acute respiratory syndrome in elderly humans. J Virol 79:5833–5838

Roberts A, Vogel L, Guarner J, Hayes N, Murphy B, Zaki S, Subbarao K (2005b) Severe acute respiratory syndrome coronavirus infection of golden Syrian hamsters. J Virol 79:503–511

Roberts A, Wood J, Subbarao K, Ferguson M, Wood D, Cherian T (2006) Animal models and antibody assays for evaluating candidate SARS vaccines: summary of a technical meeting 25–26 August 2005, London, UK. Vaccine 24:7056–7065

Roberts A, Deming D, Paddock CD, Cheng A, Yount B, Vogel L, Herman BD, Sheahan T, Heise M, Genrich GL, Zaki SR, Baric R, Subbarao K (2007) A mouse-adapted SARS-coronavirus causes disease and mortality in BALB/c mice. PLoS Pathog 3:e5

Rockx B, Sheahan T, Donaldson E, Harkema J, Sims A, Heise M, Pickles R, Cameron M, Kelvin D, Baric R (2007) Synthetic reconstruction of zoonotic and early human severe acute respiratory syndrome coronavirus isolates that produce fatal disease in aged mice. J Virol 81:7410–7423

Rockx B, Corti D, Donaldson E, Sheahan T, Stadler K, Lanzavecchia A, Baric R (2008) Structural basis for potent cross-neutralizing human monoclonal antibody protection against lethal human and zoonotic SARS-CoV challenge. J Virol 82(7):3220–3235

Rockx B, Baas T, Zornetzer GA, Haagmans B, Sheahan T, Frieman M, Dyer MD, Teal TH, Proll S, van den Brand J, Baric R, Katze MG (2009) Early upregulation of acute respiratory distress syndrome-associated cytokines promotes lethal disease in an aged-mouse model of severe acute respiratory syndrome coronavirus infection. J Virol 83(14):7062–7074

Rota PA, Oberste MS, Monroe SS, Nix WA, Campagnoli R, Icenogle JP, Penaranda S, Bankamp B, Maher K, Chen MH, Tong S, Tamin A, Lowe L, Frace M, DeRisi JL, Chen Q, Wang D, Erdman DD, Peret TC, Burns C, Ksiazek TG, Rollin PE, Sanchez A, Liffick S, Holloway B, Limor J, McCaustland K, Olsen-Rasmussen M, Fouchier R, Gunther S, Osterhaus AD, Drosten C, Pallansch MA, Anderson LJ, Bellini WJ (2003) Characterization of a novel coronavirus associated with severe acute respiratory syndrome. Science 300:1394–1399

Rudd BD, Schaller MA, Smit JJ, Kunkel SL, Neupane R, Kelley L, Berlin AA, Lukacs NW (2007) MyD88-mediated instructive signals in dendritic cells regulate pulmonary immune responses during respiratory virus infection. J Immunol 178:5820–5827

See RH, Zakhartchouk AN, Petric M, Lawrence DJ, Mok CP, Hogan RJ, Rowe T, Zitzow LA, Karunakaran KP, Hitt MM, Graham FL, Prevec L, Mahony JB, Sharon C, Auperin TC, Rini JM, Tingle AJ, Scheifele DW, Skowronski DM, Patrick DM, Voss TG, Babiuk LA, Gauldie J, Roper RL, Brunham RC, Finlay BB (2006) Comparative evaluation of two severe acute respiratory syndrome (SARS) vaccine candidates in mice challenged with SARS coronavirus. J Gen Virol 87:641–650

See RH, Petric M, Lawrence DJ, Mok CP, Rowe T, Zitzow LA, Karunakaran KP, Voss TG, Brunham RC, Gauldie J, Finlay BB, Roper RL (2008) Severe acute respiratory syndrome vaccine efficacy in ferrets: whole killed virus and adenovirus-vectored vaccines. J Gen Virol 89:2136–2146

Sheahan T, Morrison TE, Funkhouser W, Uematsu S, Akira S, Baric RS, Heise MT (2008) MyD88 is required for protection from lethal infection with a mouse-adapted SARS-CoV. PLoS Pathog 4(12):e1000240

Sheahan T, Rockx B, Donaldson E, Corti D, Baric R (2008a) Pathways of cross-species transmission of synthetically reconstructed zoonotic severe acute respiratory syndrome coronavirus. J Virol 82:8721–8732

Sheahan T, Rockx B, Donaldson E, Sims A, Pickles R, Corti D, Baric R (2008b) Mechanisms of zoonotic severe acute respiratory syndrome coronavirus host range expansion in human airway epithelium. J Virol 82:2274–2285

Siegal FP, Kadowaki N, Shodell M, Fitzgerald-Bocarsly PA, Shah K, Ho S, Antonenko S, Liu YJ (1999) The nature of the principal type 1 interferon-producing cells in human blood. Science 284:1835–1837

Sims AC, Baric RS, Yount B, Burkett SE, Collins PL, Pickles RJ (2005) Severe acute respiratory syndrome coronavirus infection of human ciliated airway epithelia: role of ciliated cells in viral spread in the conducting airways of the lungs. J Virol 79:15511–15524

Spruth M, Kistner O, Savidis-Dacho H, Hitter E, Crowe B, Gerencer M, Bruhl P, Grillberger L, Reiter M, Tauer C, Mundt W, Barrett PN (2006) A double-inactivated whole virus candidate SARS coronavirus vaccine stimulates neutralising and protective antibody responses. Vaccine 24:652–661

Stockman LJ, Bellamy R, Garner P (2006) SARS: systematic review of treatment effects. PLoS Med 3:e343

Subbarao K, Roberts A (2006) Is there an ideal animal model for SARS? Trends Microbiol 14:299–303

Subbarao K, McAuliffe J, Vogel L, Fahle G, Fischer S, Tatti K, Packard M, Shieh WJ, Zaki S, Murphy B (2004) Prior infection and passive transfer of neutralizing antibody prevent replication of severe acute respiratory syndrome coronavirus in the respiratory tract of mice. J Virol 78:3572–3577

Sui J, Li W, Murakami A, Tamin A, Matthews LJ, Wong SK, Moore MJ, Tallarico AS, Olurinde M, Choe H, Anderson LJ, Bellini WJ, Farzan M, Marasco WA (2004) Potent neutralization of severe acute respiratory syndrome (SARS) coronavirus by a human mAb to S1 protein that blocks receptor association. Proc Natl Acad Sci USA 101:2536–2541

Sui J, Li W, Roberts A, Matthews LJ, Murakami A, Vogel L, Wong SK, Subbarao K, Farzan M, Marasco WA (2005) Evaluation of human monoclonal antibody 80R for immunoprophylaxis of severe acute respiratory syndrome by an animal study, epitope mapping, and analysis of spike variants. J Virol 79:5900–5906

Tang L, Zhu Q, Qin E, Yu M, Ding Z, Shi H, Cheng X, Wang C, Chang G, Zhu Q, Fang F, Chang H, Li S, Zhang X, Chen X, Yu J, Wang J, Chen Z (2004) Inactivated SARS-CoV vaccine prepared from whole virus induces a high level of neutralizing antibodies in BALB/c mice. DNA Cell Biol 23:391–394

Thompson JM, Whitmore AC, Konopka JL, Collier ML, Richmond EM, Davis NL, Staats HF, Johnston RE (2006) Mucosal and systemic adjuvant activity of alphavirus replicon particles. Proc Natl Acad Sci USA 103:3722–3727

Tipnis SR, Hooper NM, Hyde R, Karran E, Christie G, Turner AJ (2000) A human homolog of angiotensin-converting enzyme. Cloning and functional expression as a captopril-insensitive carboxypeptidase. J Biol Chem 275:33238–33243

Towner JS, Pourrut X, Albarino CG, Nkogue CN, Bird BH, Grard G, Ksiazek TG, Gonzalez JP, Nichol ST, Leroy EM (2007) Marburg virus infection detected in a common african bat. PLoS ONE 2:e764

Tse GM, To KF, Chan PK, Lo AW, Ng KC, Wu A, Lee N, Wong HC, Mak SM, Chan KF, Hui DS, Sung JJ, Ng HK (2004) Pulmonary pathological features in coronavirus associated severe acute respiratory syndrome (SARS). J Clin Pathol 57:260–265

Tseng CT, Perrone LA, Zhu H, Makino S, Peters CJ (2005) Severe acute respiratory syndrome and the innate immune responses: modulation of effector cell function without productive infection. J Immunol 174:7977–7985

Tsunetsugu-Yokota Y, Ato M, Takahashi Y, Hashimoto S, Kaji T, Kuraoka M, Yamamoto K, Mitsuki YY, Yamamoto T, Oshima M, Ohnishi K, Takemori T (2007) Formalin-treated UV-inactivated SARS coronavirus vaccine retains its immunogenicity and promotes Th2-type immune responses. Jpn J Infect Dis 60:106–112

Tumpey TM, Basler CF, Aguilar PV, Zeng H, Solorzano A, Swayne DE, Cox NJ, Katz JM, Taubenberger JK, Palese P, Garcia-Sastre A (2005) Characterization of the reconstructed 1918 Spanish influenza pandemic virus. Science 310:77–80

Turner AJ, Hooper NM (2002) The angiotensin-converting enzyme gene family: genomics and pharmacology. Trends Pharmacol Sci 23:177–183

Uehara A, Fujimoto Y, Fukase K, Takada H (2007) Various human epithelial cells express functional Toll-like receptors, NOD1 and NOD2 to produce anti-microbial peptides, but not proinflammatory cytokines. Mol Immunol 44:3100–3111

Vallejo AN (2005) CD28 extinction in human T cells: altered functions and the program of T-cell senescence. Immunol Rev 205:158–169

Varga SM, Wang X, Welsh RM, Braciale TJ (2001) Immunopathology in RSV infection is mediated by a discrete oligoclonal subset of antigen-specific CD4(+) T cells. Immunity 15:637–646

Vasto S, Malavolta M, Pawelec G (2006) Age and immunity. Immun Ageing 3:2

Vennema H, de Groot RJ, Harbour DA, Dalderup M, Gruffydd-Jones T, Horzinek MC, Spaan WJ (1990) Early death after feline infectious peritonitis virus challenge due to recombinant vaccinia virus immunization. J Virol 64:1407–1409

Vickers C, Hales P, Kaushik V, Dick L, Gavin J, Tang J, Godbout K, Parsons T, Baronas E, Hsieh F, Acton S, Patane M, Nichols A, Tummino P (2002) Hydrolysis of biological peptides by human angiotensin-converting enzyme-related carboxypeptidase. J Biol Chem 277:14838–14843

Vogel LN, Roberts A, Paddock CD, Genrich GL, Lamirande EW, Kapadia SU, Rose JK, Zaki SR, Subbarao K (2007) Utility of the aged BALB/c mouse model to demonstrate prevention and control strategies for severe acute respiratory syndrome coronavirus (SARS-CoV). Vaccine 25:2173–2179

Wang CQ, Udupa KB, Xiao H, Lipschitz DA (1995) Effect of age on marrow macrophage number and function. Aging 7:379–384

Wang Z, Yuan Z, Matsumoto M, Hengge UR, Chang YF (2005) Immune responses with DNA vaccines encoded different gene fragments of severe acute respiratory syndrome coronavirus in BALB/c mice. Biochem Biophys Res Commun 327:130–135

Wang S, Wei M, Han Y, Zhang K, He L, Yang Z, Su B, Zhang Z, Hu Y, Hui W (2008) Roles of TNF-alpha gene polymorphisms in the occurrence and progress of SARS-Cov infection: a case-control study. BMC Infect Dis 8:27

Weingartl H, Czub M, Czub S, Neufeld J, Marszal P, Gren J, Smith G, Jones S, Proulx R, Deschambault Y, Grudeski E, Andonov A, He R, Li Y, Copps J, Grolla A, Dick D, Berry J, Ganske S, Manning L, Cao J (2004) Immunization with modified vaccinia virus Ankara-based recombinant vaccine against severe acute respiratory syndrome is associated with enhanced hepatitis in ferrets. J Virol 78:12672–12676

Werle B, Fromantin C, Alexandre A, Kohli E, Pothier P (1999) Dose-dependent effects of IL-12 treatment to immune response induced after immunization with a recombinant respiratory syncytial virus (RSV) fusion protein fragment. Vaccine 17:2983–2990

WHO (2008a) Global polio eradication initiative. WHO, Geneva

WHO (2008b) Measles fact sheet. WHO, Geneva

WHO (2008c) Polio fact sheet. WHO, Geneva

Wolfe ND, Dunavan CP, Diamond J (2007) Origins of major human infectious diseases. Nature 447:279–283

Wong CK, Lam CW, Wu AK, Ip WK, Lee NL, Chan IH, Lit LC, Hui DS, Chan MH, Chung SS, Sung JJ (2004) Plasma inflammatory cytokines and chemokines in severe acute respiratory syndrome. Clin Exp Immunol 136:95–103

Woo PC, Lau SK, Yuen KY (2006) Infectious diseases emerging from Chinese wet-markets: zoonotic origins of severe respiratory viral infections. Curr Opin Infect Dis 19:401–407

Xiong S, Wang YF, Zhang MY, Liu XJ, Zhang CH, Liu SS, Qian CW, Li JX, Lu JH, Wan ZY, Zheng HY, Yan XG, Meng MJ, Fan JL (2004) Immunogenicity of SARS inactivated vaccine in BALB/c mice. Immunol Lett 95:139–143

Yang ZY, Kong WP, Huang Y, Roberts A, Murphy BR, Subbarao K, Nabel GJ (2004) A DNA vaccine induces SARS coronavirus neutralization and protective immunity in mice. Nature 428:561–564

Yang L, Peng H, Zhu Z, Li G, Huang Z, Zhao Z, Koup RA, Bailer RT, Wu C (2007) Persistent memory CD4+ and CD8+ T-cell responses in recovered severe acute respiratory syndrome (SARS) patients to SARS coronavirus M antigen. J Gen Virol 88:2740–2748

Yasui F, Kai C, Kitabatake M, Inoue S, Yoneda M, Yokochi S, Kase R, Sekiguchi S, Morita K, Hishima T, Suzuki H, Karamatsu K, Yasutomi Y, Shida H, Kidokoro M, Mizuno K, Matsushima K, Kohara M (2008) Prior immunization with severe acute respiratory syndrome

(SARS)-associated coronavirus (SARS-CoV) nucleocapsid protein causes severe pneumonia in mice infected with SARS-CoV. J Immunol 181:6337–6348

Ye J, Zhang B, Xu J, Chang Q, McNutt MA, Korteweg C, Gong E, Gu J (2007) Molecular pathology in the lungs of severe acute respiratory syndrome patients. Am J Pathol 170:538–545

Yount B, Roberts RS, Sims AC, Deming D, Frieman MB, Sparks J, Denison MR, Davis N, Baric RS (2005) Severe acute respiratory syndrome coronavirus group-specific open reading frames encode nonessential functions for replication in cell cultures and mice. J Virol 79: 14909–14922

Zhang CH, Lu JH, Wang YF, Zheng HY, Xiong S, Zhang MY, Liu XJ, Li JX, Wan ZY, Yan XG, Qi SY, Cui Z, Zhang B (2005) Immune responses in Balb/c mice induced by a candidate SARS-CoV inactivated vaccine prepared from F69 strain. Vaccine 23:3196–3201

Zhou J, Wang W, Zhong Q, Hou W, Yang Z, Xiao SY, Zhu R, Tang Z, Wang Y, Xian Q, Tang H, Wen L (2005) Immunogenicity, safety, and protective efficacy of an inactivated SARS-associated coronavirus vaccine in rhesus monkeys. Vaccine 23:3202–3209

Zhou L, Ni B, Luo D, Zhao G, Jia Z, Zhang L, Lin Z, Wang L, Zhang S, Xing L, Li J, Liang Y, Shi X, Zhao T, Zhou L, Wu Y, Wang X (2007a) Inhibition of infection caused by severe acute respiratory syndrome-associated coronavirus by equine neutralizing antibody in aged mice. Int Immunopharmacol 7:392–400

Zhou S, Kurt-Jones EA, Fitzgerald KA, Wang JP, Cerny AM, Chan M, Finberg RW (2007b) Role of MyD88 in route-dependent susceptibility to vesicular stomatitis virus infection. J Immunol 178:5173–5181

Zhu MS, Pan Y, Chen HQ, Shen Y, Wang XC, Sun YJ, Tao KH (2004) Induction of SARS-nucleoprotein-specific immune response by use of DNA vaccine. Immunol Lett 92:237–243

Zhu Z, Chakraborti S, He Y, Roberts A, Sheahan T, Xiao X, Hensley LE, Prabakaran P, Rockx B, Sidorov IA, Corti D, Vogel L, Feng Y, Kim JO, Wang LF, Baric R, Lanzavecchia A, Curtis KM, Nabel GJ, Subbarao K, Jiang S, Dimitrov DS (2007) Potent cross-reactive neutralization of SARS coronavirus isolates by human monoclonal antibodies. Proc Natl Acad Sci USA 104:12123–12128

Zust R, Cervantes-Barragan L, Kuri T, Blakqori G, Weber F, Ludewig B, Thiel V (2007) Coronavirus non-structural protein 1 is a major pathogenicity factor: implications for the rational design of coronavirus vaccines. PLoS Pathog 3:e109

Chapter 14
Modulation of Host Cell Death by SARS Coronavirus Proteins

Claudia Diemer, Martha Schneider, Hermann M. Schätzl, and Sabine Gilch

Abstract Both types of cell death, namely necrosis and apoptosis, are found in organs of SARS coronavirus (CoV) infected patients. The gastrointestinal tract, however, although also a target for SARS-CoV replication, is obviously not affected by cell death mechanisms. Such differences in cell death induction are paralleled by in-vitro studies. In a colon-derived cell line (Caco-2), proapoptotic proteins were down- and antiapoptotic proteins were upregulated during SARS-CoV infection. By contrast, in SARS-CoV infected Vero E6 cells, apoptosis was induced via the p38 MAPK and caspase dependent pathways. Both apoptotic pathways, although mostly the intrinsic signal transduction, can be targeted by structural as well as accessory proteins of SARS-CoV. The fact that all structural and most of the accessory proteins of SARS-CoV are implicated in apoptotic scenarios indicates the fundamental role of apoptosis in the SARS-CoV life cycle. Interestingly, at least for the nucleocapsid protein of SARS-CoV, a cell-type specific manipulation of apoptosis was confirmed.

14.1 Cell Death During SARS-CoV Infection

SARS manifests predominantly as a viral pneumonia with diffuse alveolar damage. Although the exact mechanisms of SARS pathogenesis are not known, the lung damage in patients with SARS appears to be due to apoptosis and necrosis, both induced directly by viral replication as well as indirectly by production of immune mediators. SARS-CoV has also been found in various extrapulmonary tissues. Infected tissues from autopsied or biopsied SARS-CoV patients, varying in histopathological findings with regard to necrosis, apoptosis or no tissue damage, are

H.M. Schätzl (✉)
Institute of Virology, Technische Universität München, Trogerstr. 30, 81675 Munich, Germany
e-mail: schaetzl@virologie.med.tum.de

summarized in Table 14.1 (Guo et al. 2008; Zhang et al. 2003). Whereas apoptosis- and/or necrosis-induced tissue damage is found in various SARS-CoV infected organs such as lung, liver, brain, and immune cells, SARS-CoV replication in the gastrointestinal tract apparently does not cause histopathological changes.

Moreover, differences of SARS-CoV pathology in infected organs are paralleled by in-vitro studies employing SARS-CoV infected cell lines. It was reported that SARS-CoV infection of Vero E6 (African green monkey kidney) cells induces apoptosis via the p38 MAPK and caspase dependent pathways (Bordi et al. 2006; Mizutani et al. 2004b; Ren et al. 2005; Yan et al. 2004). By contrast, in SARS-CoV infected Caco-2 cells (colon carcinoma), proapoptotic proteins were down- and antiapoptotic factors were upregulated (Cinatl et al. 2004). These data indicate that the modulation of apoptosis by SARS-CoV seems to be crucial for the cell-type specific phenotype of infection and might subsequently account for the differences of SARS-CoV pathology. Notably, no induction of necrosis was observed in SARS-CoV infected cell lines. Necrosis is frequently induced by immunological mediators and is characterized as nongenetically regulated cell death. Apoptosis, in contrast, is genetically highly regulated and the molecular mechanism is classically divided into two major apoptotic pathways (Green 2000), called extrinsic and intrinsic pathway. Extrinsic signals are transmitted by members of the tumor necrosis factor (TNF) superfamiliy. Ligand binding to TNF family death receptors recruits adaptors and initiator procaspases-8 and/or -10 to form the death-inducing signaling complex (DISC). The DISC formation results in the autocatalytic activation of initiator caspases. Once initiator caspases are activated they directly cleave effector caspases (3, 6, 7) to convert them into their active form. Now effector caspases are able to degrade their targets such as cytoskeletal proteins or nuclear lamins to facilitate cell death.

Intrinsic signals are propagated to mitochondria by the Bcl-2 protein family. There are pro-survival members of this family (Bcl-2, Bcl-xL, Bcl-w, Mcl-1,

Table 14.1 Organs and cell types infected by SARS-CoV

	Organs					
	Lung	Kidney	Liver	Brain	Intestine	Immune system
Detection of SARS-CoV	ISH, EM, RT-PCR	ISH, EM	RT-PCR	EM, RT-PCR	ISH, EM, RT-PCR	EM, ISH
Cell types	Pneumocyte, endothelial and epithelial cells, macrophage, lymphocyte	Epithelial cells of renal tubules	Hepatocyte	Neurons of cortex and hypothalamus	Epithelial cells of mucosa	Monocyte, lymphocyte, macrophage in lymph node, spleen
Tissue damage	+	+	+	+	−	+
Cell death	Apoptosis, necrosis	Necrosis	Apoptosis	Necrosis	No tissue damage	Necrosis

+ positive; − negative; ISH: In situ hybridization; RT-PCR: Reverse-transcription polymerase chain reaction; EM: Electronic microscopy and A1) which oppose proapoptotic members (Bax, Bak, Bok, Bad, Bid, Bik, Puma, and Noxa). Moreover, cross-talk between the intrinsic and extrinsic pathway exists by caspase-8 mediated cleavage of Bid. The intrinsic pathway is highly regulated by the interactions of pro- and antiapoptotic Bcl-2 proteins. Monitoring of intrinsic signals leads, for example, to binding of Bad to the pro-survival Bcl-2 proteins and results in release of Bax and Bak from these proteins. Once Bax and Bak are released they oligomerize and insert into the mitochondrial membrane causing an efflux of cytochrome c. Cytochrome c oligomerizes with Apaf1 and recruits initiator procaspase-9 to form the apoptosome, finally resulting in proteolytic activation of caspase-9. Caspase-9 then initiates the caspase cascade by cleaving effector caspases (3, 6, 7) which subsequently degrade their target proteins.

A counterpart to the intrinsic apoptotic pathway is the pro-survival PI3K-Akt signal transduction. PI3K activates several downstream effectors such as the serine–threonine kinase Akt which regulates cell growth, cell cycle and cell survival (Cantrell 2001). Activated Akt phosphorylates a number of proapoptotic proteins including Bad and caspase-9. Thereby, the proapoptotic proteins are inactivated (Kulik et al. 1997). Similar to several other viruses, SARS-CoV also promotes PI3K-Akt signal transduction to establish persistent infection in Vero cell lines (Mizutani et al. 2005). Moreover it was shown that the activation of PI3K-Akt is differentiation state-specific in intestinal cells (Gauthier et al. 2001) and inhibits FAS-induced apoptosis in human intestinal epithelial cells (Abreu et al. 2001). In several studies it was shown that SARS-CoV can both induce and inhibit apoptosis in a cell-type specific manner. Activation of PI3K-Akt signal transduction in specific cell lines could offer an explanation; the mechanism for cell-type specific modulation of apoptosis, however, needs to be elucidated. Although it was shown by over-expression of Bcl-2 (Bordi et al. 2006) or by treatment with caspase inhibitors (Ren et al. 2005) that induction of apoptosis by SARS-CoV does not favor viral replication, SARS-CoV might induce apoptosis upon effective replication, potentially to evade immune response or to enable spread to other target organs. Alternatively, SARS-CoV may establish persistent infection by inhibition of apoptosis in cell lines derived from the intestine.

In summary, these data indicate a host cell specific modulation of cell death by SARS-CoV and an important role of apoptosis in SARS-CoV pathogenesis. Although the role of apoptosis for the severe clinical outcome of SARS-CoV infection is not entirely clear, the fact that all structural proteins and several of the accessory proteins investigated so far can induce apoptosis indicates that apoptotic cell death is important for the SARS-CoV life-cycle.

14.2 Induction of Host Cell Death by SARS-CoV Structural Proteins

The SARS-CoV virion consists of four main structural proteins, namely the nucleocapid (N), envelope (E), membrane (M), and spike (S) proteins. N provides the structural basis for the helical nucleocapsid by complexing the viral RNA genome. The nucleocapsid is surrounded by an envelope containing the other three structural proteins E, S and M. Beside the primary functions of the structural proteins, namely to constitute the virion scaffold and to execute viral morphogenesis, these proteins also modulate host cell death pathways such as apoptosis.

14.2.1 E Protein

Similar to the E protein of murine hepatitis virus (MHV), SARS-CoV E can induce apoptosis under certain conditions. It was demonstrated that adenoviral-mediated over-expression of E promotes cell death in the human T-cell line Jurkat upon serum deprivation (Yang et al. 2005). In contrast, Vero E6 viability was not affected by E over-expression (Chow et al. 2005); however, in this study cells were cultivated in the presence of growth factors and the role of starvation was not addressed. It is of particular interest that T-cell death can be induced by a structural protein of SARS-CoV, since lymphopenia is one clinical hallmark of patients suffering from SARS that occurs in up to 100% of the patients (Booth et al. 2003; Yang et al. 2004). Thus, expression of E in infected lymphocytes could contribute to this remarkable cell death. On the molecular level, E induces cell death by sequestering the antiapoptotic Bcl-xL protein to ER membranes (Yang et al. 2005), thereby probably preventing its incorporation into the mitochondrial membrane. Consequently, over-expression of Bcl-xL attenuates the proapoptotic effect of E (Yang et al. 2005). Interestingly, in Vero E6 cells infected with SARS-CoV apoptosis can also be prevented by over-expression of Bcl-2 (Bordi et al. 2006).

14.2.2 M Protein

The intrinsic apoptotic pathway is also activated by the M protein. Initially, induction of apoptosis was shown in the absence of growth factors in human pulmonary fibroblast (HPF) cells. In these cells apoptosis was also induced by N over-expression. In contrast, using a baculoviral expression system in insect cells, only M but not N expression led to cell death (Lai et al. 2006). Upon co-expression of M and N in serum-depleted HPF cells, apoptosis was increased compared to cultures expressing either M or N alone (Zhao et al. 2006). This could be due to the observed interaction between these two structural proteins (He et al. 2004). Furthermore, HEK293 cells and a *Drosophila* model over-expressing M exhibited an apoptotic phenotype (Chan et al. 2007). M expression in HEK293 cells induced cytochrome *c* release (Chan et al. 2007), leading to activation of the intrinsic apoptotic pathway. In addition, serine phosphorylation of Akt was reduced in HEK cells expressing M, indicating that the transduction of survival signals could be prevented thereby. This is in line with alterations observed in SARS-CoV infected Vero E6 cells (Mizutani et al. 2004a). Here, Akt was dephosphorylated 18 h postinfection (p.i.). At this time point of infection, M is detectable in SARS-CoV infected Vero E6 cells (Mizutani et al. 2004b), and it might be possible that M expression leads to Akt dephosphorylation.

14.2.3 S Protein

Vero E6 cells were employed to study the proapoptotic potential of the SARS-CoV structural proteins upon adenoviral expression (Chow et al. 2005). In this experimental set-up, apoptosis could only be activated by S, in particular the S2 domain (Chow et al. 2005). Induction of apoptosis by S appears to be unique for SARS-CoV S since neither MHV (An et al. 1999) nor infectious bronchitis virus (IBV) (Liu et al. 2001) S proteins induced apoptosis. In a follow-up study, microarray analysis of S2-transduced Vero E6 cells was performed (Yeung et al. 2008). Here, expressions of several genes involved in the extrinsic apoptotic pathway were downregulated, whereas the upregulation of CYCS and of Mdm-2 involved in the regulation of p53 degradation provides a hint to an activation of the intrinsic apoptotic pathway. The antiapoptotic proteins Mcl-1, Bcl-xL and Bcl-2 were downregulated by S2 over-expression (Yeung et al. 2008), in line with a downregulation of Bcl-2 in SARS-CoV infected Vero E6 cells (Ren et al. 2005). However, in contrast to infected cells (Mizutani et al. 2004b), components of the MAPK pathway were also downregulated (Yeung et al. 2008). Since S expression triggers ER stress response (Chan et al. 2006), prolonged ER stress in S expressing cells might finally lead to apoptotic cell death.

14.2.4 N Protein

The most compelling evidence for the proapoptotic potential of SARS-CoV structural proteins is provided for the N protein. Several studies performed by different groups indicated that N leads to apoptotic cell death in the absence of growth factors in COS-1 (Surjit et al. 2004; Zhang et al. 2007), HPF (Zhao et al. 2006), or in Vero E6 and A549 (human lung carcinoma) cells, but not Caco-2 (human colon carcinoma) or N2a (murine neuroblastoma) cells (Diemer et al. 2008). In addition, the human hepatoma cell lines Hep-G2 (Zhang et al. 2007) and Huh-7 (Surjit et al. 2004; Zhang et al. 2007) did not undergo apoptosis upon N expression. In COS-1 cells, induction of apoptosis was independent of Bax and p53. Bcl-2 expression as well as Akt phosphorylation were reduced, whereas the MAPK pathway and caspases were activated. It was speculated that apoptosis was initiated via an integrin dependent pathway (Surjit et al. 2004). In addition, mitochondrial cytochrome *c* release was observed (Zhang et al. 2007). Interestingly, N-induced apoptosis appears to be cell-type specific and among the investigated cell lines only COS-1, Vero E6 and A549 clearly exhibited apoptosis (Diemer et al. 2008; Surjit et al. 2004; Zhang et al. 2007). The fact that A549 cells, which are similar to Caco-2 and Huh-7 cells derived from a carcinoma, undergo apoptosis, argues against an artificial effect of cell-type specificity due to different immortalization protocols, as discussed by Surjit et al. (2004) for Huh-7 and COS-1. Diemer et al. (2008) demonstrated that upon N over-expression and SARS-CoV infection N acts not only as an inducer but finally also as a substrate of effector caspases. The significance of this finding is supported by a proteomics approach, by which N also was identified to be a substrate of caspases-3 and -6 (Ying et al. 2004). Interestingly, the N cleavage by caspases indicative for the induction of apoptosis correlated with the localization of N to the nucleus and the efficiency of viral replication (Diemer et al. 2008). In Vero E6 cells N was partially located to the nucleus, cleaved by caspases and high viral titers were produced. By contrast, in Caco-2 cells N was located only in the cytoplasm, was not cleaved and virus production was about 100-fold lower than in Vero E6 cells. Furthermore, activation of the proapoptotic protein Bad correlated with the nuclear localization of N. Thus, nuclear localization of N appears to be a factor that critically influences induction of apoptosis via the intrinsic pathway and there might also be a correlation with the replication efficiency (Diemer et al. 2008; illustrated in Fig. 14.1).

Furthermore, TGF-β (transforming growth factor beta) signaling which plays a pivotal role in cell growth, differentiation, apoptosis, and tissue remodeling is affected by N expression. In TGF-β stimulated cells, binding of N to Smad3, a downstream target of the TGF-β signaling pathway, promoted Smad3–p300 complex formation. Thereby, Smad3/Smad4 heterocomplex formation which leads to induction of apoptosis was inhibited. This was observed in HEK293, MEF and human peripheral lung epithelial (HPL1) cells (Zhao et al. 2008). The interaction of N and Smad3 was able to compete for the nuclear localization of N, resulting in attenuation of apoptosis, and could therefore describe a control mechanism of cell death during SARS-CoV infection. The observed interference of N expression with TGF-β signaling on the one hand inhibits apoptosis, and on the other hand increases the expression of plasminogen activator inhibitor-1 (PAI-1) which promotes fibrosis. This provides an interesting molecular link to the lung pathology observed in SARS patients, since pulmonary fibrosis is associated with lung failure (Nicholls et al. 2003, 2006).

14.3 Accessory Proteins

The SARS-CoV genome harbors at the 3′ end some unique, group-specific genes which encode the eight putative accessory proteins 3a, 3b, 6, 7a, 7b, 8a, 8b, and 9b (Marra et al. 2003). Their exact contributions to viral replication or pathogenesis in the natural host have not been entirely elucidated. Yount et al. have demonstrated that the deletion of SARS-CoV ORF3a, 3b, 6, 7a, and 7b did not dramatically

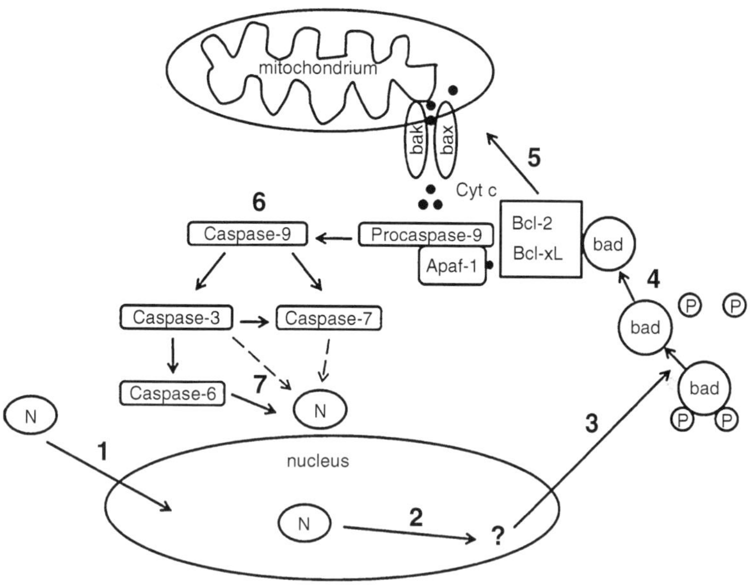

Fig. 14.1 Hypothetical model for correlation of nuclear localization and caspase-mediated cleavage of N in lytic SARS-CoV infection (adapted from Diemer et al. 2008). N translocates into the nucleus (*1*) and may activate gene expression or interact with nuclear components there (*2*), resulting in the dephosphorylation of Bad (*3*). In this form Bad is enabled to interact with Bcl-2 and Bcl-xL (*4*). This interaction releases Bax and Bak from pro-survival Bcl-2 proteins, enabling insertion of Bax and Bak in the mitochondrial membrane. The resulting membrane permeabilization facilitates cytochrome *c* efflux (*5*) which provokes the activation of procaspase-9 by formation of apoptosome. Once caspase-9 is activated the caspase cascade occurs and effector caspases (i.e., 3, 6, and 7) are activated (*6*). Finally, N is cleaved by caspase-6 (*7*), but cleavage of N may additionally be caused by caspase-3 or -7

influence replication efficiencies in cell culture or in a murine model of SARS-CoV infection. However, in the same study it was observed that upon deletion of ORF3a, virus titers were reduced to a certain degree (Yount et al. 2005). Akerström et al. also observed a significant reduction in the yield of progeny virus when Vero E6 cells expressing siRNA against ORF3 or 7 were infected with SARS-CoV (Akerstrom et al. 2007). Even if the applied siRNAs caused knockdown of the complete ORF3, ORF7 and ORF8, encoding for 3a, 3b, 7a, 7b, 8a, and 8b proteins, respectively, a partial participation in viral replication cannot be excluded for some of them. In addition, over-expression of the 8a protein in infected Vero E6 cells can enhance viral replication and, in infected HuH-7 cells, virus-mediated cytopathic effects (Chen et al. 2007). As the mechanisms of viral pathogenesis are still unclear, the ability of accessory proteins to modulate cell death is of special interest.

14.3.1 Protein 3a

The largest of the accessory proteins consists of 274 amino acids and is expressed in transfected (Huang et al. 2006b; Law et al. 2005; Tan et al. 2005; Yuan et al. 2005a) and infected cells (Lu et al. 2006; Yu et al. 2004; Zeng et al. 2004). In both cases the protein was shown to be associated with intracellular membranes and the plasma membrane (Ito et al. 2005; Tan et al. 2004b; Yuan et al. 2005a), to be released into the surrounding media (Huang et al. 2006b), and it appears to be a structural protein of the SARS-CoV (Ito et al. 2005; Shen et al. 2005). By transient expression in Vero E6 cells, 3a induced the extrinsic apoptotic pathway via caspase-8, whereas no induction of the intrinsic apoptotic pathway could be observed (Law et al. 2005). However, Padhan et al. demonstrated that protein 3a additionally can activate the intrinsic apoptotic pathway. Although activation of caspase-8 was confirmed, activation of caspase-9 and increased Bax-oligomerization, characteristic for the intrinsic apoptotic pathway, were demonstrated. Treatment of transfected cells with a p38 MAPK inhibitor decreased the level of 3a-induced apoptosis, indicating upstream activation of p38 MAPK. Protein 3a seems to be potently proapoptotic as it is apparently able to induce and link both apoptotic pathways, as cleavage of Bid was also detected (Padhan et al. 2008). Although apoptosis can be the result of p38 MAPK activation, there exist alternative ways for the induction of apoptosis by protein 3a. Transient over-expression in HEK293, Cos-7, and Vero E6 cells led to decreased levels of cyclin D3 protein and mRNA, resulting in growth inhibition and cell cycle arrest at the G0/G1 phase (Yuan et al. 2007). Subsequently, apoptosis was induced to eliminate these cells. Additionally, the induction of the ER stress-inducible proapoptotic kinase JNK by transient expression of 3a in HEK293 cells might lead to apoptosis (Kanzawa et al. 2006).

14.3.2 Protein 3b

Protein 3b was shown to be expressed upon transient transfection in several cell lines (Khan et al. 2006; Kopecky-Bromberg et al. 2007; Yuan et al. 2005b) with a predominant nucleolar localization (Yuan et al. 2005c). In infected Vero E6 cells the protein could be detected also in the cytoplasm (Chan et al. 2005) and protein 3b is likely to be expressed during infection in vivo. In Cos-7 and Vero E6 cells, overexpression of protein 3b caused apoptosis (Yuan et al. 2005b; Khan et al. 2006). As seen for protein 3a, transient over-expression of 3b deregulates cell cycle progression of Cos-7, HEK293 and Vero E6 cells and causes growth inhibition and cell cycle arrest at the G0/G1 phase. Whereas the induction of cell cycle arrest was comparable in all used cell lines, the ability to induce apoptosis was significant in Cos-7 cells, but almost undetectable in HEK293 and Vero E6 cells (Yuan et al. 2005b). In addition, when Vero E6 cells were transiently transfected with protein 3b, increased levels of lactate dehydrogenase (LDH), a key marker of cells undergoing necrosis, were released. Therefore, the authors concluded that the accessory protein 3b is able to induce necrosis (Khan et al. 2006).

14.3.3 Protein 6

ORF6 encodes a small 63-residue protein which is expressed upon transient transfection in several cell lines. It was also detected in infected Caco-2 and Vero E6 cells and in SARS patients. It localizes to the ER or Golgi membranes and to a perinuclear region (Frieman et al. 2007; Geng et al. 2005; Huang et al. 2007; Kopecky-Bromberg et al. 2007; Pewe et al. 2005). Furthermore, it was found incorporated into virus particles and is released from transfected and infected cells (Huang et al. 2007). Recently, it was shown that protein 6 is able to induce apoptosis in several cell lines, e.g., Cos-7, HEK293, and Vero E6, via a caspase-3 dependent pathway (Ye et al. 2008). However, the mechanisms leading to apoptosis are unclear.

14.3.4 Protein 7a

Protein 7a consists of 122 amino acids and is probably the best studied SARS-CoV accessory protein. It was the first accessory protein for which proapoptotic abilities were demonstrated (Tan et al. 2004a). In several transfected as well as infected cell lines, expression of 7a was confirmed (Fielding et al. 2004; Kanzawa et al. 2006; Kopecky-Bromberg et al. 2006; Nelson et al. 2005; Tan et al. 2004a; Yuan et al. 2006). Notably, it was detected in lung tissues from SARS-CoV infected patients (Chen et al. 2005). Protein 7a also appears to be a structural component of the virion (Huang et al. 2006a). With regard to the cellular localization, discrepancies exist as to whether it localizes to the ER, the ER–Golgi intermediate compartment, or the *trans*-Golgi network (Fielding et al. 2004; Kopecky-Bromberg et al. 2006; Nelson et al. 2005). Tan et al. however proposed that it is synthesized in the ER and shuttles rapidly to the Golgi or to mitochondria (Tan et al. 2007). The induction of apoptosis by 7a has been observed in several studies (Kopecky-Bromberg et al. 2006; Tan et al. 2004a, 2007; Yuan et al. 2006). Its proapoptotic properties appear to be common for many cell types as over-expression of 7a induced the apoptotic pathway in cell lines originating from different tissues such as lung (A549 cell line), kidney (Cos-7, Vero E6, HEK293), cervix (HeLa), and liver (Hep-G2) (Tan et al. 2004a). Moreover, contribution to the modulation of apoptosis in the viral context has been confirmed. (Schaecher et al. 2007) demonstrated that a recombinant virus with deletion of ORF7a did not show alterations in replication kinetics or viral progeny but the mode of DNA fragmentation induced in infected Vero E6 cells was significantly altered as compared to cells infected with the wild-type virus. Furthermore, in this study a population of SARS-CoV infected cells which exhibited DNA fragmentation but no active caspase-3 was observed at 48 h and 72 h postinfection. This population was not detected if cells were infected with the respective ORF7a deletion mutant. Therefore, these cells might represent a population in a late stage of apoptosis in which active caspase-3 is undetectable or, as DNA fragmentation can also be observed during necrosis, might alternatively represent a population of necrotic cells (Schaecher et al. 2007). The induction of cell cycle arrest at the G0/G1 phase appears to be a common feature of several accessory proteins and it was shown that transient over-expression of 7a in HEK293, Cos-7, and Vero E6 cells resulted in a deregulated cell cycle (Yuan et al. 2006). Kopecky-Bromberg et al. observed growth inhibition of 7a-expressing HEK293 cells and demonstrated that 7a blocks gene expression at the level of translation. The influences of 7a on the host translation machinery together with induction of cell cycle arrest in G0/G1 phase could subsequently activate apoptosis. Despite activation of p38 MAPK, the inhibition of this kinase could not block induction of apoptosis (Kopecky-Bromberg et al. 2006). Whereas in this study no activation of JNK was found, Kanzawa et al. showed the activation of JNK in HEK293 cells overexpressing 7a (Kanzawa et al. 2006). These discrepancies concerning involved mediators or effectors imply that the protein 7a probably uses alternative or several ways to induce apoptosis. Tan et al. recently demonstrated that 7a induces cell death by directly interacting with the antiapoptotic Bcl-xL, thereby preventing the interaction with proapoptotic Bcl-2 proteins. Consequently, 7a-induced apoptosis could be abrogated by over-expression of Bcl-xL (Tan et al. 2007). A further hint for the putative role of 7a in cell death scenarios was recently obtained in yeast-2-hybrid screens in which the proteins BiP and Bik were found as interactors of 7a (Diemer, Schneider, Schätzl and Gilch, personal communication).

14.3.5 Proteins 8a/8b

ORFs 8a and 8b emerged at late stages of the SARS-CoV epidemic by a 29-nt deletion in the single ORF8 (Guan et al. 2003). The cytoplasmic expression of both accessory proteins was confirmed in virus-infected Vero E6 cells (Keng et al. 2006) and transfected cells (Chen et al. 2007; Law et al. 2006). Protein 8a mainly localizes to the cytoplasm or mitochondrial membranes of transfected HEK293 cells (Chen et al. 2007; Keng et al. 2006), whereas for 8b only a cytoplasmic localization was reported (Chen et al. 2007; Keng et al. 2006). Transient expression of 8a leads to induction of apoptosis in Huh-7 cells in a caspase-3 dependent way. In stably transfected Huh-7 cells, Chen et al. observed an enhanced production of reactive oxygen species (ROS), an increase of cellular oxygen consumption, and a hyperpolarization of the mitochondrial membrane. As the mitochondrial homeostasis is of special importance for induction of the intrinsic apoptotic pathway and consequently for cell survival, these changes might lead to apoptosis (Chen et al. 2007). However, accessory proteins are not always proapoptotic, they can also show antiapoptotic abilities. Along this line, Keng et al. demonstrated that the protein level of E was significantly decreased in Vero E6 cells cotransfected with SARS-CoV E and 8b. As the E protein is able to induce apoptosis in T cells, downregulation of this protein by 8b might prevent apoptosis in these cells (Keng et al. 2006).

14.4 Conclusion

As described in this chapter, SARS-CoV structural and accessory proteins are potent inducers and modulators of apoptosis in vitro. Structural proteins mainly trigger the intrinsic apoptotic pathway, with p38 MAPK and PI3K/Akt pathways regulating cell death. Expression of accessory proteins leads to apoptosis by more diverse routes, involving various possibilities even for the same protein to induce cell death. However, one has to note that these observations were mostly obtained by cell culture experiments and might differ from effects in vivo. Whether the proteins are able to induce necrosis remains to be elucidated, although necrosis is assumed to be the underlying mechanism for the extensive immune overreaction and massive inflammation observed during clinical SARS-CoV infection. Induction of cell death might furthermore be modulated by the interplay of viral and host cell proteins, as exemplified for the N protein. In addition, it has to be assumed that one structural or accessory protein may not be solely responsible for induction or modulation of cell death but rather that the orchestrated activity of all viral proteins expressed at a certain time point finally results in the exquisite virulence and pathogenesis throughout infection with SARS-CoV.

References

Abreu MT, Arnold ET, Chow JY, Barrett KE (2001) Phosphatidylinositol 3-kinase-dependent pathways oppose Fas-induced apoptosis and limit chloride secretion in human intestinal epithelial cells. Implications for inflammatory diarrheal states. J Biol Chem 276:47563–47574

Akerstrom S, Mirazimi A, Tan YJ (2007) Inhibition of SARS-CoV replication cycle by small interference RNAs silencing specific SARS proteins, 7a/7b, 3a/3b and S. Antiviral Res 73: 219–227

An S, Chen CJ, Yu X, Leibowitz JL, Makino S (1999) Induction of apoptosis in murine coronavirus-infected cultured cells and demonstration of E protein as an apoptosis inducer. J Virol 73:7853–7859

Booth CM, Matukas LM, Tomlinson GA, Rachlis AR, Rose DB, Dwosh HA, Walmsley SL, Mazzulli T, Avendano M, Derkach P, Ephtimios IE, Kitai I, Mederski BD, Shadowitz SB, Gold WL, Hawryluck LA, Rea E, Chenkin JS, Cescon DW, Poutanen SM, Detsky AS (2003) Clinical features and short-term outcomes of 144 patients with SARS in the greater Toronto area. JAMA 289:2801–2809

Bordi L, Castilletti C, Falasca L, Ciccosanti F, Calcaterra S, Rozera G, Di Caro A, Zaniratti S, Rinaldi A, Ippolito G, Piacentini M, Capobianchi MR (2006) Bcl-2 inhibits the caspase-dependent apoptosis induced by SARS-CoV without affecting virus replication kinetics. Arch Virol 151: 369–377

Cantrell DA (2001) Phosphoinositide 3-kinase signalling pathways. J Cell Sci 114:1439–1445

Chan WS, Wu C, Chow SC, Cheung T, To KF, Leung WK, Chan PK, Lee KC, Ng HK, Au DM, Lo AW (2005) Coronaviral hypothetical and structural proteins were found in the intestinal surface enterocytes and pneumocytes of severe acute respiratory syndrome (SARS). Mod Pathol 18:1432–1439

Chan CP, Siu KL, Chin KT, Yuen KY, Zheng B, Jin DY (2006) Modulation of the unfolded protein response by the severe acute respiratory syndrome coronavirus spike protein. J Virol 80:9279–9287

Chan CM, Ma CW, Chan WY, Chan HY (2007) The SARS-Coronavirus Membrane protein induces apoptosis through modulating the Akt survival pathway. Arch Biochem Biophys 459:197–207

Chen YY, Shuang B, Tan YX, Meng MJ, Han P, Mo XN, Song QS, Qiu XY, Luo X, Gan QN, Zhang X, Zheng Y, Liu SA, Wang XN, Zhong NS, Ma DL (2005) The protein X4 of severe acute respiratory syndrome-associated coronavirus is expressed on both virus-infected cells and lung tissue of severe acute respiratory syndrome patients and inhibits growth of Balb/c 3T3 cell line. Chin Med J 118:267–274

Chen CY, Ping YH, Lee HC, Chen KH, Lee YM, Chan YJ, Lien TC, Jap TS, Lin CH, Kao LS, Chen YM (2007) Open reading frame 8a of the human severe acute respiratory syndrome coronavirus not only promotes viral replication but also induces apoptosis. J Infect Dis 196:405–415

Chow KY, Yeung YS, Hon CC, Zeng F, Law KM, Leung FC (2005) Adenovirus-mediated expression of the C-terminal domain of SARS-CoV spike protein is sufficient to induce apoptosis in Vero E6 cells. FEBS Lett 579:6699–6704

Cinatl J Jr, Hoever G, Morgenstern B, Preiser W, Vogel JU, Hofmann WK, Bauer G, Michaelis M, Rabenau HF, Doerr HW (2004) Infection of cultured intestinal epithelial cells with severe acute respiratory syndrome coronavirus. Cell Mol Life Sci 61:2100–2112

Diemer C, Schneider M, Seebach J, Quaas J, Frosner G, Schatzl HM, Gilch S (2008) Cell type-specific cleavage of nucleocapsid protein by effector caspases during SARS coronavirus infection. J Mol Biol 376:23–34

Fielding BC, Tan YJ, Shuo S, Tan TH, Ooi EE, Lim SG, Hong W, Goh PY (2004) Characterization of a unique group-specific protein (U122) of the severe acute respiratory syndrome coronavirus. J Virol 78:7311–7318

Frieman M, Yount B, Heise M, Kopecky-Bromberg SA, Palese P, Baric RS (2007) Severe acute respiratory syndrome coronavirus ORF6 antagonizes STAT1 function by sequestering nuclear import factors on the rough endoplasmic reticulum/Golgi membrane. J Virol 81:9812–9824

Gauthier R, Harnois C, Drolet JF, Reed JC, Vezina A, Vachon PH (2001) Human intestinal epithelial cell survival: differentiation state-specific control mechanisms. Am J Physiol Cell Physiol 280:C1540–C1554

Geng H, Liu YM, Chan WS, Lo AW, Au DM, Waye MM, Ho YY (2005) The putative protein 6 of the severe acute respiratory syndrome-associated coronavirus: expression and functional characterization. FEBS Lett 579:6763–6768

Green DR (2000) Apoptotic pathways: paper wraps stone blunts scissors. Cell 102:1–4

Guan Y, Zheng BJ, He YQ, Liu XL, Zhuang ZX, Cheung CL, Luo SW, Li PH, Zhang LJ, Guan YJ, Butt KM, Wong KL, Chan KW, Lim W, Shortridge KF, Yuen KY, Peiris JS, Poon LL (2003) Isolation and characterization of viruses related to the SARS coronavirus from animals in southern China. Science 302:276–278

Guo Y, Korteweg C, McNutt MA, Gu J (2008) Pathogenetic mechanisms of severe acute respiratory syndrome. Virus Res 133:4–12

He R, Leeson A, Ballantine M, Andonov A, Baker L, Dobie F, Li Y, Bastien N, Feldmann H, Strocher U, Theriault S, Cutts T, Cao J, Booth TF, Plummer FA, Tyler S, Li X (2004) Characterization of protein-protein interactions between the nucleocapsid protein and membrane protein of the SARS coronavirus. Virus Res 105:121–125

Huang C, Ito N, Tseng CT, Makino S (2006a) Severe acute respiratory syndrome coronavirus 7a accessory protein is a viral structural protein. J Virol 80:7287–7294

Huang C, Narayanan K, Ito N, Peters CJ, Makino S (2006b) Severe acute respiratory syndrome coronavirus 3a protein is released in membranous structures from 3a protein-expressing cells and infected cells. J Virol 80:210–217

Huang C, Peters CJ, Makino S (2007) Severe acute respiratory syndrome coronavirus accessory protein 6 is a virion-associated protein and is released from 6 protein-expressing cells. J Virol 81:5423–5426

Ito N, Mossel EC, Narayanan K, Popov VL, Huang C, Inoue T, Peters CJ, Makino S (2005) Severe acute respiratory syndrome coronavirus 3a protein is a viral structural protein. J Virol 79:3182–3186

Kanzawa N, Nishigaki K, Hayashi T, Ishii Y, Furukawa S, Niiro A, Yasui F, Kohara M, Morita K, Matsushima K, Le MQ, Masuda T, Kannagi M (2006) Augmentation of chemokine production by severe acute respiratory syndrome coronavirus 3a/X1 and 7a/X4 proteins through NF-kappaB activation. FEBS Lett 580:6807–6812

Keng CT, Choi YW, Welkers MR, Chan DZ, Shen S, Gee LS, Hong W, Tan YJ (2006) The human severe acute respiratory syndrome coronavirus (SARS-CoV) 8b protein is distinct from its counterpart in animal SARS-CoV and down-regulates the expression of the envelope protein in infected cells. Virology 354(1):132–142

Khan S, Fielding BC, Tan TH, Chou CF, Shen S, Lim SG, Hong W, Tan YJ (2006) Overexpression of severe acute respiratory syndrome coronavirus 3b protein induces both apoptosis and necrosis in Vero E6 cells. Virus Res 122(1–2):20–27

Kopecky-Bromberg SA, Martinez-Sobrido L, Palese P (2006) 7a protein of severe acute respiratory syndrome coronavirus inhibits cellular protein synthesis and activates p38 mitogen-activated protein kinase. J Virol 80:785–793

Kopecky-Bromberg SA, Martinez-Sobrido L, Frieman M, Baric RA, Palese P (2007) Severe acute respiratory syndrome coronavirus open reading frame (ORF) 3b, ORF 6, and nucleocapsid proteins function as interferon antagonists. J Virol 81:548–557

Kulik G, Klippel A, Weber MJ (1997) Antiapoptotic signalling by the insulin-like growth factor I receptor, phosphatidylinositol 3-kinase, and Akt. Mol Cell Biol 17:1595–1606

Lai CW, Chan ZR, Yang DG, Lo WH, Lai YK, Chang MD, Hu YC (2006) Accelerated induction of apoptosis in insect cells by baculovirus-expressed SARS-CoV membrane protein. FEBS Lett 580:3829–3834

Law PT, Wong CH, Au TC, Chuck CP, Kong SK, Chan PK, To KF, Lo AW, Chan JY, Suen YK, Chan HY, Fung KP, Waye MM, Sung JJ, Lo YM, Tsui SK (2005) The 3a protein of severe acute respiratory syndrome-associated coronavirus induces apoptosis in Vero E6 cells. J Gen Virol 86:1921–1930

Law PY, Liu YM, Geng H, Kwan KH, Waye MM, Ho YY (2006) Expression and functional characterization of the putative protein 8b of the severe acute respiratory syndrome-associated coronavirus. FEBS Lett 580:3643–3648

Liu C, Xu HY, Liu DX (2001) Induction of caspase-dependent apoptosis in cultured cells by the avian coronavirus infectious bronchitis virus. J Virol 75:6402–6409

Lu W, Zheng BJ, Xu K, Schwarz W, Du L, Wong CK, Chen J, Duan S, Deubel V, Sun B (2006) Severe acute respiratory syndrome-associated coronavirus 3a protein forms an ion channel and modulates virus release. Proc Natl Acad Sci USA 103:12540–12545

Marra MA, Jones SJ, Astell CR, Holt RA, Brooks-Wilson A, Butterfield YS, Khattra J, Asano JK, Barber SA, Chan SY, Cloutier A, Coughlin SM, Freeman D, Girn N, Griffith OL, Leach SR, Mayo M, McDonald H, Montgomery SB, Pandoh PK, Petrescu AS, Robertson AG, Schein JE, Siddiqui A, Smailus DE, Stott JM, Yang GS, Plummer F, Andonov A, Artsob H, Bastien N, Bernard K, Booth TF, Bowness D, Czub M, Drebot M, Fernando L, Flick R, Garbutt M, Gray M, Grolla A, Jones S, Feldmann H, Meyers A, Kabani A, Li Y, Normand S, Stroher U, Tipples GA, Tyler S, Vogrig R, Ward D, Watson B, Brunham RC, Krajden M, Petric M, Skowronski DM, Upton C, Roper RL (2003) The genome sequence of the SARS-associated coronavirus. Science 300:1399–1404

Mizutani T, Fukushi S, Saijo M, Kurane I, Morikawa S (2004a) Importance of Akt signaling pathway for apoptosis in SARS-CoV-infected Vero E6 cells. Virology 327:169–174

Mizutani T, Fukushi S, Saijo M, Kurane I, Morikawa S (2004b) Phosphorylation of p38 MAPK and its downstream targets in SARS coronavirus-infected cells. Biochem Biophys Res Commun 319:1228–1234

Mizutani T, Fukushi S, Saijo M, Kurane I, Morikawa S (2005) JNK and PI3k/Akt signaling pathways are required for establishing persistent SARS-CoV infection in Vero E6 cells. Biochim Biophys Acta 1741:4–10

Nelson CA, Pekosz A, Lee CA, Diamond MS, Fremont DH (2005) Structure and intracellular targeting of the SARS-coronavirus Orf7a accessory protein. Structure 13:75–85

Nicholls JM, Poon LL, Lee KC, Ng WF, Lai ST, Leung CY, Chu CM, Hui PK, Mak KL, Lim W, Yan KW, Chan KH, Tsang NC, Guan Y, Yuen KY, Peiris JS (2003) Lung pathology of fatal severe acute respiratory syndrome. Lancet 361:1773–1778

Nicholls JM, Butany J, Poon LL, Chan KH, Beh SL, Poutanen S, Peiris JS, Wong M (2006) Time course and cellular localization of SARS-CoV nucleoprotein and RNA in lungs from fatal cases of SARS. PLoS Med 3:e27

Padhan K, Minakshi R, Towheed MA, Jameel S (2008) Severe acute respiratory syndrome coronavirus 3a protein activates the mitochondrial death pathway through p38 MAP kinase activation. J Gen Virol 89:1960–1969

Pewe L, Zhou H, Netland J, Tangudu C, Olivares H, Shi L, Look D, Gallagher T, Perlman S (2005) A severe acute respiratory syndrome-associated coronavirus-specific protein enhances virulence of an attenuated murine coronavirus. J Virol 79:11335–11342

Ren L, Yang R, Guo L, Qu J, Wang J, Hung T (2005) Apoptosis induced by the SARS-associated coronavirus in Vero cells is replication-dependent and involves caspase. DNA Cell Biol 24:496–502

Schaecher SR, Touchette E, Schriewer J, Buller RM, Pekosz A (2007) Severe acute respiratory syndrome coronavirus gene 7 products contribute to virus-induced apoptosis. J Virol 81:11054–11068

Shen S, Lin PS, Chao YC, Zhang A, Yang X, Lim SG, Hong W, Tan YJ (2005) The severe acute respiratory syndrome coronavirus 3a is a novel structural protein. Biochem Biophys Res Commun 330:286–292

Surjit M, Liu B, Jameel S, Chow VT, Lal SK (2004) The SARS coronavirus nucleocapsid protein induces actin reorganization and apoptosis in COS-1 cells in the absence of growth factors. Biochem J 383:13–18

Tan YJ, Fielding BC, Goh PY, Shen S, Tan TH, Lim SG, Hong W (2004a) Overexpression of 7a, a protein specifically encoded by the severe acute respiratory syndrome coronavirus, induces apoptosis via a caspase-dependent pathway. J Virol 78:14043–14047

Tan YJ, Teng E, Shen S, Tan TH, Goh PY, Fielding BC, Ooi EE, Tan HC, Lim SG, Hong W (2004b) A novel severe acute respiratory syndrome coronavirus protein, U274, is transported to the cell surface and undergoes endocytosis. J Virol 78:6723–6734

Tan YJ, Tham PY, Chan DZ, Chou CF, Shen S, Fielding BC, Tan TH, Lim SG, Hong W (2005) The severe acute respiratory syndrome coronavirus 3a protein up-regulates expression of fibrinogen in lung epithelial cells. J Virol 79:10083–10087

Tan YX, Tan TH, Lee MJ, Tham PY, Gunalan V, Druce J, Birch C, Catton M, Fu NY, Yu VC, Tan YJ (2007) Induction of apoptosis by the severe acute respiratory syndrome coronavirus 7a protein is dependent on its interaction with the Bcl-XL protein. J Virol 81:6346–6355

Yan H, Xiao G, Zhang J, Hu Y, Yuan F, Cole DK, Zheng C, Gao GF (2004) SARS coronavirus induces apoptosis in Vero E6 cells. J Med Virol 73:323–331

Yang M, Li CK, Li K, Hon KL, Ng MH, Chan PK, Fok TF (2004) Hematological findings in SARS patients and possible mechanisms (review). Int J Mol Med 14:311–315

Yang Y, Xiong Z, Zhang S, Yan Y, Nguyen J, Ng B, Lu H, Brendese J, Yang F, Wang H, Yang XF (2005) Bcl-xL inhibits T-cell apoptosis induced by expression of SARS coronavirus E protein in the absence of growth factors. Biochem J 392:135–143

Ye Z, Wong CK, Li P, Xie Y (2008) A SARS-CoV protein, ORF-6, induces caspase-3 mediated, ER stress and JNK-dependent apoptosis. Biochim Biophys Acta 1780(12):1383–1387

Yeung YS, Yip CW, Hon CC, Chow KY, Ma IC, Zeng F, Leung FC (2008) Transcriptional profiling of Vero E6 cells over-expressing SARS-CoV S2 subunit: insights on viral regulation of apoptosis and proliferation. Virology 371:32–43

Ying W, Hao Y, Zhang Y, Peng W, Qin E, Cai Y, Wei K, Wang J, Chang G, Sun W, Dai S, Li X, Zhu Y, Li J, Wu S, Guo L, Dai J, Wang J, Wan P, Chen T, Du C, Li D, Wan J, Kuai X, Li W, Shi R, Wei H, Cao C, Yu M, Liu H, Dong F, Wang D, Zhang X, Qian X, Zhu Q, He F (2004) Proteomic analysis on structural proteins of Severe Acute Respiratory Syndrome coronavirus. Proteomics 4:492–504

Yount B, Roberts RS, Sims AC, Deming D, Frieman MB, Sparks J, Denison MR, Davis N, Baric RS (2005) Severe acute respiratory syndrome coronavirus group-specific open reading frames encode nonessential functions for replication in cell cultures and mice. J Virol 79: 14909–14922

Yu CJ, Chen YC, Hsiao CH, Kuo TC, Chang SC, Lu CY, Wei WC, Lee CH, Huang LM, Chang MF, Ho HN, Lee FJ (2004) Identification of a novel protein 3a from severe acute respiratory syndrome coronavirus. FEBS Lett 565:111–116

Yuan X, Li J, Shan Y, Yang Z, Zhao Z, Chen B, Yao Z, Dong B, Wang S, Chen J, Cong Y (2005a) Subcellular localization and membrane association of SARS-CoV 3a protein. Virus Res 109:191–202

Yuan X, Shan Y, Zhao Z, Chen J, Cong Y (2005b) G0/G1 arrest and apoptosis induced by SARS-CoV 3b protein in transfected cells. Virol J 2:66

Yuan X, Yao Z, Shan Y, Chen B, Yang Z, Wu J, Zhao Z, Chen J, Cong Y (2005c) Nucleolar localization of non-structural protein 3b, a protein specifically encoded by the severe acute respiratory syndrome coronavirus. Virus Res 114:70–79

Yuan X, Wu J, Shan Y, Yao Z, Dong B, Chen B, Zhao Z, Wang S, Chen J, Cong Y (2006) SARS coronavirus 7a protein blocks cell cycle progression at G0/G1 phase via the cyclin D3/pRb pathway. Virology 346:74–85

Yuan X, Yao Z, Wu J, Zhou Y, Shan Y, Dong B, Zhao Z, Hua P, Chen J, Cong Y (2007) G1 phase cell cycle arrest induced by SARS-CoV 3a protein via the cyclin D3/pRb pathway. Am J Respir Cell Mol Biol 37:9–19

Zeng R, Yang RF, Shi MD, Jiang MR, Xie YH, Ruan HQ, Jiang XS, Shi L, Zhou H, Zhang L, Wu XD, Lin Y, Ji YY, Xiong L, Jin Y, Dai EH, Wang XY, Si BY, Wang J, Wang HX, Wang CE, Gan YH, Li YC, Cao JT, Zuo JP, Shan SF, Xie E, Chen SH, Jiang ZQ, Zhang X, Wang Y, Pei G, Sun B, Wu JR (2004) Characterization of the 3a protein of SARS-associated coronavirus in infected vero E6 cells and SARS patients. J Mol Biol 341:271–279

Zhang QL, Ding YQ, He L, Wang W, Zhang JH, Wang HJ, Cai JJ, Geng J, Lu YD, Luo YL (2003) Detection of cell apoptosis in the pathological tissues of patients with SARS and its significance. Di Yi Jun Yi Da Xue Xue Bao 23:770–773

Zhang L, Wei L, Jiang D, Wang J, Cong X, Fei R (2007) SARS-CoV nucleocapsid protein induced apoptosis of COS-1 mediated by the mitochondrial pathway. Artif Cells Blood Substit Immobil Biotechnol 35:237–253

Zhao G, Shi SQ, Yang Y, Peng JP (2006) M and N proteins of SARS coronavirus induce apoptosis in HPF cells. Cell Biol Toxicol 22:313–322

Zhao X, Nicholls JM, Chen YG (2008) Severe acute respiratory syndrome-associated coronavirus nucleocapsid protein interacts with Smad3 and modulates transforming growth factor-beta signaling. J Biol Chem 283:3272–3280

Chapter 15
SARS Coronavirus and Lung Fibrosis

Wei Zuo, Xingang Zhao, and Ye-Guang Chen

Abstract Severe acute respiratory syndrome (SARS) is an acute infectious disease with significant mortality. A novel coronavirus (SARS-CoV) has been shown to be the causative agent of SARS. The typical clinical feature associated with SARS is diffuse alveolar damage in lung, and lung fibrosis is evident in patients who died from this disease. The mechanisms by which SARS-CoV infection causes lung fibrosis are not fully understood, but transforming growth factor-β (TGF-β) and angiotensin-converting enzyme 2 (ACE2)-mediated lung fibrosis are among the most documented ones. The activation of the TGF-β/Smad pathway is critical to lung fibrosis. SARS-CoV infection not only enhances the expression of TGF-β, but also facilitates its signaling activity. The SARS-CoV receptor ACE2 is a negative regulator of lung fibrosis, and SARS-CoV infection decreases ACE2 expression. Therefore, SARS-CoV infection may lead to lung fibrosis through multiple signaling pathways and TGF-β activation is one of the major contributors.

15.1 Introduction

Severe acute respiratory syndrome (SARS) is an acute infectious disease with significant morbidity and mortality (Wang and Chang 2004). An intense, cooperative worldwide effort led to the identification of the etiological agent as a novel SARS coronavirus (SARS-CoV) (Guan et al. 2003; Peiris et al. 2003; Rota et al. 2003) and the subsequent complete sequencing of the viral genome. SARS-CoV belongs to a family of large, positive, single-stranded RNA viruses. The SARS-CoV genome is 29.7 kb in length, and encodes 14 putative open reading frames

Y.-G. Chen (✉)
Department of Biological Sciences and Biotechnology, Tsinghua University, Beijing 100084, China
e-mail: ygchen@tsinghua.edu.cn

S.K. Lal (ed.), *Molecular Biology of the SARS-Coronavirus*,
DOI 10.1007/978-3-642-03683-5_15, © Springer-Verlag Berlin Heidelberg 2010

generating 28 potential proteins, the functions of many of which are not known (Marra et al. 2003). Sequence comparison with other known coronaviruses revealed a similar organization of *SARS-CoV* genes to typical coronaviruses (Chen et al. 2006).

Coronaviruses are known to cause up to 30% of common colds in humans but their infection leads only to lower respiratory tract diseases in livestock and poultry. However, SARS-CoV infection results in severe and even fatal lung disease in humans. At 1–2 weeks after the onset of the disease, the majority of patients can recover but one-third of patients develop severe symptoms. Many patients in the latter group deteriorate into acute respiratory distress syndrome (ARDS) with high mortality. During the convalescence, patients gradually shed virus but this process may take a long time. However, no chronic infection by SARS-CoV has been documented in humans. The virus can be found in respiratory tract secretions, lung tissue, serum, and stool (Tse et al. 2004).

According to the data from autopsy of fatal cases, the major pathological characteristic in SARS patients is diffuse alveolar damage (DAD) (Chan et al. 2003). The mechanism of DAD is believed to be endothelial and alveolar epithelial injury due to both direct viral effects and other indirect factors (Nicholls et al. 2003). During the early phase of SARS development (7–10 days), the lungs demonstrate the features of DAD including extensive edema, hyaline membrane formation, fluid and cellular exudation, collapse of alveoli, and desquamation of alveolar epithelial cells. At the same time, fibrous tissue could be detected in alveolar spaces. This phase is referred as the acute stage. In the medium phase of SARS development (10–14 days), the lungs display fibrous organization including interstitial and airspace fibrosis, reparative fibroblastic proliferation and type II pneumocytic hyperplasia. The fibrous organization becomes more extensive with time. This phase is referred as the proliferative organizing stage. In the late phase of SARS (2–3 weeks), referred as the fibrotic stage, the lungs exhibit dense septal and alveolar fibrosis. The time duration is counted from onset of the symptoms, and the pathological process does not always evolve through all the three stages, but may cease or recover at any phase (Chan et al. 2003; Cheung et al. 2004; Gu and Korteweg 2007; Guo et al. 2008; Ketai et al. 2006; Ng et al. 2006; Nicholls et al. 2003; Tse et al. 2004).

The extent of lung fibrosis is positively correlated with the duration of the SARS disease. Clinical data showed that fibrous organization is more common in patients in the late phase than in patients in the early or medium phases. In contrast, DAD more likely happens in patients in the early or medium phases than in patients in the late phase. Thus lung fibrosis can also display without DAD (Hwang et al. 2005). Importantly, lung fibrosis was even observed in SARS patients who had recovered and been discharged from hospital. Furthermore, the prevalence rate of lung fibrosis in cured SARS patients at 9 months after discharge is about 21% (42/200). The traditional therapy for lung fibrosis (glucocorticoid and hormone treatment) has no effect on the SARS-induced lung fibrosis regardless of the dosage, method, or length of drug usage. Interestingly, individuals who had lung fibrosis showed some spontaneous recovery (Xie et al. 2005).

15.2 SARS-CoV-Mediated Lung Fibrosis

Lung fibrosis is a pathological consequence of acute and chronic interstitial lung diseases. Lung fibrosis is characterized by failed alveolar re-epithelialization, fibroblast persistence, and excessive deposition of collagen and other extracellular matrix (ECM) components as well as destruction of the normal lung architecture (Sime and O'Reilly 2001). Progression of lung fibrosis results in the widening of interstitial matrix, eventual compression and destruction of normal lung parenchyma, and resultant damage to the capillaries leading to ventilatory insufficiency (Razzaque and Taguchi 2003). The etiology of lung fibrosis is uncertain: smoking, viral infection, drug exposure, and genetic predisposition may contribute to the fibrotic process. Chronic inflammation was regarded as the main reason for lung fibrosis and it can result in epithelial injury and fibroblast activation. However, recent studies have suggested that alveolar epithelial injury and formation of active myofibroblast foci are the main reasons for most lung fibrotic processes (Chapman 2004; Gharaee-Kermani et al. 2007; Kalluri and Neilson 2003; Razzaque and Taguchi 2003; Scotton and Chambers 2007).

Once damage takes place in lung tissue, a set of growth factors and cytokines, including monocyte chemoattractant protein-1 (MCP-1), transforming growth factor-β1 (TGF-β1), tumor necrosis factor-α (TNF-α), interleukin-1β (IL-1β), and interleukin-6 (IL-6), are overexpressed and released by the cells. Type II alveolar epithelial cells are one of the major sources of these fibrogenic factors. These factors in turn stimulate hyperproliferation of type II alveolar epithelial cells, recruit fibroblasts to the fibrotic loci, and induce transdifferentiation/activation of fibroblasts into myofibroblasts. The myofibroblasts are responsible for the excessive accumulation of ECM in the basement membranes and interstitial tissues, which finally leads to the loss of alveolar function, especially gas exchange between alveoli and capillaries (Razzaque and Taguchi 2003).

The excessive accumulation of ECM may result from increased synthesis or decreased degradation of ECM or both. The expression and deposition of collagen are the major events of ECM accumulation. Various types of collagens, such as type I, type III, and type VI, have been found in fibrotic lesions of the lungs. Many factors are involved in the ECM degradation, amongst which the plasminogen activator/plasmin system plays an important role. The activity of plasminogen is regulated by a physiologic inhibitor, plasminogen activator inhibitor 1 (PAI-1). In fibrotic lungs of patients, PAI-1 expression is increased. The deletion of the *PAI-1* gene can reduce the susceptibility of mouse to lung fibrosis induced by various stimuli whereas PAI-1 protein overexpression enhances it, indicating that PAI-1 is critical in the development of lung fibrosis (Izuhara et al. 2008). In addition to plasminogen, other proteolytic enzymes including matrix metalloproteinases (MMP) and a disintegrin and metalloproteinase domain (ADAM) also contribute to ECM degradation. The tissue inhibitors of metalloproteinases (TIMP) accelerate EMC deposition by neutralizing MMP activities (Gill and Parks 2008). In fibrotic lungs of patients, a wider distribution of TIMP compared with MMP has been described, suggesting that in lung fibrosis the total ECM degradation is reduced (Gill and Parks 2008; Izuhara et al. 2008; Liu 2008; Razzaque and Taguchi 2003).

15.3 TGF-β and SARS-Induced Lung Fibrosis

15.3.1 TGF-β Signal Transduction

TGF-β is a secreted cytokine which belongs to a superfamily of structure-related growth factors including TGF-β, activin, bone morphogenetic proteins (BMPs), and others. TGF-β signaling is initiated via its binding to the transmembrane receptors (Heldin et al. 1997; Hu et al. 2006; Massague and Chen 2000; Shi and Massague 2003). Ligand binding results in the formation of heterocomplexes between type I and type II receptors, in which the type II receptor phosphorylates and activates the type I receptor. The activated type I receptor then phoshorylates downstream signaling mediators – R-Smad proteins – which form a heteromeric complex with common Smad–Smad4 and are accumulated in the nucleus to orchestrate the expression of target genes in collaboration with other DNA-binding factors or transcription cofactors. In addition to the canonical Smad pathway, TGF-β has been reported to activate mitogen-activated protein kinases (ERK, p38 and JNK), phosphoinositide-3 kinase (PI3K)/Akt, p21-activated kinase (PAK), and others (Derynck and Akhurst 2007; Moustakas and Heldin 2005).

The TGF-β superfamily members play multifunctional regulatory roles in cell growth, differentiation, death, and migration and matrix remodeling (Derynck and Miyazono 2008; Hu et al. 2006; Massague and Chen 2000). Dysregulation of TGF-β signaling has been associated with a variety of human diseases, such as cancer, tissue fibrosis, and cardiovascular disorders (Derynck and Miyazono 2008; Massague and Chen 2000; ten Dijke and Arthur 2007).

15.3.2 TGF-β and Lung Fibrosis

Transgenic studies on an animal model reveal the importance of TGF-β activation in the process of lung fibrosis. Using replication-deficient adenovirus vectors to transfer the cDNA of TGF-β1 to rat lung, the effect of TGF-β1 protein in the respiratory tract was directly tested. Transient overexpression of active TGF-β1 resulted in severe interstitial and pleural fibrosis (Sime et al. 1997). Soluble TGF-β type II receptor consisting of the extracellular domain, which can bind to TGF-β and function as a TGF-β antagonist, has been shown to inhibit lung fibrosis in an animal model (Lee et al. 2001).

TGF-β activation is the hallmark event in lung fibrotic processes. In fibrotic lung tissue, the levels of TGF-β mRNA as well as TGF-β protein are increased

(Rube et al. 2000). TGF-β stimulates the production of ECM proteins, enhances the secretion of protease inhibitors (PAI-1 and TIMP), and reduces the secretion of proteases, thus leading to deposition of ECM proteins (Liu and Matsuura 2005). Blocking the activity of TGF-β or disrupting TGF-β signaling cascades in animal models could inhibit matrix production and prevent the fibrotic process (Liu et al. 2005; Rube et al. 2000; Willis et al. 2005).

TGF-β1 can also promote lung fibrosis through induction of myofibroblast expansion. As mentioned above, myofibroblasts are the key effector cells in lung fibrogenesis. Myofibroblasts localize to fibrotic foci and other sites of active fibrosis, and contribute to most ECM production. TGF-β can directly promote resident lung fibroblasts to differentiate into myofibroblasts. Furthermore, TGF-β induces the epithelial cells to undergo transition to form fibroblasts and hence myofibroblasts, a process termed epithelial–mesenchymal transition (EMT) (Derynck and Akhurst 2007). It has been shown that alveolar epithelial cells can undergo EMT in response to TGF-β both in vitro and in animal models (Kim et al. 2006; Willis et al. 2005). In biopsy samples, both epithelial and mesenchymal markers were found to be coexpressed in a cell, suggesting the existence of the EMT state in fibrotic lungs (Burkett et al. 2007; Kim et al. 2006; Willis et al. 2005; Wu et al. 2007). Thus, TGF-β can induce lung fibrosis through multiple mechanisms.

15.3.3 The Role of TGF-β in SARS-Induced Lung Fibrosis

Consistent with the promoting effect of TGF-β in tissue fibrosis, viral infection (such as HIV, HCV, and HBV infection) has been reported to upregulate TGF-β expression in various tissues (Lin et al. 2007; Ray et al. 2003). In the case of SARS-CoV infection, the serum level of TGF-β1 was elevated during the early phase of SARS (Pang et al. 2003). A high level of TGF-β was also observed in SARS-CoV-infected lung cells (including alveolar epithelial cells, bronchial epithelial cells, and monocytes/macrophages), but not in uninfected lung cells (Baas et al. 2006; He et al. 2006). The virus-induced high level of serum and in situ TGF-β ligand leads to hyperactivation of the TGF-β pathway, thus promoting the lung fibrosis. However, the mechanism by which SARS-CoV infection modulates the tissue level of TGF-β is unknown.

Importantly, SARS-CoV not only modulates TGF-β expression, but also directly regulates the signal transduction process of the TGF-β pathway through viral nucleocapsid (N) protein. The SARS-CoV N protein is a 46 kDa viral RNA-binding protein and shares little homology with the counterparts of other known coronaviruses (He et al. 2004). Our recent study demonstrated that N protein can associate with Smad3 and interferes with the formation of the complex between Smad3 and Smad4 (Zhao et al. 2008). However, N protein has no effect either on TGF-β-stimulated phosphorylation or on the nuclear accumulation of Smad3. Furthermore, it could promote the interaction between Smad3 and the transcriptional coactivator p300 to activate downstream genes. In lung epithelial and fibroblast cells, overexpressing N protein enhances the TGF-β-induced expression of PAI-1 and collagen I, and this enhancement is independent of Smad4. Interestingly, N protein attenuates Smad4-dependent apoptosis, which is in agreement with the fact that N protein was not found in the apoptotic cells (Zhao et al. 2008). Thus, SARS-CoV can promote the lung fibrosis process through its N protein.

15.4 ACE2/Angiotensin II and SARS-Induced Lung Fibrosis

The renin–angiotensin II system is an important regulator of blood pressure homeostasis. Angiotensin-converting enzyme (ACE) plays a key role in generating angiotensin II (ANG-II) from ANG-I which has no biological function. ACE2 is a recently identified homologue of ACE, which decreases ANG-II level by cleaving ANG-II. Therefore, ACE2 plays an opposite role to ACE in ANG-II generation. In respiratory tract, ACE2 is expressed on the surface of tracheobronchial and alveolar epithelium (Kuba et al. 2006a).

ACE2 has been reported to have a protective role in lung fibrosis. In lung biopsy specimens from patients with lung fibrosis, the ACE2 mRNA and enzyme activity are decreased by 92% ($p < 0.01$) and 74% ($p < 0.05$), respectively (Li et al. 2008). In a mouse model of bleomycin-induced lung fibrosis, intratracheal administration of ACE2-specific small interfering RNAs increased the accumulation of lung collagen, while recombinant human ACE2 decreased the accumulation of lung collagen (Imai et al. 2008; Kuba et al. 2006a, 2006b).

It is suggested that ACE2 protects against lung fibrosis through negative regulation of the local ANG-II level. In addition to its role in vasoconstriction regulation, ANG-II also contributes to many fibrotic diseases including lung, renal, hepatic, and cardiac fibrosis (Watanabe et al. 2005). ANG-II is mainly produced by fibroblasts, as well as activated macrophages. ANG-II signals via two receptors AGTR1 and AGTR2. ANG-II could stimulate the production and secretion of TGF-β cytokine in lung tissue, and this process may be mediated by AGTR1 (Molteni et al. 2007). As TGF-β itself can also regulate the level of ANG-II, it is believed that an "autocrine loop" between ANG-II and TGF-β exists in lung tissue. Therefore, both ACE inhibitor and AGTR antagonist can block lung fibrosis in experimental animals (Uhal et al. 2007). In a mouse model of bleomycin-induced lung fibrosis, the antagonist of angiotensin II receptor reduces the lung fibrosis score. This effect is accompanied with a reduction of the TGF-β level (Otsuka et al. 2004; Waseda et al. 2008).

In addition to regulating the TGF-β ligand level, ANG-II also increases the intracellular Smad2 and Smad4 protein levels and facilitates nuclear translocation of phosphorylated Smad2 (Hao et al. 2000). It has been recently reported that ANG-II activates Smad signaling in a TGF-β-independent and MAPK-dependent manner (Su et al. 2006). ANG-II also induces the EMT process. This effect can be blocked by a TGF-β antagonist Smad7 (Rodriguez-Vita et al. 2004). Altogether, ANG-II enhances TGF-β/Smad signaling and the interplay between ANG-II and TGF-β may promote lung fibrosis.

In addition to enhancing TGF-β expression, ANG-II has been reported to upregulate connective tissue growth factor (CTGF) (Rodriguez-Vita et al. 2005). CTGF may promote deposition of ECM and lung fibrosis via the MEK/Erk pathway (Ponticos et al. 2003). ANG-II can also directly upregulate the level of α-SMA, a key protein in pathogenesis of lung fibrosis (Molteni et al. 2007), as well as production of ECM (Rodriguez-Vita et al. 2005).

Several studies have verified ACE2 as an essential receptor for SARS-CoV infection . The spike protein of SARS-CoV could bind to ACE2 on the cell surface, and this binding promotes viral entry into the cell and induces cell–cell fusion into multinucleated cells (syncytia formation) (Li et al. 2003). ACE2 proteins are expressed by alveolar epithelial cells, the primary targets of SARS-CoV in lung. Thus, ACE2 contributes to SARS disease both as protector of lung fibrosis and as the cellular receptor for viral infection (Imai et al. 2005). Interestingly, SARS-CoV infection and the spike protein of SARS-CoV reduce ACE2 expression level during viral infection (Kuba et al. 2005). As discussed above, the decreased ACE2 expression results in an increased ANG-II level and leads to lung fibrosis and lung failure. Accordingly, the in-vivo administration of the spike protein did not affect the severity of lung failure in *Ace2* knockout mice (Kuba et al. 2005). Furthermore, administration of spike protein to mice led to a significant increase in ANG-II level in the lung tissue. Inhibition of ANG-II receptor can rescue the effect of the spike protein (Imai et al. 2007; Kuba et al. 2005). Altogether, infection with SARS-CoV results in ACE2 downregulation through the binding of SARS-CoV spike protein to ACE2, and this spike protein-mediated ACE2 downregulation is responsible for the pathogenesis of lung fibrosis by upregulating ANG-II and activating TGF-β signaling.

15.5 Other Mechanisms of SARS-Mediated Lung Fibrosis

Besides TGF-β and ACE2, other mechanisms may contribute to SARS-CoV-mediated lung fibrosis. MCP-1 is a chemokine that promote lung fibrosis. In patients with lung fibrosis, the MCP-1 level is usually upregulated (Emad and Emad 2007; Wynn 2008). SARS-CoV infection upregulates the MCP-1 level in patients. The elevation of the MCP-1 level happens at least 2 weeks after disease onset. Corticosteroid treatment reduces MCP-1 concentrations at 5–8 days after treatment (Wong et al. 2004). MAPK p38 is a signal transducer that responds to extracellular stimulation including viral infection, and in turn regulates cell differentiation and ECM production. Hyperactivation of p38 was detected in patients of lung fibrosis, and phosphorylation of p38 is responsible for pulmonary myofibroblast activation and α-SMA expression (Hu et al. 2006; Yoshida et al. 2002). SARS-CoV infection has been reported to induce p38 phosphorylation in lung epithelial and other cells, which lead to actin reorganization (Mizutani et al. 2006; Surjit et al. 2004). Thus, p38 may be involved in SARS-CoV-mediated lung fibrosis.

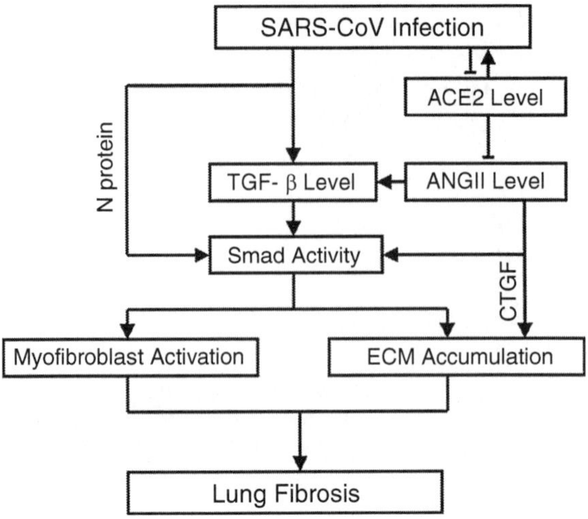

Fig. 15.1 The possible molecular mechanisms of SARS-CoV-mediated lung fibrosis. SARS-CoV infection upregulates TGF-β expression while downregulating the level of the SARS-CoV receptor ACE2. The decreased ACE2 expression leads to a high ANG-II level, which further enhances the TGF-β level. TGF-β activates Smad proteins via receptor-mediated phosphorylation and stimulates Smad-dependent gene transcription, which is facilitated by N protein of SARS-CoV and ANG-II. Both the TGF-β/Smad and the ACE2/ANG-II/CTGF pathways contribute to myofibroblast activation and ECM accumulation, collectively leading to the final lung fibrosis

15.6 Conclusion

Fibrosis usually brings irreversible damage to the lung. Lung fibrosis is widely observed in patients who died from SARS. However, the mechanisms by which SARS-CoV infection leads to lung fibrosis remain poorly understood. We have discussed two important mechanisms for SARS-mediated fibrosis: activation of TGF-β signaling and degradation of ACE2 (Fig. 15.1). A major challenge will be how to use what we have learned to prevent SARS-CoV-evoked lung fibrosis.

Acknowledgments The research in Y.G.C's laboratory is supported by grants from the National Natural Science Foundation of China (30430360, 30671033), 973 Program (2004CB720002, 2006CB943401, 2006CB910102), and 863 Program (2006AA02Z172).

References

Baas T, Taubenberger JK, Chong PY, Chui P, Katze MG (2006) SARS-CoV virus-host interactions and comparative etiologies of acute respiratory distress syndrome as determined by transcriptional and cytokine profiling of formalin-fixed paraffin-embedded tissues. J Interferon Cytokine Res 26:309–317

Burkett PR, Noth I, Husain AN (2007) Evidence of epithelial/mesenchymal transition in idiopathic pulmonary fibrosis. Lab Invest 87:319a

Chan KS, Zheng JP, Mok YW, Li YM, Liu YN, Chu CM, Ip MS (2003) SARS: prognosis, outcome and sequelae. Respirology 8:S36–S40

Chapman HA (2004) Disorders of lung matrix remodeling. J Clin Invest 113:148–157

Chen SA, Luo HB, Chen LL, Chen J, Shen JH, Zhu WL, Chen KX, Shen X, Jiang HL (2006) An overall picture of SARS coronavirus (SARS-CoV) genome-encoded major proteins: Structures, functions and drug development. Curr Pharm Des 12:4539–4553

Cheung OY, Chan JWM, Ng CK, Koo CK (2004) The spectrum of pathological changes in severe acute respiratory syndrome (SARS). Histopathology 45:119–124

Derynck R, Akhurst RJ (2007) Differentiation plasticity regulated by TGF-beta family proteins in development and disease. Nat Cell Biol 9:1000–1004

Derynck R, Miyazono K (2008) The TGF-beta family. Cold Spring Harbor Laboratory, Cold Spring Harbor

Emad A, Emad V (2007) Elevated levels of MCP-1, MIP-alpha and MIP-1 beta in the bronchoalveolar lavage (BAL) fluid of patients with mustard gas-induced pulmonary fibrosis. Toxicology 240:60–69

Gharaee-Kermani M, Gyetko MR, Hu B, Phan SH (2007) New insights into the pathogenesis and treatment of idiopathic pulmonary fibrosis: a potential role for stem cells in the lung parenchyma and implications for therapy. Pharm Res 24:819–841

Gill SE, Parks WC (2008) Metalloproteinases and their inhibitors: regulators of wound healing. Int J Biochem Cell Biol 40:1334–1347

Gu J, Korteweg C (2007) Pathology and pathogenesis of severe acute respiratory syndrome. Am J Pathol 170:1136–1147

Guan Y, Zheng BJ, He YQ, Liu XL, Zhuang ZX, Cheung CL, Luo SW, Li PH, Zhang LJ, Guan YJ et al (2003) Isolation and characterization of viruses related to the SARS coronavirus from animals in southern China. Science 302:276–278

Guo Y, Korteweg C, McNutt MA, Gu J (2008) Pathogenetic mechanisms of severe acute respiratory syndrome. Virus Res 133:4–12

Hao JM, Wang BQ, Jones SC, Jassal DS, Dixon IMC (2000) Interaction between angiotensin II and Smad proteins in fibroblasts in failing heart and in vitro. Am J Physiol Heart Circ Physiol 279:H3020–H3030

He RT, Dobie F, Ballantine M, Leeson A, Li Y, Bastien N, Cutts T, Andonov A, Cao JX, Booth TF et al (2004) Analysis of multimerization of the SARS coronavirus nucleocapsid protein. Biochem Biophys Res Commun 316:476–483

He L, Ding Y, Zhang Q, Che X, He Y, Shen H, Wang H, Li Z, Zhao L, Geng J et al (2006) Expression of elevated levels of pro-inflammatory cytokines in SARS-CoV-infected ACE2(+) cells in SARS patients: relation to the acute lung injury and pathogenesis of SARS. J Pathol 210:288–297

Heldin CH, Miyazono K, ten Dijke P (1997) TGF-beta signalling from cell membrane to nucleus through SMAD proteins. Nature 390:465–471

Hu YB, Peng JW, Feng DY, Chu L, Li XA, Jin ZY, Lin Z, Zeng QF (2006) Role of extracellular signal-regulated kinase, p38 kinase, and activator protein-1 in transforming growth factor-beta 1-induced alpha smooth muscle actin expression in human fetal lung fibroblasts in vitro. Lung 184:33–42

Hwang DM, Chamberlain DW, Poutanen SM, Low DE, Asa SL, Butany J (2005) Pulmonary pathology of severe acute respiratory syndrome in Toronto. Mod Pathol 18:1–10

Imai Y, Kuba K, Rao S, Huan Y, Guo F, Guan B, Yang P, Sarao R, Wada T, Leong-Poi H et al (2005) Angiotensin-converting enzyme 2 protects from severe acute lung failure. Nature 436:112–116

Imai Y, Kuba K, Penninger JM (2007) Angiotensin-converting enzyme 2 in acute respiratory distress syndrome. Cell Mol Life Sci 64:2006–2012

Imai Y, Kuba K, Penninger JM (2008) The discovery of angiotensin-converting enzyme 2 and its role in acute lung injury in mice. Exp Physiol 93:543–548

Izuhara Y, Takahashi S, Nangaku M, Takizawa SY, Ishida H, Kurokawa K, de Strihou CV, Hirayama N, Miyata T (2008) Inhibition of plasminogen activator inhibitor-1: its mechanism and effectiveness on coagulation and fibrosis. Arteriosclerosis Thrombosis Vasc Biol 28: 672–677

Kalluri R, Neilson EG (2003) Epithelial-mesenchymal transition and its implications for fibrosis. J Clin Invest 112:1776–1784

Ketai L, Paul NS, Wong KTT (2006) Radiology of severe acute respiratory syndrome (SARS) – The emerging pothologic-radiologic correlates of an emerging disease. J Thoracic Imag 21: 276–283

Kim KK, Kugler MC, Wolters PJ, Robillard L, Galvez MG, Brumwell AN, Sheppard D, Chapman HA (2006) Alveolar epithelial cell mesenchymal transition develops in vivo during pulmonary fibrosis and is regulated by the extracellular matrix. Proc Natl Acad Sci U S A 103: 13180–13185

Kuba K, Imai Y, Rao SA, Gao H, Guo F, Guan B, Huan Y, Yang P, Zhang YL, Deng W et al (2005) A crucial role of angiotensin converting enzyme 2 (ACE2) in SARS coronavirus-induced lung injury. Nat Med 11:875–879

Kuba K, Imai Y, Penninger JM (2006a) Angiotensin-converting enzyme 2 in lung diseases. Curr Opin Pharmacol 6:271–276

Kuba K, Imai Y, Rao S, Jiang CY, Penninger JM (2006b) Lessons from SARS: control of acute lung failure by the SARS receptor ACE2. J Mol Med 84:814–820

Lee CG, Homer R, Zhou Z, Lanone Z, Wang XM, Koteliansky V, Shipley JM, Gotwals P, Noble P, Chen QS et al (2001) Interleukin-13 induces tissue fibrosis by selectively stimulating and activating transforming growth factor beta(1). J Exp Med 194:809–821

Liu F, Matsuura I (2005) Inhibition of Smad antiproliferative function by CDK phosphorylation. Cell Cycle 4:63–66

Li WH, Moore MJ, Vasilieva N, Sui JH, Wong SK, Berne MA, Somasundaran M, Sullivan JL, Luzuriaga K, Greenough TC et al (2003) Angiotensin-converting enzyme 2 is a functional receptor for the SARS coronavirus. Nature 426:450–454

Li XP, Molina-Molina M, Abdul-Hafez A, Uhal V, Xaubet A, Uhal BD (2008) Angiotensin converting enzyme-2 is protective but downregulated in human and experimental lung fibrosis. Am J Physiol Lung Cell Mol Physiol 295:L178–L185

Lin WY, Weinberg E, Kim KA, Peng LF, Kim SS, Brockman M, Lopez-Marra H, De Sa Borges CB, Hyppolite G, Shao RX, Chung RT (2007) Hiv and GP120 enhance HCV replication and upregulate TGF-beta 1. Hepatology 46:447a

Liu RM (2008) Oxidative stress, plasminogen activator inhibitor 1, and lung fibrosis. Antioxidants Redox Signal 10:303–319

Liu RM, Vayalil P, Wang SQ, Ballinger C, Gauldie J, Postlethwait E (2005) TGF-beta1 induces concomitant lung fibrosis, decreased GSH and ascorbate concentrations, and increased PAI-1 gene expression. Free Radical Biol Med 39:S37

Marra MA, Jones SJ, Astell CR, Holt RA, Brooks-Wilson A, Butterfield YS, Khattra J, Asano JK, Barber SA, Chan SY et al (2003) The Genome sequence of the SARS-associated coronavirus. Science 300:1399–1404

Massague J, Chen YG (2000) Controlling TGF-beta signaling. Genes Dev 14:627–644

Mizutani T, Fukushi S, Ishii K, Sasaki Y, Kenri T, Saijo M, Kanaji Y, Shirota K, Kurane I, Morikawa S (2006) Mechanisms of establishment of persistent SARS-CoV-infected cells. Biochem Biophys Res Commun 347:261–265

Molteni A, Wolfe LF, Ward WF, Ts'ao CH, Molteni LB, Veno P, Fish BL, Taylor JM, Quintanilla N, Herndon B, Moulder JE (2007) Effect of an angiotensin II receptor blocker and two angiotensin converting enzyme inhibitors on transforming growth factor-beta (TGF-beta) and alpha-actomyosin (alpha SMA), important mediators of radiation-induced pneumopathy and lung fibrosis. Curr Pharm Des 13:1307–1316

Moustakas A, Heldin CH (2005) Non-Smad TGF-beta signals. J Cell Sci 118:3573–3584

Ng WF, To KF, Lam WWL, Ng TK, Lee KC (2006) The comparative pathology of severe acute respiratory syndrome and avian influenza A subtype H5N1 – a review. Hum Pathol 37:381–390

Nicholls J, Dong XP, Jiang G, Peiris M (2003) SARS: clinical virology and pathogenesis. Respirology 8:S6–S8

Otsuka M, Takahashi H, Shiratori M, Chiba H, Abe S (2004) Reduction of bleomycin induced lung fibrosis by candesartan cilexetil, an angiotension II type 1 receptor antagonist. Thorax 59: 31–37

Pang BS, Wang Z, Zhang LM, Tong ZH, Xu LL, Huang XX, Guo WJ, Zhu M, Wang C, Li XW et al (2003) Dynamic changes in blood cytokine levels as clinical indicators in severe acute respiratory syndrome. Chin Med J 116:1283–1287

Peiris JSM, Lai ST, Poon LLM, Guan Y, Yam LYC, Lim W, Nicholls J, Yee WKS, Yan WW, Cheung MT et al (2003) Coronavirus as a possible cause of severe acute respiratory syndrome. Lancet 361:1319–1325

Ponticos M, Holmes AM, Rajkumar V, Wen XS, Leask A, Abraham DJ, Black CM (2003) Connective tissue growth factor (CTGF) mediates extracellular matrix deposition in bleomycin-induced lung fibrosis via the MEK/ERK signaling pathways. Arthritis Rheum 48:S156

Ray S, Broor SL, Vaishnav Y, Sarkar C, Girish R, Dar L, Seth P, Broor S (2003) Transforming growth factor beta in hepatitis C virus infection: in vivo and in vitro findings. J Gastroenterol Hepatol 18:393–403

Razzaque MS, Taguchi T (2003) Pulmonary fibrosis: Cellular and molecular events. Pathol Int 53:133–145

Rodriguez-Vita J, Sanchez-Lopez E, Esteban V, Ruperez M, Lopez A, Egido J, Ruiz-Ortega M (2004) Angiotensin II activates the Smad signalling pathway: Role in vascular fibrosis. J Hypertens 22:S351–S352

Rodriguez-Vita J, Sanchez-Lopez E, Esteban V, Ruperez M, Egido J, Ruiz-Ortega M (2005) Angiotensin II activates the Smad pathway in vascular smooth muscle cells by a transforming growth factor-beta-independent mechanism. Circulation 111:2509–2517

Rota PA, Oberste MS, Monroe SS, Nix WA, Campagnoli R, Icenogle JP, Penaranda S, Bankamp B, Maher K, Chen MH et al (2003) Characterization of a novel coronavirus associated with severe acute respiratory syndrome. Science 300:1394–1399

Rube CE, Uthe D, Schmid KW, Richter KD, Wessel J, Schuck A, Willich N, Rube C (2000) Dose-dependent induction of transforming growth factor beta (TGF-beta) in the lung tissue of fibrosis-prone mice after thoracic irradiation. Int J Radiat Oncol Biol Phys 47:1033–1042

Scotton CJ, Chambers RC (2007) Molecular targets in pulmonary fibrosis – the myofibroblast in focus. Chest 132:1311–1321

Shi Y, Massague J (2003) Mechanisms of TGF-beta signaling from cell membrane to the nucleus. Cell 113:685–700

Sime PJ, O'Reilly KMA (2001) Fibrosis of the lung and other tissues: new concepts in pathogenesis and treatment. Clin Immunol 99:308–319

Sime PJ, Xing Z, Graham FL, Csaky KG, Gauldie J (1997) Adenovector-mediated gene transfer of active transforming growth factor-beta 1 induces prolonged severe fibrosis in rat lung. J Clin Invest 100:768–776

Su Z, Zimpelmann J, Burns KD (2006) Angiotensin-(1–7) inhibits angiotensin II-stimulated phosphorylation of MAP kinases in proximal tubular cells. Kidney Int 69:2212–2218

Surjit M, Liu BP, Jameel S, Chow VTK, Lal SK (2004) The SARS coronavirus nucleocapsid protein induces actin reorganization and apoptosis in COS-1 cells in the absence of growth factors. Biochem J 383:13–18

ten Dijke P, Arthur HM (2007) Extracellular control of TGFbeta signalling in vascular development and disease. Nat Rev Mol Cell Biol 8:857–869

Tse GMK, To KF, Chan PKS, Lo AWI, Ng KC, Wu A, Lee N, Wong HC, Mak SM, Chan KF et al (2004) Pulmonary pathological features in coronavirus associated severe acute respiratory syndrome (SARS). J Clin Pathol 57:260–265

Uhal BD, Kim JK, Li XP, Molina-Molina M (2007) Angiotensin-TGF-beta 1 crosstalk in human idiopathic pulmonary fibrosis: autocrine mechanisms in myofibroblasts and macrophages. Curr Pharm Des 13:1247–1256

Wang JT, Chang SC (2004) Severe acute respiratory syndrome. Curr Opin Infect Dis 17:143–148

Waseda Y, Yasui M, Nishizawa Y, Inuzuka K, Takato H, Ichikawa Y, Tagami A, Fujimura M, Nakao S (2008) Angiotensin II type 2 receptor antagonist reduces bleomycin-induced pulmonary fibrosis in mice. Respir Res 9:43

Watanabe T, Barker TA, Berk BC (2005) Angiotensin II and the endothelium – diverse signals and effects. Hypertension 45:163–169

Willis BC, Liebler JM, Luby-Phelps K, Nicholson AG, Crandall ED, du Bois RM, Borok Z (2005) Induction of epithelial-mesenchymal transition in alveolar epithelial cells by transforming growth factor-ss 1 – potential role in idiopathic pulmonary fibrosis. Am J Pathol 166:1321–1332

Wong CK, Lam CWK, Wu AKL, Ip WK, Lee NLS, Chan IHS, Lit LCW, Hui DSC, Chan MHM, Chung SSC, Sung JJY (2004) Plasma inflammatory cytokines and chemokines in severe acute respiratory syndrome. Clin Exp Immunol 136:95–103

Wu Z, Yang LL, Cai L, Zhang M, Cheng X, Yang X, Xu J (2007) Detection of epithelial to mesenchymal transition in airways of a bleomycin induced pulmonary fibrosis model derived from an alpha-smooth muscle actin-Cre transgenic mouse. Respir Res 8:1

Wynn TA (2008) Cellular and molecular mechanisms of fibrosis. J Pathol 214:199–210

Xie LX, Liu YN, Fan BX, Xiao YY, Tian Q, Chen LG, Zhao H, Chen WJ (2005) Dynamic changes of serum SARS-Coronavirus IgG, pulmonary function and radiography in patients recovering from SARS after hospital discharge. Respir Res 6:5

Yoshida K, Kuwano K, Hagimoto N, Watanabe K, Matsuba T, Fujita M, Inoshima I, Hara N (2002) MAP kinase activation and apoptosis in lung tissues from patients with idiopathic pulmonary fibrosis. J Pathol 198:388–396

Zhao X, Nicholls JM, Chen YG (2008) Severe acute respiratory syndrome-associated coronavirus nucleocapsid protein interacts with Smad3 and modulates transforming growth factor-beta signaling. J Biol Chem 283:3272–3280

Chapter 16
Host Immune Responses to SARS Coronavirus in Humans

Chris Ka-fai Li and Xiaoning Xu

Abstract The severe acute respiratory syndrome (SARS) is a newly identified infectious disease caused by a novel zoonotic coronavirus (SARS-CoV) with unknown animal reservoirs. The risk of SARS reemergence in humans remains high due to the large animal reservoirs of SARS-CoV-like coronavirus and the genome instability of RNA coronaviruses. An epidemic in 2003 affected more than 8,000 patients in 29 countries, with 10% mortality.

SARS infection is transmitted by air droplets. Clinical and laboratory manifestations include fever, chills, rigor, myalgia, malaise, diarrhea, cough, dyspnoea, pneumonia, lymphopenia, neutrophilia, thrombocytopenia, and elevated serum lactate dehydrogenase, alanine aminotransferase, and creatine kinase activities. Health care workers are a high-risk group, and advanced age is strongly associated with disease severity. Treatment has been empirical, and there is no licensed SARS vaccine for humans so far. However, presence of long-lived neutralizing antibodies and memory T- and B-lymphocytes in convalescent SARS patients raises hope for active immunization. Furthermore, results from preclinical SARS vaccines expressing spike protein to elicit neutralizing antibodies and cellular responses that are protective in mouse and nonhuman primate models are encouraging.

Very little is known of the early events in viral clearance and the onset of innate and inflammatory responses during the SARS infection. Regulation of the innate immune response is associated with the development of adaptive immunity and disease severity in SARS infection. Notably, SARS-CoV has evolved evasive strategies to suppress antiviral type I interferon responses in infected cells. In addition, inflammatory responses are characterized by upregulation of proinflammatory cytokines/chemokines such as IL-6, IP-10, and MCP-1 in tissues and serum, and massive infiltrations of inflammatory cells such as macrophages in infected tissues.

C. Ka-fai Li (✉)
MRC Human Immunology Unit, The Weatherhall Institute of Molecular Medicine, John Radcliffe Hospital, University of Oxford, Oxford, United Kingdom OX3 9DS
e-mail: chris.li@imm.ox.ac.uk

S.K. Lal (ed.), *Molecular Biology of the SARS-Coronavirus*,
DOI 10.1007/978-3-642-03683-5_16,

Due to the lack of animal models that mimic the clinical manifestations of human SARS infection for mechanistic study and vaccine evaluation, development of a safe prophylactic SARS vaccine for human use remains a huge challenge. This chapter is written to summarize and highlight the latest clinical, serological, and immunological parameters relevant to the pathogenesis and protective immunity of SARS infection in humans.

16.1 Introduction

The severe acute respiratory syndrome (SARS) is a newly identified infectious disease caused by a novel zoonotic coronavirus (SARS-CoV) with unknown animal reservoirs. SARS emerged first in the Guangdong province of China in late 2002 and had spread to 29 countries by July 2003. During the 2002 and 2003 outbreak, more than 8,000 cases with 9.6% mortality were reported from 29 countries (WHO, http//www.who.int.csr/sars/country/table2003_09_23/en/). Health care workers are the most vulnerable group, and advanced age (>60) is strongly associated with disease severity (reviewed in Cheng et al. 2007; Zhao 2007). SARS is transmitted by air droplets, and transmission efficiency is associated with severely ill patients and those with rapid clinical deterioration. The mean incubation period is 4.6 days with a variance of 15.9 days, and the infectious period is 7 days after illness onset. Clinical diagnosis of SARS is made by X-ray and CT radiography, and rapid definitive laboratory diagnosis is mainly by virus isolation and RT-PCR from respiratory and/or stool specimens.

The clinical spectrum of the outcome of SARS infection is highly variable, from mild flu-like symptoms to severe pneumonia (Tsang et al. 2003; Lee et al. 2003; Poutanen et al. 2003; Drosten et al. 2003; Ksiazek et al. 2003; Peiris et al. 2003). Typically, the disease is manifested by high fever (>38°C), chills, rigor, myalgia, malaise, diarrhea, cough, dyspnea, pneumonia, and rapidly progressing radiographic changes. Upper respiratory tract symptoms are not prominent, but watery diarrhea is common in most patients. About 10–15% of patients fail to respond to treatment and may progress to acute respiratory distress syndrome (ARDS), which is the major cause of death among fatal SARS cases.

The respiratory system, especially the lower lung, is the major site of SARS-CoV replication, although the virus can also be found in urine and fecal samples from SARS patients (Peiris et al. 2003). In autopsy samples, SARS-CoV can be detected in intestine, liver, kidney, brain, spleen, and lymph nodes, as well as lung samples, by immunohistochemical and in-situ-hybridization techniques (Nicholls et al. 2003, 2006; To et al. 2004; Gu et al. 2005). In the lungs, diffuse alveolar damage (DAD) with fibrosis is the prominent pathological feature of SARS infection, suggestive of active tissue injury and repair. Other pathological changes include massive infiltrations of inflammatory cells, predominantly macrophages, and enlargement of pneumocytes in the lungs. Lymphoid depletion is also common in the spleen and other lymphoid organs.

SARS-CoV is the largest RNA virus and is classified in group 2 of the Coronavirus family (Nicholls et al. 2006; Marra et al. 2003; Stadler et al. 2003). The virion is roughly 90–120 nm in diameter. It is an enveloped positive-strand RNA virus and contains a lipid bilayer surrounding a helical nucleocapsid structure that protects the genome. The genome size is about 30 kb in length and encodes 15 open reading frames (Orfs). Major Orfs encode four structural proteins: spike (S) protein (180/90 kDa), envelope (E) protein (8 kDa), membrane (M) protein (23 kDa), and nucleocapsid (N) protein (50–60 kDa). Replicase proteins encoded in the 5′-most two-thirds of the SARS-CoV genome are essential for polyprotein processing, replicase complex formation, and efficient virus replication. M protein and other Orfs, such as Orf 3a, Orf 3b, Orf 6 and Orf 7a, are important in apoptosis and antagonism of type I interferon responses in infected cells (Chan et al. 2007; Law et al. 2005b; Tan et al. 2004; Freundt et al. 2009). N protein is essential in the formation of the helical nucleocapsid during virion assembly. Recently, N protein was found to modulate TGF-β signaling to block apoptosis of SARS-CoV-infected host cells, with important implications in tissue fibrosis (Zhao et al. 2008).

Entry of SARS-CoV into the host target cell is mediated by surface spike protein, which binds and fuses with the host cell. The mechanism of membrane fusion of SARS-CoV is very similar to that of class I fusion proteins, such as HIV gp160 and influenza hemagglutinin (reviewed in Kielian and Rey (2006)). The N-terminal half of the S protein (S1) contains the receptor-binding domain (RBD), and the C-terminal half (S2) is the membrane-anchored subunit that encodes a putative fusion peptide and two heptad repeat regions (HR1 and HR2) (Liu et al. 2004; Bosch et al. 2004). Entry of SARS-CoV into the host target cell involves receptor binding and a conformational change in the S protein, followed by cathepsin L-mediated proteolysis within endosomes (Simmons et al. 2005; Tripet et al. 2004; Bosch et al. 2003; Hofmann and Pohlmann 2004).

The functional receptor for the spike protein of SARS-CoV is a metallopeptidase, angiotensin-converting enzyme 2 (ACE2) (Li et al. 2003). It is expressed on human lung alveolar epithelial cells (type I and type II pneumocytes), enterocytes of the small intestine, brush borders of the proximal tubular cells of the kidney, endothelial cells of arteries and veins, and arterial smooth muscle cells in several organs (Hamming et al. 2004). SARS-CoV infection of ACE2-expressing cells is dependent on the activity of the proteolytic enzyme cathepsin L, which is pH-sensitive (Simmons et al. 2005; Huang et al. 2006). Hence, tissue tropism and viral replication of SARS-CoV in these organs could be explained by the distribution of the ACE2 receptor and cathepsin L activity in tissues. In addition, SARS infection could be enhanced by C-type lectins such as DC-SIGN and L-SIGN (Yang et al. 2004a; Jeffers et al. 2004). It is interesting to note that another newly identified human coronavirus, NL63, which only causes minor colds, also uses ACE2 as a receptor, and the infection is independent of cathepsin L (Hofmann et al. 2005; Huang et al. 2006). The protective function of ACE2 in acute lung injury is regulated by the balance of angiotensin and angiotensin-converting enzyme (ACE) in the renin–angiotensin pathway. Tissue injury caused by angiotensin II could be negatively regulated by ACE2 (Imai et al. 2005). In an animal model of acute lung injury, acute lung failure could be induced by injection of the SARS S protein, which downregulates ACE2; moreover, such effect could be attenuated by blocking the renin–angiotensin pathway (Kuba et al. 2005).

Although a huge global public health initiative by the WHO successfully contained the SARS outbreak in 2003, the risk of reemergence of SARS in humans remains high due to the large number of animal reservoirs that harbor SARS-CoV-like coronavirus. Phylogenetic study has shown that SARS-CoV is a zoonotic virus that crossed the species barrier and evolved in palm civets and humans, with the observation that the SARS-CoV became increasingly virulent following human-to-human transmission (reviewed in Zhao (2007)). However, failure to isolate SARS-CoV from wild or farmed civets from nonepidemic areas argued against civets being the natural reservoir of the virus (Shi and Hu 2008). Recent sequence analysis of SARS-CoV-like viruses from other wild animals suggests that wild bats are the natural reservoir of SARS-CoV (Lau et al. 2005; Li et al. 2005). If this is the case, it would be difficult to control the further spread of this virus to the human population, as bats are asymptomatic carriers of many human pathogens, such as Hendra and Nipha paramyxoviruses (Murray et al. 1995; Chua et al. 2000). Given the highly infectious properties and complex public health impact of SARS infection, there is an urgent need to develop effective treatments and prophylactic vaccines to control and prevent any future SARS outbreak.

So far, there has been no consensus regarding whether any treatment, especially the use of steroids and convalescent plasma therapy, could benefit SARS patients (Stockman et al. 2006). Moreover, the role played by the host immunity against SARS-CoV in viral clearance or tissue damage during the disease progression is not clear. High initial viral load is shown to be independently associated with severity of the disease and may be influenced by host immune responses (Chu et al. 2004). However, recent studies have suggested that type I IFN plays a key role in the switch from innate to adaptive immunity during the acute phase of SARS; indeed, patients with poor outcomes showed type I IFN-mediated immunopathological events and deficient adaptive immune responses (Cameron et al. 2007).

Several studies have shown that most recovered SARS patients have higher and sustainable antibody responses compared to those observed in fatal cases (Temperton et al. 2005; Ho et al. 2005). Taken together, this suggests that antibody responses are likely to play an important role in determining the ultimate disease outcome of SARS infection. Several forms of possible vaccines, such as attenuated or inactivated SARS-CoV, DNA, and viral vector-based vaccines have been evaluated in a number of animal models, including nonhuman primates (Roberts et al. 2008). Neutralizing antibodies to S protein are the major components of protective immunity (Yang et al. 2004b; Zhu et al. 2007). However, these animal models, including nonhuman primates, lack the severe clinical features observed in humans (Hogan 2006). Hence, it is difficult to evaluate whether these vaccines will prevent the disease in humans. Progress is further complicated by concerns over the safety of coronavirus vaccines; indeed, immune-mediated enhancement of pathology has been reported in other animal coronavirus infections (reviewed in Perlman and Dandekar (2005)) as well as in animals vaccinated with modified vaccine virus encoding SARS-CoV S protein (Weingartl et al. 2004). In addition, some variants of SARS-CoV are resistant to antibody neutralization, and the infection is enhanced by the antibodies (Yang et al. 2005b; Kam et al. 2007). As the inflammatory response is strongly implicated in the pathogenesis of SARS, a full understanding of the mechanism of protective immunity against SARS-CoV in humans is critical in the development of a safe prophylactic vaccine for use in humans.

16.2 Immune Responses

Host responses to viral infection include both immune activation and programmed cell death. Resistance to many viral infections is predominantly mediated by adaptive immunity comprising both humoral and cell-mediated immunity (reviewed in Zinkernagel and Hengartner (2004)). The antiviral effect of humoral immunity is mediated by the generation of neutralizing antibodies capable of blocking virus entry/infection of the target cells. The effector components of cell-mediated immunity, CD4 and CD8 T cells, mediate their antiviral effect through the secretion of cytokines/effector molecules and the killing of virus-infected target cells. Recently, there has been increasing evidence that the development of the adaptive immunity is governed by the initial interaction between invading microorganisms and the innate immune system. SARS-CoV virus has evolved a way to evade innate antiviral type I interferon responses of host cells in order to prolong viral replication and survival (reviewed in Frieman and Baric (2008)). Recent studies have also suggested that dysregulated type I and II interferon responses may culminate in a failure of the switch from hyperinnate immunity to protective adaptive immune responses in SARS patients with a poor outcome (Cameron et al. 2007).

16.3 Innate Immunity

Proteomic analysis of plasma from SARS patients shows activation of innate immune responses by SARS-CoV, including increased acute phase proteins such as serum amyloid A and MBL (Chen et al. 2004). MBL could bind to SARS-CoV and inhibit virus infectivity in vitro by blocking the carbohydrate-recognition domains of the S protein (Ip et al. 2005). Moreover, low MBL serum levels and haplotypes associated with MBL deficiency have been observed in SARS patients. Reduction of NK ($CD3^-CD56^+CD16^+$) cells and $CD158b^+$ NK cells is detected during the course of SARS infection, and such reduction is associated with disease severity (Anonymous 2004; He et al. 2005). However, the significance of these changes in SARS pathogenesis is not clear. In a C57BL/6 mouse model of SARS infection, NK and NKT cells are not required for viral clearance in beige and $CD1^{-/-}$ mice (Glass et al. 2004).

A drastic elevation of cytokines and chemokines (known as a cytokine storm) in the tissues and serum is a classical feature in acute SARS infection (Ng et al. 2004b; Wong et al. 2004; Zhang et al. 2004; Cheung et al. 2005; Law et al. 2005a; Huang et al. 2005; Tseng et al. 2005; Li et al. 2008). Although there is considerable variation in the literature with regards to the stage of sample collection and method of measurement, the general pattern is the same: during the first two weeks of illness onset, levels of IL1β, IL-6, IL-12, IFN-γ, IL-8, IP-10, and MCP-1 are markedly upregulated; a subsequent reduction in the level of these cytokines is associated with recovery from SARS pneumonia. These cytokines are predominantly proinflammatory cytokines and represent mainly the innate immune responses such as monocytes/macrophages and NK cells. The predictive value of cytokines in disease outcome is still limited. IL-6 has been strongly associated with radiographic score, and increased Th2 cytokines are observed in patients with a fatal outcome (Chien et al. 2006; Li et al. 2008). The presence of immunosuppressive molecules such as IL-10, TGF-β, and prostaglandin E_2 (PGE_2) in the serum of SARS patients may provide an alternative explanation for prolonged and severe clinical outcomes (Huang et al. 2005; Tseng et al. 2005; Li et al. 2008). Furthermore, dysregulation of type I and II interferon responses was associated with defective adaptive immunity and a severe disease outcome for SARS patients (Cameron et al. 2007).

The cellular source of proinflammatory cytokines/chemokines is identified by immunocytochemical methods and is validated by an in vitro infection model. IP-10 is detected in the infected pneumocytes and alveolar macrophages of infected lungs by immunohistochemical methods (Jiang et al. 2005). Under in vitro conditions, SARS-CoV infection could induce monocytes/macrophages, dendritic cells, and lung epithelial cells such as A549 to produce IL-1β, IL-6, IL8, IP-10, and MCP-1 (Cheung et al. 2005; Law et al. 2005a; Yen et al. 2006). Among all the primary cells tested, only type II pneumocytes could support SARS-CoV replication and release of viral particles in vitro (Mossel et al. 2008). Infections of purified monocytes, monocyte-derived macrophages/dendritic cells, and peripheral blood mononuclear cells (PBMC) by SARS-CoV are all abortive, without the production of mature viral particles, although production of chemokines is not impaired (Yilla et al. 2005; Cheung et al. 2005).

The role of the cytokine storm in the pathogenesis of SARS or similar human H5N1 infections is very complex, and the circular cause and consequence relationship of cytokines makes the interpretation very difficult. Attempts have been made to elucidate the mechanism with both in vitro and in vivo models. Using a polarized lung epithelial cell line, Calu-3, supernatants from the apical and basolateral domains of infected epithelial cells are potent in suppressing the phagocytic functions of macrophages, the maturation of dendritic cells, and T-cell priming by dendritic cells (Yoshikawa et al. 2008). Moreover, this inhibition is caused by IL-6 and IL-8 from SARS-infected epithelial cells (Yoshikawa et al. 2009). Recently, oxidized phospholipids (OxPC) resulting from severe inflammation have been detected in infected lung samples from patients with fatal SARS and

H5N1 infections. These oxidized phospholipids are shown to stimulate IL-6 production from lung macrophages via the TLR4–TRIF–TRAF6 pathway in vitro (Sheahan et al. 2008) and in an animal model of acute lung injury. Furthermore, such stimulation occurred independently of MyD88, an adaptor protein in the classical pathway, which is triggered by lipo-polysaccharide (LPS) (Imai et al. 2008). However, in a separate study using a recombinant mouse-adapted SARS-CoV, mice deficient in MyD88 were defective in inflammatory cell recruitment and more susceptible to lethal SARS-CoV (rMA15) infection (Sheahan et al. 2008). Compared to wild-type mice, MyD88$^{-/-}$ mice have a more enhanced pulmonary pathology, and most die by day 6 postinfection. The pathology is associated with increased viral loads, reduction of proinflammatory cytokines, and mononuclear infiltrates in the infected lungs. These data suggest that innate and inflammatory responses mediated by MyD88 are important for protective immunity in lethal mouse-adapted SARS (rMA15) infection. As excessive inflammatory cytokines have been implicated in severe lung disease such as SARS and H5N1 infection, the importance of the pathogenic or protective role of proinflammatory cytokines and chemokines in disease progression needs to be clarified for future treatment strategies.

16.4 Viral Evasive Strategies

Inhibition of antiviral type I interferons (IFN) is a characteristic of SARS infection, as well as other group 2 coronaviruses, such as mouse hepatitis virus (reviewed in Frieman and Baric (2008)). It has been found that production of type I IFN is impaired in cells infected with SARS-CoV, but pretreatment of cells with IFN prevents growth of SARS-CoV (Spiegel et al. 2005; Cinatl et al. 2003). This suggests that the virus has evolved strategies to overcome the antiviral IFN response in infected host cells. Type I IFNs are rarely detected in acute SARS patients, and SARS-CoV is sensitive to the antiviral effect of pegylated IFN-α in a cynomolgus monkey model (Haagmans et al. 2004). Data from in vitro experiments show that suppression of the type I IFN response in SARS-CoV infected cells is due to the inactivation of interferon regulatory factor 3 (IRF-3), which is a latent cytoplasmic transcriptional factor that regulates IFN transcription (Spiegel et al. 2005). Other accessory proteins of SARS-CoV also function as potent IFN antagonists via various mechanisms. N protein inhibits interferon synthesis, whereas Orf 3b and Orf 6 proteins inhibit both IFN synthesis and signaling. Orf 6 protein also inhibits nuclear translocation of STAT1 (Kopecky-Bromberg et al. 2006). Recently, Orf 3b has been shown to be a shuttling protein and inhibits induction of type I IFN induced by retinoic acid-inducible gene 1 and the mitochondrial signaling protein (Freundt et al. 2009). Moreover, M protein inhibits type I IFN production by impeding the formation of the TRAF3–TANK–TBK1/IKKϵ complex (Siu et al. 2009). In addition, NSP1 protein of SARS-CoV could suppress host

IFN-β gene expression by promoting host mRNA degradation and inhibiting translation (Narayanan et al. 2008). Papain-like protease (PLpro), a polyprotein from SARS-CoV replicase proteins, inhibits the phosphorylation and nuclear translocation of IRF-3, thereby disrupting the activation of type I IFN responses through either the Toll-like receptor 3 or the retinoic acid-inducible gene I/melanoma differentiation-associated gene 5 pathway. Such inhibition of the IFN response is independent of the protease activity of PLpro (Devaraj et al. 2007). Besides, SARS-CoV infected cells activate both PKR and PERK, resulting in a sustained eIF2alpha phosphorylation. However, viral replication and viral-induced apoptosis are unaffected (Krahling et al. 2009).

16.5 T Cell Responses

Lymphopenia is a common hematological feature in acute SARS infection, as with other acute viral infections such as measles, Ebola, and respiratory syncytial virus (Schneider-Schaulies et al. 2001; Formenty et al. 1999; Openshaw et al. 2001). Lymphopenia is observed in most SARS patients (>98%) in different studies, and most of them have reduced T-helper (CD4) and T-cytotoxic/suppressor (CD8) cell counts during the early phase of their illness; these reach their lowest values on day 5 and 7 from disease onset (fever), and restore gradually during the recovery phase (Wong et al. 2003; Lam et al. 2004; He et al. 2005). Low CD4 and CD8 lymphocyte counts at presentation are associated with adverse outcomes. B-lymphocyte count and the ratio of T-helper to T-cytotoxic/suppressor lymphocytes (CD4:CD8) remain normal and stable. Lymphopenia as a result of enhanced cell death during the acute SARS infection is attributed to the activation of cytoplasmic caspase 3 in both CD4 and CD8 lymphocytes (Chen et al. 2006). Transfection of SARS E protein in Jurkat T cells could also induce T-cell apoptosis by inactivating antiapoptotic protein Bcl-xL (Yang et al. 2005a). However, very little is known of whether apoptosis of bystander uninfected T cells is present in acute SARS infection.

There is little data available regarding the kinetics and correlation of SARS-specific T-cell responses in viral clearance and disease severity during the acute phase of primary SARS infection. Most of the information on the antigen-specific T-cell response in humans is from retrospective studies using PBMC samples from convalescent SARS patients (Chen et al. 2005; Zhou et al. 2006; Peng et al. 2006a, 2006b; Yang et al. 2006, 2007; Poccia et al. 2006; Li et al. 2008). However, antigen-specific responses from convalescent samples provide information on the memory response of T lymphocytes after primary SARS infection; this is important in the generation of the design of prophylactic SARS vaccines. Recent advances have provided more sensitive techniques to detect and enumerate low frequency antigen-specific T cells by ELISPOT assay, and to identify their surface, maturation, and functional phenotypes by multiparameter flow cytometry at a single cell level.

Using a matrix of overlapping peptides spanning the entire SARS-CoV proteome and an IFN-γ Elispot assay, it was shown that 50% of 128 convalescent SARS patients from endemic areas of the SARS outbreak retain memory T cells against SARS-CoV protein in their peripheral blood one year after SARS infection (Li et al. 2008). The memory response is polyclonal, and SARS-specific T-cell responses are directed against different SARS-CoV proteins and against multiple epitopes (Fig. 16.1). Despite the large genome size of SARS-CoV, only 3% of the proteome

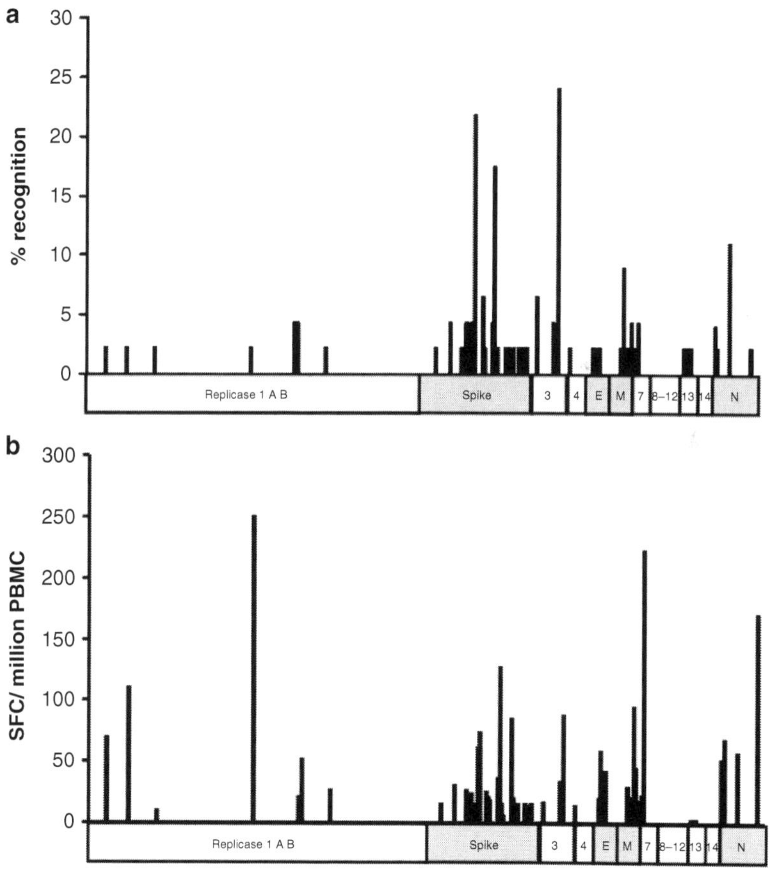

Fig. 16.1 Distribution of peptide recognition and magnitude of SARS-specific T-cell responses across the entire expressed SARS-CoV genome in convalescent SARS patients, as determined by immediate IFN-γ release ELISPOT assay. Freshly isolated PBMC from convalescent SARS samples were stimulated overnight with overlapping peptides from different proteins across the SARS genome, and the IFN-γ response was measured by ELISPOT assay. The individual 1,843 overlapping peptides of different proteins are represented on the x-axis. (**a**) The y-axis represents the percentage recognition of the study subjects to the individual peptides. (**b**) The y-axis represents the average magnitude of response by the study subjects to individual peptides

is immunogenic. Eight of fourteen SARS Orfs are able to stimulate T-cell responses: replicase, S, Orf 3, Orf 4, E, M, Orf 13 and N. Most responses are focused on structural proteins (S, E, M, and N). S protein induces the most dominant T-cell responses, the majority of which are CD4 responses, suggesting that these may provide help for the B cells to produce neutralizing antibodies. In fact, there is a strong association between the CD4 memory response and neutralizing antibody against S protein in recovered SARS patients. However, overall the SARS-specific CD8 response predominates over CD4 in terms of frequency and magnitude of responses. In terms of quality of T cells by polyfunctional status, most SARS-specific CD4 cells produce mainly one cytokine, and they tend to be of a central memory phenotype ($CD27^+$ and $CD45RO^+$). Most SARS-specific CD8 responses are IFN-γ positive, but a significant proportion of the responding cells could also produce TNF-α and degranulate (based upon the mobilization of CD107a to the cell membrane). Observations in the mild–moderate and severe groups show that the frequency of SARS-specific memory CD4 producing IFN-γ, IL2, and TNF-α is significantly higher in the severe group than in the mild–moderate group. The influence of preexisting cross-reactive T cells from other human coronaviruses on the outcome of SARS infection is unlikely, as most published human SARS epitopes do not share amino acid sequence homology with other known human coronaviruses such as NL63, OC43, 229E, and HKU1.

Data from other independent studies also show that SARS-specific T lymphocytes against S, M, E, and N proteins are detected in convalescent SARS samples from one to four years postinfection using overlapping peptides against individual structural proteins, rather than a genome-wide approach (Peng et al. 2006a, 2006b; Yang et al. 2006, 2007). Both CD4 and CD8 responses are detected for all proteins, and they are IFN-γ positive. Moreover, the S protein response is predominantly CD4-dependent (Yang et al. 2006; Libraty et al. 2007), and N-specific T cells could be detected two years postinfection in the absence of antigen stimulation (Peng et al. 2006a and b). In the absence of antigen and reexposure, memory T cells specific to SARS-CoV further declined and could be detected in a small proportion of convalescent patients (Fan et al. 2009). Using a panel of TCR-specific antibodies, effector/memory Vγ9Vδ2 cells were found in convalescent SARS patients (Poccia et al. 2006). These cells are associated with higher anti-SARS IgG levels. Stimulated Vγ9Vδ2 cells display IFN-γ-dependent anti-SARS-CoV activity and are able to kill SARS-CoV infected target cells. In addition, using a prediction algorithm to identify putative cytotoxic T-lymphocyte (CTL) epitopes, S-specific CTL in convalescent SARS patients show a differentiated effector phenotype, which is characterized by $CD45RA^+CCR7^-CD62L^-CCR5^+CD44^+$ (Chen et al. 2005). In a separate study, the CTL epitopes identified could be used to induce CTL responses in A2 transgenic mice after DNA immunization (Zhou et al. 2006). Association of specific HLA alleles with susceptibility to SARS infection and disease severity has been proposed in several studies (Lin et al. 2003; Ng et al. 2004a). However, due to diverse polymorphism of HLA alleles and the small number of samples tested, the results are inconclusive.

16.6 Antibody and B-Cell Responses

Antibodies to SARS-CoV are present in patients after primary SARS infection. Using immunofluorescence assays and ELISA against N protein, serum IgG could be detected as early as four days after illness onset. Serum IgG, IgM, and IgA responses to SARS-CoV are present around the same time, with most patients seroconverted by day 14 after illness onset (Hsueh et al. 2004). In a 36-month follow-up study of 56 convalescent SARS patients, SARS-specific IgG and neutralizing antibodies peaked at month 4 and diminished thereafter (Liu et al. 2006; Cao et al. 2007). IgG and neutralizing antibodies are only detectable in 70% and 20% of convalescent patients respectively by month 36. There is no significant difference in the kinetics of specific antibodies with regard to disease severity, use of steroids, and type and number of coexisting conditions. In a large study of 623 SARS patients, antibodies were able to neutralize pseudotyped viruses bearing S proteins from four different SARS-CoV strains, suggesting that these antibodies are cross-reactive (Nie et al. 2004). Among structural proteins, only the S protein elicits neutralizing antibody (Buchholz et al. 2004). The major immunodominant epitope in S protein lies between amino acids 441 and 700 (Lu et al. 2004).

Memory B cells could be isolated from convalescent SARS patients and immortalized with Epstein–Barr virus to generate human monoclonal antibodies with sufficient affinities to protect from SARS infection (Traggiai et al. 2004). These antibodies could protect mice from viral replication in the lungs after nasal virus challenge. Randomized placebo-controlled trials evaluating passive postexposure prophylaxis in at-risk groups have not been undertaken; however, a retrospective analysis of outcomes in a limited number of SARS patients who continue to deteriorate but receive plasma therapy suggests that passive immunization may shorten hospital stays without obvious adverse effects in patients (Soo et al. 2004; Cheng et al. 2005). Currently, only hyperimmune globulin produced from plasma from convalescent patients and equine plasma produced by immunization with inactivated SARS-CoV are available for prophylactic trials in humans (Lu et al. 2005; Zhang et al. 2005). A human monoclonal IgG1 produced from a single-chain variable region fragment against the S1 domain from two nonimmune human antibody libraries has also been produced (Sui et al. 2004). One of the single-chain variable region fragments, 80R, blocks spike–ACE2 receptor interactions through binding to the S1 domain. Passive immunization of ferrets, hamsters, and mice with human mAbs is effective in suppressing viral replication in the lungs (ter Meulen et al. 2004; Traggiai et al. 2004; Roberts et al. 2006). Recently, potent cross-reactive monoclonal antibodies against highly conserved sites within the S protein, which can neutralize zoonotic or epidemic SARS-CoV, have been reported (He et al. 2006; Zhu et al. 2007). These broadly cross-reactive neutralizing antibodies may be useful for treatment options.

The role of antibodies in antibody-dependent enhancement (ADE) in SARS vaccine has been controversial and extensively debated (Perlman and Dandekar 2005; Chen and Subbarao 2007). However, most of the evidence is taken from limited data from in vitro and animal experiments. The interpretation of these findings with regard to implications for human disease is not clear. Concerns over the safety of SARS vaccines came from a previous animal study in which ADE was observed in domestic cats after vaccination against feline infectious peritonitis coronavirus (FIPV) (Vennema et al. 1990). Moreover, ferrets that are immunized with MVA vaccine expressing S protein suffer hepatitis whilst failing to develop protection (Weingartl et al. 2004). In addition, some variants of SARS-CoV are resistant to antibody neutralization, but the infection is enhanced by antibodies against a different variant (Yang et al. 2005b). Under in vitro conditions, entry into human B-cell lines by SARS-CoV virus could be carried out in an FcγRII-dependent and ACE2-independent fashion, indicating that ADE of virus entry is an alternative cell entry mechanism of SARS-CoV (Kam et al. 2007). However, there is no direct evidence that patients with SARS have had previous exposure to a related virus. Likewise, there is no evidence of enhanced disease in the lungs of animals that are infected with SARS-CoV following passive transfer of antibodies against SARS-CoV induced by infection or immunization. Also, macrophages are not productively infected by SARS-CoV (Subbarao et al. 2004; Bisht et al. 2004; Stadler et al. 2005; Yang et al. 2004b; Greenough et al. 2005; Kam et al. 2007). Hence, the potential for ADE following SARS vaccination is low. However, due to the lack of animal models that mimic the clinical disease in humans, care must be taken in the interpretation of the safety data of SARS vaccines.

16.7 Vaccines

SARS infection begins at the mucosal surface of the respiratory system, and the virus infects the lung epithelial cells via the surface S protein. An effective SARS vaccine is required to elicit neutralizing antibodies that abolish the binding of virus to the surface of host target cells and/or to eliminate the virus at the early stages of infection. Hope for active immunization with SARS vaccine stems from the findings that natural human infection with SARS-CoV leads to long-lived neutralizing antibody responses and immune sera capable of cross-neutralizing pseudotype viruses bearing S proteins from diverse but highly related human SARS-CoV isolates (Nie et al. 2004). In addition, long-lived memory T lymphocytes against different SARS-CoV structural proteins are present in the peripheral blood of convalescent SARS patients (Li et al. 2008; Peng et al. 2006; Fan et al. 2009).

Numerous immunization strategies with various vaccine expression vectors, routes of immunization, and adjuvants have been used to develop SARS vaccines (reviewed in Chen and Subbarao 2007; Roberts et al. 2008). They include inactivated whole virus vaccine, DNA immunization, recombinant viral vector (modified vaccinia Ankara (MVA), adenovirus, baculovirus, rhabdovirus, parainfluenza virus, Venezuelan equine encephalitis virus, measles virus), recombinant bacterial vector (*Salmonella enteritica*, *Lactobacillus casei*), recombinant protein, adjuvants (alum, ODN CpG, Protollin, MALP-2), and different combinations of prime-boost strategies.

In most viral vectors, expression of the target gene in eukaryotic cells is codon optimized, and the immunogenicity is improved by using immunostimulatory sequences encoding IL-2 and internal ribosomal entry side (IRES).

Among all the structural proteins that are potential vaccine targets, S protein is the only protein that could induce neutralizing antibody and protect animals from challenge infection regardless of the method of antigen delivery (Callendret et al. 2007; Yang et al. 2004b; Kobinger et al. 2007; See et al. 2006; Bukreyev et al. 2004; Hu et al. 2007; Du et al. 2008). S protein protects hamsters from challenge when expressed in an attenuated parainfluenza type 3 vector (Buchholz et al. 2004). Mucosal immunization of African green monkeys with this parainfluenza type 3–S protein chimeric virus elicits neutralizing antibody and protection from viral replication in both the upper and lower respiratory tract after challenge with live SARS-CoV (Bukreyev et al. 2004). S protein-encoding DNA vaccines and modified vaccinia Ankara (MVA) expressing the S protein stimulate neutralizing antibody production and provide protection from live virus challenge in mice (Yang et al. 2004b). The protective mechanism is mediated by antibody and not by T cells.

Taken together, both neutralizing antibody (B-cell) responses and T-cell responses to the S protein are dominant in patients recovered from SARS, suggesting that an S protein-based vaccine may be sufficient to induce protective immunity by prophylactic vaccination. Moreover, there is a good correlation between neutralizing antibody and T-cell responses (Li et al. 2008). In humans, only two independent Phase 1 clinical trials of SARS vaccines have been published. Inactivated SARS-CoV vaccine administered via two intramuscular immunizations was reported to be well tolerated and immunogenic in 36 healthy adult volunteers in China (Lin et al. 2007). All subjects had sero-converted by day 42 postvaccination. The geometric mean titer of neutralizing antibody peaked two weeks after the second vaccination, but decreased four weeks later. In a separate DNA vaccine trial where S antigen was used as the target, 80% of the subjects had sero-converted after receiving 2–3 immunizations (Martin et al. 2008). The neutralizing antibody response peaked between weeks 8 and 12, and 50% of the subjects remained positive at week 32 postvaccination. In addition, the DNA vaccine induced 100% spike-specific T-lymphocyte response in all subjects from week 2 postvaccination. The spike-specific T-cell response was mainly CD4-dependent. The peak T-cell response occurred between weeks 8 and 12, and was maintained throughout the 32-week trial period.

16.8 Summary and Perspectives

Following the initial outbreaks of SARS which occurred in hospital-based clusters, the virus was rapidly isolated, sequenced, and characterized as a result of an international public health effort. This knowledge was quickly transformed into diagnostic reagents and tools to control community outbreaks of SARS infection.

Despite the wealth of active scientific research and information, the mechanisms of viral clearance, immune correlates of protection, and the immunopathogenesis of SARS infection remain unclear. The risk of SARS-CoV to humans is still high due to the large number of animal reservoirs of SARS-CoV-like coronaviruses and the genome instability of RNA coronaviruses. Currently available animal models are inadequate, and development of models that mimic the clinical symptoms of human SARS infection is urgently required for mechanistic study and vaccine evaluation. Serial analysis of acute SARS samples to analyze innate, antibody, and T-cell responses would be useful to develop treatment strategies and prophylactic SARS vaccines.

Acknowledgements This project was supported by Medical Research Council UK, Beijing Municipal Government, and the European Commission Euro-Asian SARS-DTV Network (SP22-CT-2004-511064). Expert editorial assistance and critical reading by Dr. Sarah Bangs is greatly appreciated.

References

Anonymous (2004) The involvement of natural killer cells in the pathogenesis of severe acute respiratory syndrome. Am J Clin Pathol 121(4):507–511

Bisht H, Roberts A et al (2004) Severe acute respiratory syndrome coronavirus spike protein expressed by attenuated vaccinia virus protectively immunizes mice. Proc Natl Acad Sci USA 101(17):6641–6646

Bosch BJ, van der Zee R et al (2003) The coronavirus spike protein is a class I virus fusion protein: structural and functional characterization of the fusion core complex. J Virol 77(16):8801–8811

Bosch BJ, Martina BE et al (2004) Severe acute respiratory syndrome coronavirus (SARS-CoV) infection inhibition using spike protein heptad repeat-derived peptides. Proc Natl Acad Sci USA 101(22):8455–8460

Buchholz UJ, Bukreyev A et al (2004) Contributions of the structural proteins of severe acute respiratory syndrome coronavirus to protective immunity. Proc Natl Acad Sci USA 101(26):9804–9809

Bukreyev A, Lamirande EW et al (2004) Mucosal immunisation of African green monkeys (Cercopithecus aethiops) with an attenuated parainfluenza virus expressing the SARS coronavirus spike protein for the prevention of SARS. Lancet 363(9427):2122–2127

Callendret B, Lorin V et al (2007) Heterologous viral RNA export elements improve expression of severe acute respiratory syndrome (SARS) coronavirus spike protein and protective efficacy of DNA vaccines against SARS. Virology 363(2):288–302

Cameron MJ, Ran L et al (2007) Interferon-mediated immunopathological events are associated with atypical innate and adaptive immune responses in patients with severe acute respiratory syndrome. J Virol 81(16):8692–8706

Cao WC, Liu W et al (2007) Disappearance of antibodies to SARS-associated coronavirus after recovery. N Engl J Med 357(11):1162–1163

Chan CM, Ma CW et al (2007) The SARS-Coronavirus membrane protein induces apoptosis through modulating the Akt survival pathway. Arch Biochem Biophys 459(2):197–207

Chen J, Subbarao K (2007) The immunobiology of SARS. Annu Rev Immunol 25:443–472

Chen JH, Chang YW et al (2004) Plasma proteome of severe acute respiratory syndrome analyzed by two-dimensional gel electrophoresis and mass spectrometry. Proc Natl Acad Sci USA 101(49):17039–17044

Chen H, Hou J et al (2005) Response of memory CD8+ T cells to severe acute respiratory syndrome (SARS) coronavirus in recovered SARS patients and healthy individuals. J Immunol 175(1):591–598

Chen RF, Chang JC et al (2006) Role of vascular cell adhesion molecules and leukocyte apoptosis in the lymphopenia and thrombocytopenia of patients with severe acute respiratory syndrome (SARS). Microbes Infect 8(1):122–127

Cheng Y, Wong R et al (2005) Use of convalescent plasma therapy in SARS patients in Hong Kong. Eur J Clin Microbiol Infect Dis 24(1):44–46

Cheng VC, Lau SK et al (2007) Severe acute respiratory syndrome coronavirus as an agent of emerging and reemerging infection. Clin Microbiol Rev 20(4):660–694

Cheung CY, Poon LL et al (2005) Cytokine responses in severe acute respiratory syndrome coronavirus-infected macrophages in vitro: possible relevance to pathogenesis. J Virol 79 (12):7819–7826

Chien JY, Hsueh PR et al (2006) Temporal changes in cytokine/chemokine profiles and pulmonary involvement in severe acute respiratory syndrome. Respirology 11(6):715–722

Chu CM, Poon LL et al (2004) Initial viral load and the outcomes of SARS. CMAJ 171(11):1349–1352

Chua KB, Bellini WJ et al (2000) Nipah virus: a recently emergent deadly paramyxovirus. Science 288(5470):1432–1435

Cinatl J, Morgenstern B et al (2003) Treatment of SARS with human interferons. Lancet 362 (9380):293–294

Devaraj SG, Wang N et al (2007) Regulation of IRF-3-dependent innate immunity by the papain-like protease domain of the severe acute respiratory syndrome coronavirus. J Biol Chem 282 (44):32208–32221

Drosten C, Gunther S et al (2003) Identification of a novel coronavirus in patients with severe acute respiratory syndrome. N Engl J Med 348(20):1967–1976

Du L, He Y et al (2008) Development of subunit vaccines against severe acute respiratory syndrome. Drugs Today 44(1):63–73

Fan YY, Huang ZT et al (2009) Characterization of SARS-CoV-specific memory T cells from recovered individuals 4 years after infection. Arch Virol 154(7):1093–1099

Formenty P, Hatz C et al (1999) Human infection due to Ebola virus, subtype Cote d'Ivoire: clinical and biologic presentation. J Infect Dis 179(Suppl 1):S48–S53

Freundt EC, Yu L et al (2009) Molecular determinants for subcellular localization of the severe acute respiratory syndrome coronavirus open reading frame 3b protein. J Virol 83(13):6631–6640

Frieman M, Baric R (2008) Mechanisms of severe acute respiratory syndrome pathogenesis and innate immunomodulation. Microbiol Mol Biol Rev 72(4):672–685

Glass WG, Subbarao K et al (2004) Mechanisms of host defense following severe acute respiratory syndrome-coronavirus (SARS-CoV) pulmonary infection of mice. J Immunol 173(6):4030–4039

Greenough TC, Babcock GJ et al (2005) Development and characterization of a severe acute respiratory syndrome-associated coronavirus-neutralizing human monoclonal antibody that provides effective immunoprophylaxis in mice. J Infect Dis 191(4):507–514

Gu J, Gong E et al (2005) Multiple organ infection and the pathogenesis of SARS. J Exp Med 202 (3):415–424

Haagmans BL, Kuiken T et al (2004) Pegylated interferon-alpha protects type 1 pneumocytes against SARS coronavirus infection in macaques. Nat Med 10(3):290–293

Hamming I, Timens W et al (2004) Tissue distribution of ACE2 protein, the functional receptor for SARS coronavirus. A first step in understanding SARS pathogenesis. J Pathol 203(2):631–637

He Z, Zhao C et al (2005) Effects of severe acute respiratory syndrome (SARS) coronavirus infection on peripheral blood lymphocytes and their subsets. Int J Infect Dis 9(6):323–330

He Y, Li J et al (2006) Cross-neutralization of human and palm civet severe acute respiratory syndrome coronaviruses by antibodies targeting the receptor-binding domain of spike protein. J Immunol 176(10):6085–6092

Ho MS, Chen WJ et al (2005) Neutralizing antibody response and SARS severity. Emerg Infect Dis 11(11):1730–1737

Hofmann H, Pohlmann S (2004) Cellular entry of the SARS coronavirus. Trends Microbiol 12 (10):466–472

Hofmann H, Pyrc K et al (2005) Human coronavirus NL63 employs the severe acute respiratory syndrome coronavirus receptor for cellular entry. Proc Natl Acad Sci USA 102(22):7988–7993

Hogan RJ (2006) Are nonhuman primates good models for SARS? PLoS Med 3(9):e411; author reply e415

Hsueh PR, Huang LM et al (2004) Chronological evolution of IgM, IgA, IgG and neutralisation antibodies after infection with SARS-associated coronavirus. Clin Microbiol Infect 10(12): 1062–1066

Hu MC, Jones T et al (2007) Intranasal Protollin-formulated recombinant SARS S-protein elicits respiratory and serum neutralizing antibodies and protection in mice. Vaccine 25 (34):6334–6340

Huang KJ, Su IJ et al (2005) An interferon-gamma-related cytokine storm in SARS patients. J Med Virol 75(2):185–194

Huang IC, Bosch BJ et al (2006) SARS coronavirus, but not human coronavirus NL63, utilizes cathepsin L to infect ACE2-expressing cells. J Biol Chem 281(6):3198–3203

Imai Y, Kuba K et al (2005) Angiotensin-converting enzyme 2 protects from severe acute lung failure. Nature 436(7047):112–116

Imai Y, Kuba K et al (2008) Identification of oxidative stress and Toll-like receptor 4 signaling as a key pathway of acute lung injury. Cell 133(2):235–249

Ip WK, Chan KH et al (2005) Mannose-binding lectin in severe acute respiratory syndrome coronavirus infection. J Infect Dis 191(10):1697–1704

Jeffers SA, Tusell SM et al (2004) CD209L (L-SIGN) is a receptor for severe acute respiratory syndrome coronavirus. Proc Natl Acad Sci USA 101(44):15748–15753

Jiang Y, Xu J et al (2005) Characterization of cytokine/chemokine profiles of severe acute respiratory syndrome. Am J Respir Crit Care Med 171(8):850–857

Kam YW, Kien F et al (2007) Antibodies against trimeric S glycoprotein protect hamsters against SARS-CoV challenge despite their capacity to mediate FcgammaRII-dependent entry into B cells in vitro. Vaccine 25(4):729–740

Kielian M, Rey FA (2006) Virus membrane-fusion proteins: more than one way to make a hairpin. Nat Rev Microbiol 4(1):67–76

Kobinger GP, Figueredo JM et al (2007) Adenovirus-based vaccine prevents pneumonia in ferrets challenged with the SARS coronavirus and stimulates robust immune responses in macaques. Vaccine 25(28):5220–5231

Kopecky-Bromberg SA, Martinez-Sobrido L et al (2006) 7a protein of severe acute respiratory syndrome coronavirus inhibits cellular protein synthesis and activates p38 mitogen-activated protein kinase. J Virol 80(2):785–793

Krahling V, Stein DA et al (2009) Severe acute respiratory syndrome coronavirus triggers apoptosis via protein kinase R but is resistant to its antiviral activity. J Virol 83(5):2298–2309

Ksiazek TG, Erdman D et al (2003) A novel coronavirus associated with severe acute respiratory syndrome. N Engl J Med 348(20):1953–1966

Kuba K, Imai Y et al (2005) A crucial role of angiotensin converting enzyme 2 (ACE2) in SARS coronavirus-induced lung injury. Nat Med 11(8):875–879

Lam CW, Chan MH et al (2004) Severe acute respiratory syndrome: clinical and laboratory manifestations. Clin Biochem Rev 25(2):121–132

Lau SK, Woo PC et al (2005) Severe acute respiratory syndrome coronavirus-like virus in Chinese horseshoe bats. Proc Natl Acad Sci USA 102(39):14040–14045

Law HK, Cheung CY et al (2005a) Chemokine up-regulation in SARS-coronavirus-infected, monocyte-derived human dendritic cells. Blood 106(7):2366–2374

Law PT, Wong CH et al (2005b) The 3a protein of severe acute respiratory syndrome-associated coronavirus induces apoptosis in Vero E6 cells. J Gen Virol 86(Pt 7):1921–1930

Lee N, Hui D et al (2003) A major outbreak of severe acute respiratory syndrome in Hong Kong. N Engl J Med 348(20):1986–1994

Li W, Moore MJ et al (2003) Angiotensin-converting enzyme 2 is a functional receptor for the SARS coronavirus. Nature 426(6965):450–454

Li W, Shi Z et al (2005) Bats are natural reservoirs of SARS-like coronaviruses. Science 310 (5748):676–679

Li CK, Wu H et al (2008) T cell responses to whole SARS coronavirus in humans. J Immunol 181 (8):5490–5500

Libraty DH, O'Neil KM et al (2007) Human CD4(+) memory T-lymphocyte responses to SARS coronavirus infection. Virology 368(2):317–321

Lin M, Tseng HK et al (2003) Association of HLA class I with severe acute respiratory syndrome coronavirus infection. BMC Med Genet 4:9

Lin JT, Zhang JS et al (2007) Safety and immunogenicity from a phase I trial of inactivated severe acute respiratory syndrome coronavirus vaccine. Antivir Ther 12(7):1107–1113

Liu S, Xiao G et al (2004) Interaction between heptad repeat 1 and 2 regions in spike protein of SARS-associated coronavirus: implications for virus fusogenic mechanism and identification of fusion inhibitors. Lancet 363(9413):938–947

Liu W, Fontanet A et al (2006) Two-year prospective study of the humoral immune response of patients with severe acute respiratory syndrome. J Infect Dis 193(6):792–795

Lu L, Manopo I et al (2004) Immunological characterization of the spike protein of the severe acute respiratory syndrome coronavirus. J Clin Microbiol 42(4):1570–1576

Lu JH, Guo ZM et al (2005) Preparation and development of equine hyperimmune globulin F(ab')2 against severe acute respiratory syndrome coronavirus. Acta Pharmacol Sin 26(12): 1479–1484

Marra MA, Jones SJ et al (2003) The genome sequence of the SARS-associated coronavirus. Science 300(5624):1399–1404

Martin JE, Louder MK et al (2008) A SARS DNA vaccine induces neutralizing antibody and cellular immune responses in healthy adults in a Phase I clinical trial. Vaccine 26(50):6338–6343

Mossel EC, Wang J et al (2008) SARS-CoV replicates in primary human alveolar type II cell cultures but not in type I-like cells. Virology 372(1):127–135

Murray K, Selleck P et al (1995) A morbillivirus that caused fatal disease in horses and humans. Science 268(5207):94–97

Narayanan K, Huang C et al (2008) Severe acute respiratory syndrome coronavirus nsp1 suppresses host gene expression, including that of type I interferon, in infected cells. J Virol 82 (9):4471–4479

National Research Project for SARS, Beijing Group (2004) The involvement of natural killer cells in the pathogenesis of severe acute respiratory syndrome

Ng MH, Lau KM et al (2004a) Association of human-leukocyte-antigen class I (B*0703) and class II (DRB1*0301) genotypes with susceptibility and resistance to the development of severe acute respiratory syndrome. J Infect Dis 190(3):515–518

Ng PC, Lam CW et al (2004b) Inflammatory cytokine profile in children with severe acute respiratory syndrome. Pediatrics 113(Pt 1):e7–e14

Nicholls JM, Poon LL et al (2003) Lung pathology of fatal severe acute respiratory syndrome. Lancet 361(9371):1773–1778

Nicholls JM, Butany J et al (2006) Time course and cellular localization of SARS-CoV nucleoprotein and RNA in lungs from fatal cases of SARS. PLoS Med 3(2):e27

Nie Y, Wang G et al (2004) Neutralizing antibodies in patients with severe acute respiratory syndrome-associated coronavirus infection. J Infect Dis 190(6):1119–1126

Openshaw PJ, Culley FJ et al (2001) Immunopathogenesis of vaccine-enhanced RSV disease. Vaccine 20(Suppl 1):S27–S31
Peiris JS, Lai ST et al (2003) Coronavirus as a possible cause of severe acute respiratory syndrome. Lancet 361(9366):1319–1325
Peng H, Yang LT et al (2006a) Human memory T cell responses to SARS-CoV E protein. Microbes Infect 8(9–10):2424–2431
Peng H, Yang LT et al (2006b) Long-lived memory T lymphocyte responses against SARS coronavirus nucleocapsid protein in SARS-recovered patients. Virology 351(2):466–475
Perlman S, Dandekar AA (2005) Immunopathogenesis of coronavirus infections: implications for SARS. Nat Rev Immunol 5(12):917–927
Poccia F, Agrati C et al (2006) Anti-severe acute respiratory syndrome coronavirus immune responses: the role played by V gamma 9V delta 2 T cells. J Infect Dis 193(9):1244–1249
Poutanen SM, Low DE et al (2003) Identification of severe acute respiratory syndrome in Canada. N Engl J Med 348(20):1995–2005
Roberts A, Thomas WD et al (2006) Therapy with a severe acute respiratory syndrome-associated coronavirus-neutralizing human monoclonal antibody reduces disease severity and viral burden in golden Syrian hamsters. J Infect Dis 193(5):685–692
Roberts A, Lamirande EW et al (2008) Animal models and vaccines for SARS-CoV infection. Virus Res 133(1):20–32
Schneider-Schaulies S, Niewiesk S et al (2001) Measles virus induced immunosuppression: targets and effector mechanisms. Curr Mol Med 1(2):163–181
See RH, Zakhartchouk AN et al (2006) Comparative evaluation of two severe acute respiratory syndrome (SARS) vaccine candidates in mice challenged with SARS coronavirus. J Gen Virol 87(Pt 3):641–650
Sheahan T, Morrison TE et al (2008) MyD88 is required for protection from lethal infection with a mouse-adapted SARS-CoV. PLoS Pathog 4(12):e1000240
Shi Z, Hu Z (2008) A review of studies on animal reservoirs of the SARS coronavirus. Virus Res 133(1):74–87
Simmons G, Gosalia DN et al (2005) Inhibitors of cathepsin L prevent severe acute respiratory syndrome coronavirus entry. Proc Natl Acad Sci USA 102(33):11876–11881
Siu KL, Kok KH et al (2009) Severe acute respiratory syndrome Coronavirus M protein inhibits type I interferon production by impeding the formation of TRAF3{middle dot}TANK{middle dot}TBK1/IKK{epsilon} complex. J Biol Chem 284(24):16202–16209
Soo YO, Cheng Y et al (2004) Retrospective comparison of convalescent plasma with continuing high-dose methylprednisolone treatment in SARS patients. Clin Microbiol Infect 10(7):676–678
Spiegel M, Pichlmair A et al (2005) Inhibition of Beta interferon induction by severe acute respiratory syndrome coronavirus suggests a two-step model for activation of interferon regulatory factor 3. J Virol 79(4):2079–2086
Stadler K, Masignani V et al (2003) SARS–beginning to understand a new virus. Nat Rev Microbiol 1(3):209–218
Stadler K, Roberts A et al (2005) SARS vaccine protective in mice. Emerg Infect Dis 11(8):1312–1314
Stockman LJ, Bellamy R et al (2006) SARS: systematic review of treatment effects. PLoS Med 3(9):e343
Subbarao K, McAuliffe J et al (2004) Prior infection and passive transfer of neutralizing antibody prevent replication of severe acute respiratory syndrome coronavirus in the respiratory tract of mice. J Virol 78(7):3572–3577
Sui J, Li W et al (2004) Potent neutralization of severe acute respiratory syndrome (SARS) coronavirus by a human mAb to S1 protein that blocks receptor association. Proc Natl Acad Sci USA 101(8):2536–2541
Tan YJ, Fielding BC et al (2004) Overexpression of 7a, a protein specifically encoded by the severe acute respiratory syndrome coronavirus, induces apoptosis via a caspase-dependent pathway. J Virol 78(24):14043–14047

Temperton NJ, Chan PK et al (2005) Longitudinally profiling neutralizing antibody response to SARS coronavirus with pseudotypes. Emerg Infect Dis 11(3):411–416

ter Meulen J, Bakker AB et al (2004) Human monoclonal antibody as prophylaxis for SARS coronavirus infection in ferrets. Lancet 363(9427):2139–2141

To KF, Tong JH et al (2004) Tissue and cellular tropism of the coronavirus associated with severe acute respiratory syndrome: an in-situ hybridization study of fatal cases. J Pathol 202(2):157–163

Traggiai E, Becker S et al (2004) An efficient method to make human monoclonal antibodies from memory B cells: potent neutralization of SARS coronavirus. Nat Med 10(8):871–875

Tripet B, Howard MW et al (2004) Structural characterization of the SARS-coronavirus spike S fusion protein core. J Biol Chem 279(20):20836–20849

Tsang KW, Ho PL et al (2003) A cluster of cases of severe acute respiratory syndrome in Hong Kong. N Engl J Med 348(20):1977–1985

Tseng CT, Perrone LA et al (2005) Severe acute respiratory syndrome and the innate immune responses: modulation of effector cell function without productive infection. J Immunol 174(12):7977–7985

Vennema H, de Groot RJ et al (1990) Early death after feline infectious peritonitis virus challenge due to recombinant vaccinia virus immunization. J Virol 64(3):1407–1409

Weingartl H, Czub M et al (2004) Immunization with modified vaccinia virus Ankara-based recombinant vaccine against severe acute respiratory syndrome is associated with enhanced hepatitis in ferrets. J Virol 78(22):12672–12676

Wong RS, Wu A et al (2003) Haematological manifestations in patients with severe acute respiratory syndrome: retrospective analysis. BMJ 326(7403):1358–1362

Wong CK, Lam CW et al (2004) Plasma inflammatory cytokines and chemokines in severe acute respiratory syndrome. Clin Exp Immunol 136(1):95–103

Yang ZY, Huang Y et al (2004a) pH-dependent entry of severe acute respiratory syndrome coronavirus is mediated by the spike glycoprotein and enhanced by dendritic cell transfer through DC-SIGN. J Virol 78(11):5642–5650

Yang ZY, Kong WP et al (2004b) A DNA vaccine induces SARS coronavirus neutralization and protective immunity in mice. Nature 428(6982):561–564

Yang Y, Xiong Z et al (2005a) Bcl-xL inhibits T-cell apoptosis induced by expression of SARS coronavirus E protein in the absence of growth factors. Biochem J 392(Pt 1):135–143

Yang ZY, Werner HC et al (2005b) Evasion of antibody neutralization in emerging severe acute respiratory syndrome coronaviruses. Proc Natl Acad Sci USA 102(3):797–801

Yang LT, Peng H et al (2006) Long-lived effector/central memory T-cell responses to severe acute respiratory syndrome coronavirus (SARS-CoV) S antigen in recovered SARS patients. Clin Immunol 120(2):171–178

Yang L, Peng H et al (2007) Persistent memory CD4+ and CD8+ T-cell responses in recovered severe acute respiratory syndrome (SARS) patients to SARS coronavirus M antigen. J Gen Virol 88(Pt 10):2740–2748

Yen YT, Liao F et al (2006) Modeling the early events of severe acute respiratory syndrome coronavirus infection in vitro. J Virol 80(6):2684–2693

Yilla M, Harcourt BH et al (2005) SARS-coronavirus replication in human peripheral monocytes/macrophages. Virus Res 107(1):93–101

Yoshikawa T, Hill T et al (2009) Severe acute respiratory syndrome (SARS) coronavirus-induced lung epithelial cytokines exacerbate SARS pathogenesis by modulating intrinsic functions of monocyte-derived macrophages and dendritic cells. J Virol 83(7):3039–3048

Zhang Y, Li J et al (2004) Analysis of serum cytokines in patients with severe acute respiratory syndrome. Infect Immun 72(8):4410–4415

Zhang Z, Xie YW et al (2005) Purification of severe acute respiratory syndrome hyperimmune globulins for intravenous injection from convalescent plasma. Transfusion 45(7):1160–1164

Zhao GP (2007) SARS molecular epidemiology: a Chinese fairy tale of controlling an emerging zoonotic disease in the genomics era. Philos Trans R Soc Lond B Biol Sci 362(1482):1063–1081

Zhao X, Nicholls JM et al (2008) Severe acute respiratory syndrome-associated coronavirus nucleocapsid protein interacts with Smad3 and modulates transforming growth factor-beta signaling. J Biol Chem 283(6):3272–3280

Zhou M, Xu D et al (2006) Screening and identification of severe acute respiratory syndrome-associated coronavirus-specific CTL epitopes. J Immunol 177(4):2138–2145

Zhu Z, Chakraborti S et al (2007) Potent cross-reactive neutralization of SARS coronavirus isolates by human monoclonal antibodies. Proc Natl Acad Sci USA 104(29):12123–12128

Zinkernagel RM, Hengartner H (2004) On immunity against infections and vaccines: credo 2004. Scand J Immunol 60(1–2):9–13

Chapter 17
The Use of Retroviral Pseudotypes for the Measurement of Antibody Responses to SARS Coronavirus

Nigel James Temperton

Abstract Neutralization assays allow for sensitive detection of functional antibody responses directed against the surface protein envelopes of many viruses. For high-containment viruses like SARS coronavirus (CoV), however, these assays are not widely applicable due to the requirement for high biosafety laboratory facilities and specially trained personnel. In order to effectively address this containment issue, retroviral pseudotypes have been used as surrogates of the live virus for neutralization assays. The pseudotype-based neutralization assay system is highly flexible, allowing for a choice of reporter systems, and is readily adaptable to newly emerging virus strains.

17.1 Introduction

The coronavirus that causes severe acute respiratory syndrome (SARS-CoV) is a relatively new human pathogen for which a vaccine will be urgently required should this virus reemerge and new outbreaks occur.

The SARS-CoV genome encodes four structural proteins, the spike (S), membrane (M), envelope (E), and nucleocapsid (N) proteins. The S protein is the major surface antigen of the virus, and the antibody response is primarily directed against this protein. Preclinical studies of SARS-CoV vaccines provide evidence that generating a strong neutralizing antibody response to SARS-CoV S may protect against SARS infection. Many methods of detecting and measuring antibody responses against SARS-CoV have been developed and include IFA, ELISA, ICT, Western blot and virus neutralization (Bermingham et al. 2004; Wu et al. 2004).

N.J. Temperton
MRC/UCL Centre for Medical Molecular Virology, University College London, Windeyer Building, 46 Cleveland Street, London W1T 4JF, United Kingdom
e-mail: nigel.temperton@ucl.ac.uk

Neutralization assays allow for sensitive detection of functional antibody responses directed against the surface protein envelopes of many viruses. For high-containment viruses like SARS-CoV and influenza H5N1, however, these assays are not widely applicable due to the requirement for high biosafety laboratory facilities and specially trained personnel. In order to effectively address this containment issue, retroviral pseudotypes have been used as alternatives to the live virus. These pseudotypes carry the retroviral RNA genome coding for marker genes (GFP, luciferase and beta-galactosidase (β-gal) are the most frequently used). The RNA genome is packaged by retroviral core proteins, but these viral cores bear foreign, often nonretroviral envelope proteins on their surface. Pseudotype infection, i.e. transfer of their genome (containing the transfer/marker gene) to target cells, depends on the function of the foreign envelope proteins and results in integration and subsequent expression of the marker gene. The function of the foreign envelope can therefore be detected/quantified by measuring marker gene expression. Using retroviral particles pseudotyped with the envelopes of high-containment viruses as "surrogate viruses" for use in neutralization assays is highly advantageous. Using a pseudotype approach, only the envelope protein(s) of the high-containment virus is required, with no possibility of recombination or virus escape. These pseudotypes undergo abortive replication and are unable to give rise to replication-competent progeny (Temperton and Wright 2009).

Safety considerations aside, for virus neutralization assays, pseudotypes are excellent serological reagents as they encode reporter genes and bear the heterologous viral envelope proteins of interest. The transfer of these reporter genes to target cells depends on the function of the viral envelope protein; therefore the titre of neutralizing antibodies against the envelope can be measured by a reduction in reporter gene transfer. These assays are also readily adaptable to a high-throughput format for use in vaccine evaluation, antiviral screening/resistance assays, monoclonal antibody screening and large sero-surveillance studies.

17.2 Construction of Retroviral Pseudotypes Bearing SARS-CoV (S) Glycoproteins

SARS pseudotypes are routinely constructed using a three-plasmid transfection approach (Fig. 17.1). Construction of retroviral pseudotypes is achieved through the concurrent introduction of a SARS spike (S) envelope gene, core retroviral genes and a marker/reporter gene into producer cells, normally HEK 293T cells as they transfect efficiently, although other cell lines may be used;

Plasmid 1: The SARS spike (S) envelope gene construct. SARS spike (S) envelope sequences (either native or codon-optimized) can be readily cloned into appropriate expression plasmids, which will permit their incorporation into the mature pseudotype. Expression plasmids that have been successfully used for this purpose include pCAGGS (Simmons et al. 2004; Temperton et al. 2005),

17 The Use of Retroviral Pseudotypes for the Measurement of Antibody Responses

Fig. 17.1 Flexible 3 plasmid transfection system for the production of SARS-CoV (S) pseudotype particles. *gag–pol module:* The retroviral structural (*gag*) and enzymatic (*pol*) proteins are expressed from a single plasmid lacking the packaging signal (removed to prevent incorporation into mature particles). *SARS spike (S) module:* Spike envelope glycoprotein expression is driven by a promoter specific to the envelope. *Transfer/reporter module:* GFP/luciferase or β-gal transfer genes are cloned between flanking LTR regions that play a role in integration and transcription of the pseudotype's RNA genome. A packaging signal (ψ) and promoter are incorporated upstream of the transfer gene to ensure it is packaged into the pseudotype particle and regulates expression once the gene is integrated

pcDNA3.1+ (Nie et al. 2004a; Zhang et al. 2006), pHCMV (Giroglou et al. 2004; Han et al. 2004) and pcDM8 (Moore et al. 2004).

Plasmid 2: The retroviral *gag–pol* construct. These are the two genes that encode the structural proteins (e.g., matrix, nucleocapsid and p7 expressed from *gag*) that comprise the core, and the enzymatic proteins (reverse transcriptase, protease and integrase expressed from *pol*) responsible for processing the structural proteins and ensuring integration of a dsDNA copy of the marker/reporter gene. Additionally in a lentiviral *gag–pol* construct, the *rev* gene is included for efficient processing.

Plasmid 3: The marker/reporter gene construct. This is the gene that is stably integrated into the target/assay cell DNA. Where it is expressed via a single plasmid, a packaging signal is engineered upstream of the gene to ensure efficient incorporation of full-length RNA copies of that gene into the pseudotype capsid. Once integrated, the gene is expressed via various *cis*-acting transcriptional elements. It should be noted that the marker/reporter gene construct is not required for pseudotype production and mature particles will be produced by a two-plasmid transfection of a *gag–pol* construct and a SARS spike (S) envelope construct. Also, pseudotypes lacking the SARS spike (S) envelope will bud but will be noninfectious.

As an alternative to using the three-plasmid transfection system for producing SARS pseudotypes, all of the necessary genes can be transfected via either one or two plasmids. Using three plasmids as detailed here, however, offers the greatest ease and flexibility (switch of plasmids, minimal cloning required), and safety (less chance of recombination) as the envelope, core and reporter genes are expressed from different plasmids. The flexibility of the system enables tailoring construction of the SARS pseudotypes depending on experimental requirements. After transfection the pseudotype virus is harvested and stored at $-80°C$ until required for neutralization assays (Fig. 17.2).

Fig. 17.2 SARS-CoV S pseudotype production, titration and neutralization. Three plasmids (*gag–pol* module, SARS spike module and transfer/reporter module) are transfected into producer cells and the resulting mature pseudotype particles are harvested 48 and 72 h post-transfection. These particles comprise a core encasing the RNA genome, surrounded by an outer matrix protein complex, with the producer cell plasma membrane displaying the SARS-CoV S envelope proteins, forming the pseudotype virus membrane. These mature pseudotype particles can subsequently be loaded onto target cells and the level of transduction/infection measured using appropriate methodologies (FACS, fluorescence microscopy, luminometry)

17.3 Neutralization Assay

A neutralization assay using SARS pseudotypes is performed routinely as follows: patient serum samples are heat-inactivated at 56°C for 30 min, serially diluted in appropriate culture medium, and mixed with SARS pseudotype virus at a 1:1 volume/volume ratio in a 96-well flat-bottomed tissue culture plate (clear plastic for GFP and β-gal, white plastic for luciferase). After incubation at 37°C for 1 h, approximately 1×10^4 target cells are added to each well. GFP, β-gal or luciferase positive cells are counted 24–48 h later, and neutralizing antibody titers calculated (Temperton et al. 2005). Similar pseudotype-based entry assays for influenza H5N1 have been evaluated in 384- and 1,536-well tissue culture plates for high-throughput screening applications (Wang et al. 2008).

17.4 Expanding the Repertoire of SARS-CoV (S) Pseudotypes

In order to achieve maximum sensitivity in serological assays for vaccine evaluation, and for seroprevalence studies, the selection of virus isolated from the same SARS outbreak, or the use of an antigenically equivalent strain, is required for

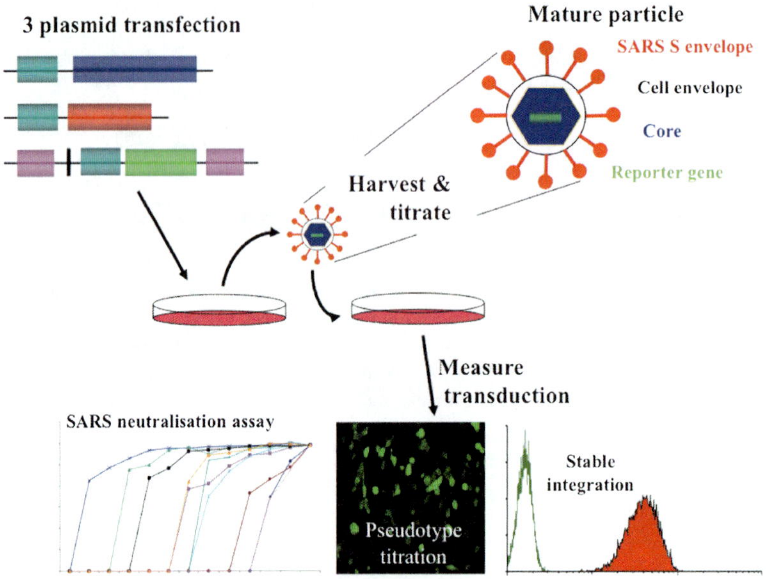

optimal antigenic match. Pseudotype viruses bearing SARS-CoV S envelopes from many different strains have been constructed and used as surrogate viruses in cell entry and neutralization assays. Hofmann et al. have constructed HIV pseudotypes using the Frankfurt strain S protein (Hofmann et al. 2004). Nie et al. have constructed HIV pseudotypes using synthetic codon-optimized S proteins from the BJ01, Frankfurt, TOR2 and TW1 strains of SARS-CoV in order to screen sera from patients with SARS (Nie et al. 2004a, 2004b). These four pseudotypes were normalized using a p24 ELISA assay. Assays were performed using human hepatoma Huh7 cells that endogenously express high levels of ACE2 receptor. Han et al. have constructed MLV pseudotypes using the Urbani strain S protein for the development of a safe neutralization assay for SARS-CoV (Han et al. 2004). Temperton et al. have constructed MLV and HIV pseudotypes using the Urbani strain S protein in order to profile neutralizing antibody responses to the S glycoprotein in sequential serum samples collected from recovered patients (Temperton et al. 2005). Assays were performed on quail QT6 cells stably over-expressing the ACE2 receptor. The FFM-1 strain of SARS was used by Giroglou et al. to construct MLV pseudotypes (Giroglou et al. 2004). Zhang et al. have constructed HIV pseudotypes using the HK-39 strain S protein to study neutralizing antibody responses in SARS-CoV infected patients. For these assays, a HEK293 cell line transduced with the SARS-CoV receptor ACE2 (293/ACE2) was used. Zhu et al. have used HIV pseudotypes bearing the S proteins from the TOR2, Urbani, GD03, SZ3 and SZ16 strains of SARS-CoV to study the in vitro neutralizing activity of human monoclonal antibodies (Zhu et al. 2007). Thus, to date at least ten different SARS-CoV spike glycoproteins have been incorporated successfully into retroviral pseudotypes and employed for a multitude of cell tropism and antibody assays.

Spike (S) proteins from other coronaviruses have also been efficiently incorporated into pseudotypes. Retroviral pseudotypes have been used to study receptor-mediated entry of the feline coronavirus, FCoV (Dye et al. 2007) and human coronaviruses NL63 and 229E (Hofmann et al. 2005). This powerful system is likely therefore to have pan-coronavirus applicability for virological and immunological studies.

If SARS-CoV were to reemerge and result in new human outbreaks it would be straightforward to update these neutralization assays to measure responses against newly emerging antigenic variants. Upon availability of viral RNA/cDNA or sequence information of the newly emergent SARS virus (from public nucleotide sequence databases such as the National Center for Biotechnology Information database held at http://www.ncbi.nlm.nih.gov/), the spike gene can be PCR-amplified or synthesized and retroviral pseudotypes prepared for use in downstream neutralization assays. With the cost of whole gene synthesis decreasing year on year, this is a viable proposition for the development of such serological assays, as access to the wild-type virus is often severely limited to a small group of laboratories (WHO, CDC, HPA etc.). These newly developed assays can be used to address the cross-clade neutralizing potential of candidate human vaccines and immuno-therapeutics (monoclonal antibodies etc.). They can also been used for the evaluation of anti-S virus entry inhibitors and could readily be adapted for sero-surveillance studies in new outbreak locations.

17.5 Pseudotype Titration and Target Cell Lines

When measuring neutralizing antibody responses against antigenically diverse SARS-CoV strains, it is necessary to accurately titre the pseudotype viruses so that virus input doses may be normalized. GFP-based pseudotypes are readily titred using flow cytometry (FACS), luciferase pseudotypes by luminometry and β-gal pseudotypes by colour reaction. The titres obtained by these methods are highly dependent on the cell type chosen for the assay, as ACE2 receptor density may differ widely between cell types. Many cell lines have been shown to be susceptible to SARS-CoV (Hattermann et al. 2005; Kaye 2006). These cell lines have been previously employed, or are predicted to be suitable target cells for use in pseudotype assays. They are BGM (buffalo green monkey kidney epithelium), COS (monkey kidney fibroblast), CV-1 (African green monkey kidney fibroblast), FRhK (rhesus monkey fetal kidney), LLC-Mk2 (rhesus monkey kidney epithelium), MA-104 (African green monkey kidney epithelium), MEK (monkey embryonic kidney), Vero (African green monkey kidney epithelium), Vero E6 (clone of Vero), HEK293 (human fetal kidney), HepG2 (human liver hepatocellular carcinoma), Huh-7 (human liver hepatocellular carconoma), Hep2 (human liver), RK-13 (rabbit kidney epithelium), POEK (porcine fetal kidney) and PS (porcine kidney).

Other methods for the titration and standardization of these retroviral pseudotypes are RT-PCR to determine virus particle number, ELISA to determine Gag levels (HIV or MLV Gag) in the retrovirus cores, and the measurement of surface envelope (SARS-CoV spike) content. Surface content may be measured by ELISA, as was recently shown for influenza H5N1 pseudotypes (Su et al. 2008).

17.6 Reporter Systems

In order for these pseudotype-based assays to have wide applicability within different laboratories, the availability of different reporter systems is highly desirable (Fig. 17.3). The HIV GFP reporter module pCSGW described recently for use in viral neutralization assays (Temperton et al. 2005, 2007; Wright et al. 2008) has been modified by PCR subcloning to express alternative reporters. These are firefly luciferase (Temperton et al. 2008; Wright et al. 2008), renilla luciferase (Capecchi et al. 2008) and β-gal. Of the three assay types, those based on luciferase reporter are the simplest to use in terms of operator time and data evaluation. However, due to the high cost of the necessary reagents (luciferase assay kits) and necessity for specialized equipment (luminometer), these assays may have limited applicability for laboratories in resource-poor regions. GFP and other fluorescence protein-based assays (RFP, YFP) do not require any additional assay reagents but do necessitate the availability of specialized equipment (fluorescence microscope with interchangeable colour filters, or a 96-well plate flow cytometry/FACS facility). The β-gal-based assays are the most cost-effective, as the required assay reagents are readily available at low cost, and specialized equipment is unnecessary.

Fig. 17.3 Reporter gene readout for SARS-CoV (S) pseudotype neutralization assays. The detection of pseudotype transduction for the quantification of neutralizing antibody responses against the SARS-CoV spike glycoprotein depends on the reporter/transfer gene incorporated into the mature particle. Options available include GFP, luciferase and β-gal. ONPG: *o*-nitrophenyl-β-D-galactopyranoside; CPRG: chlorophenol red-β-D-galactopyranoside

The expression of β-gal can be quantified by counting infected cells using a microscope following staining with X-gal (5-bromo-4-chloro-3-indolyl-β-D-galactopyranoside). However, there are two colorimetric substrates, ONPG (*o*-nitrophenyl-β-D-galactopyranoside) and CPRG (chlorophenol red-β-D-galactopyranoside) that allow the assays to be evaluated using a microplate reader. Alternatively an automated procedure using an ELISPOT reader to count X-gal stained blue cells may be used (Han et al. 2004). Finally, the availability of multiple reporters for use in these assays also lends itself to the development of multiplex assays which would facilitate the measurement of neutralizing antibody responses against surface glycoprotein envelopes derived from two (or more) different virus families (e.g., SARS-CoV and H5N1 for sero-surveillance in wet markets).

17.7 Correlation with Other Immunological Assays

Once suitable SARS-CoV pseudotype neutralization assays have been established, studies can be readily performed comparing these new surrogate assays with conventional live virus neutralization assays and other parameters of immune responses to SARS-CoV infection. Nie et al. have compared titres of neutralizing antibodies in 12 serum samples analyzed with pseudovirus and live SARS-CoV neutralization assays and found a strong correlation between the two (Nie et al. 2004a). Temperton et al. showed correlation data for neutralizing antibody titres measured by plaque reduction assay versus titres measured with a pseudotype assay

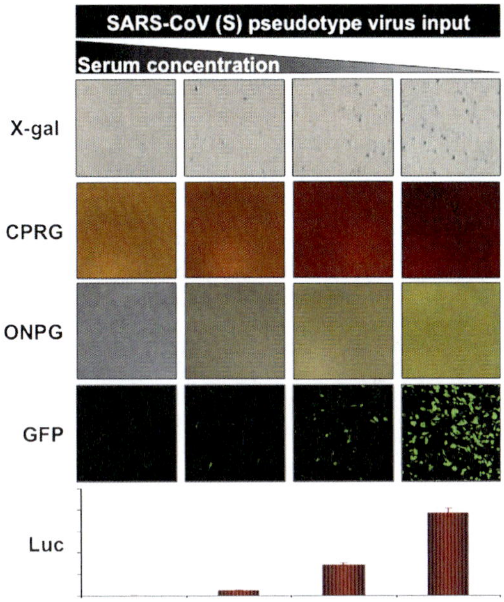

at both 50% and 90% inhibition of virus entry end-points. Correlation coefficients for live virus versus 50% and 90% pseudovirus inhibition were 0.78 and 0.69 respectively (Temperton et al. 2005). Zhang et al. studied temporal changes in N protein-specific (by quantitative ELISA) and S glycoprotein-specific neutralizing antibody responses (by SARS-CoV pseudotype assay) in infected patients who had either recovered from or succumbed to SARS-CoV infection (Zhang et al. 2006). A good correlation between strong antibody responses to N and S proteins and disease outcome of infected individuals was shown. In a large comprehensive study on 128 SARS convalescent patient sera, strong T-cell responses were shown to correlate significantly ($p < 0.05$) with higher neutralizing antibody titer as measured by an MLV(SARS) assay (Li et al. 2008).

17.8 Conclusion

In conclusion, SARS pseudotypes will enable much work that currently needs to be performed in Category III/IV laboratories to be undertaken with pseudotypes in a Category II environment. They represent a highly flexible and sensitive system for the evaluation of neutralizing antibodies against the spike surface glycoprotein.

Given that Research Councils UK and the Health Protection Agency (HPA) are currently inviting consultation on the UK national need for investment in high containment laboratories, it should be emphasized that the application of pseudotypes to highly pathogenic enveloped viruses such as SARS, H5N1 influenza A, rabies, West Nile and other high containment viruses can help to some extent to alleviate the pressure on expensive containment facilities.

Acknowledgments I would like to thank Edward Wright for invaluable assistance in the preparation of the figures, and Robin Weiss for his encouragement and mentorship.

References

Bermingham A, Heinen P, Iturriza-Gomara M, Gray J, Appleton H, Zambon MC (2004) Laboratory diagnosis of SARS. Philos Trans R Soc Lond 359:1083–1089

Capecchi B, Fasolo A, Alberini I, Baudner B, Crotta S, Temperton NJ, Montomoli E, del Giudice G, Rappuoli R (2008) Use of pseudotyped particles expressing Influenza A/Vietnam/1194/2004 hemagglutinin in neutralization assays. In Katz JM (ed) Options for the control of Influenza VI. International Medical, London, pp 303–305

Dye C, Temperton N, Siddell SG (2007) Type I feline coronavirus spike glycoprotein fails to recognize aminopeptidase N as a functional receptor on feline cell lines. J Gen Virol 88: 1753–1760

Giroglou T, Cinatl J Jr, Rabenau H, Drosten C, Schwalbe H, Doerr HW, von Laer D (2004) Retroviral vectors pseudotyped with severe acute respiratory syndrome coronavirus S protein. J Virol 78:9007–9015

Han DP, Kim HG, Kim YB, Poon LL, Cho MW (2004) Development of a safe neutralization assay for SARS-CoV and characterization of S-glycoprotein. Virology 326:140–149

Hattermann K, Muller MA, Nitsche A, Wendt S, Donoso Mantke O, Niedrig M (2005) Susceptibility of different eukaryotic cell lines to SARS-coronavirus. Arch Virol 150:1023–1031

Hofmann H, Hattermann K, Marzi A, Gramberg T, Geier M, Krumbiegel M, Kuate S, Uberla K, Niedrig M, Pohlmann S (2004) S protein of severe acute respiratory syndrome-associated coronavirus mediates entry into hepatoma cell lines and is targeted by neutralizing antibodies in infected patients. J Virol 78:6134–6142

Hofmann H, Pyrc K, van der Hoek L, Geier M, Berkhout B, Pohlmann S (2005) Human coronavirus NL63 employs the severe acute respiratory syndrome coronavirus receptor for cellular entry. Proc Natl Acad Sci USA 102:7988–7993

Kaye M (2006) SARS-associated coronavirus replication in cell lines. Emerg Infect Dis 12:128–133

Li CK, Wu H, Yan H, Ma S, Wang L, Zhang M, Tang X, Temperton NJ, Weiss RA, Brenchley JM, Douek DC, Mongkolsapaya J, Tran BH, Lin CL, Screaton GR, Hou JL, McMichael AJ, Xu XN (2008) T cell responses to whole SARS coronavirus in humans. J Immunol 181:5490–5500

Moore MJ, Dorfman T, Li W, Wong SK, Li Y, Kuhn JH, Coderre J, Vasilieva N, Han Z, Greenough TC, Farzan M, Choe H (2004) Retroviruses pseudotyped with the severe acute respiratory syndrome coronavirus spike protein efficiently infect cells expressing angiotensin-converting enzyme 2. J Virol 78:10628–10635

Nie Y, Wang G, Shi X, Zhang H, Qiu Y, He Z, Wang W, Lian G, Yin X, Du L, Ren L, Wang J, He X, Li T, Deng H, Ding M (2004a) Neutralizing antibodies in patients with severe acute respiratory syndrome-associated coronavirus infection. J Infect Dis 190:1119–1126

Nie Y, Wang P, Shi X, Wang G, Chen J, Zheng A, Wang W, Wang Z, Qu X, Luo M, Tan L, Song X, Yin X, Chen J, Ding M, Deng H (2004b) Highly infectious SARS-CoV pseudotyped virus reveals the cell tropism and its correlation with receptor expression. Biochem Biophys Res Commun 321:994–1000

Simmons G, Reeves JD, Rennekamp AJ, Amberg SM, Piefer AJ, Bates P (2004) Characterization of severe acute respiratory syndrome-associated coronavirus (SARS-CoV) spike glycoprotein-mediated viral entry. Proc Natl Acad Sci USA 101:4240–4245

Su CY, Wang SY, Shie JJ, Jeng KS, Temperton NJ, Fang JM, Wong CH, Cheng YS (2008) In vitro evaluation of neuraminidase inhibitors using the neuraminidase-dependent release assay of hemagglutinin-pseudotyped viruses. Antiviral Res 79:199–205

Temperton NJ, Wright E (2009) Retroviral pseudotypes. In: Encyclopedia of Life Sciences (ELS). Wiley, Chichester. DOI: 10.1002/9780470015902.a0021549

Temperton NJ, Chan PK, Simmons G, Zambon MC, Tedder RS, Takeuchi Y, Weiss RA (2005) Longitudinally profiling neutralizing antibody response to SARS coronavirus with pseudotypes. Emerg Infect Dis 11:411–416

Temperton NJ, Hoschler K, Major D, Nicolson C, Manvell R, Hien VM, Ha DQ, de Jong MD, Zambon M, Weiss RA (2007) A sensitive retroviral pseudotype assay for influenza H5N1 neutralizing antibodies. Influenza Other Respir Viruses 1:105–112

Temperton NJ, Capecchi B, Rappuoli R, de Jong MD, Takeuchi Y, Weiss RA (2008) Measurement of neutralizing antibody responses to influenza H5N1 clade 1 A/Viet Nam/1194/04 and clade 2 A/Indonesia/5/05 hemagglutinin using a sensitive high throughput luciferase-based retroviral pseudotype assay. In Katz JM (ed) Options for the Control of Influenza VI. International Medical Press, London, pp 94–96

Wang SY, Su CY, Temperton NJ, Jeng KS, Wong CH, Cheng YS (2008) HA-pseudotyped retroviral vectors for screening and evaluation of anti-flu inhibitors. In Katz JM (ed) Options for the Control of Influenza VI. International Medical Press, London, pp 472–474

Wright E, Temperton NJ, Marston DA, McElhinney LM, Fooks AR, Weiss RA (2008) Investigating antibody neutralization of lyssaviruses using lentiviral pseudotypes: a cross-species comparison. J Gen Virol 89:2204–2213

Wu HS, Chiu SC, Tseng TC, Lin SF, Lin JH, Hsu YH, Wang MC, Lin TL, Yang WZ, Ferng TL, Huang KH, Hsu LC, Lee LL, Yang JY, Chen HY, Su SP, Yang SY, Lin SY, Lin TH, Su IS

(2004) Serologic and molecular biologic methods for SARS-associated coronavirus infection, Taiwan. Emerg Infect Dis 10:304–310

Zhang L, Zhang F, Yu W, He T, Yu J, Yi CE, Ba L, Li W, Farzan M, Chen Z, Yuen KY, Ho D (2006) Antibody responses against SARS coronavirus are correlated with disease outcome of infected individuals. J Med Virol 78:1–8

Zhu Z, Chakraborti S, He Y, Roberts A, Sheahan T, Xiao X, Hensley LE, Prabakaran P, Rockx B, Sidorov IA, Corti D, Vogel L, Feng Y, Kim JO, Wang LF, Baric R, Lanzavecchia A, Curtis KM, Nabel GJ, Subbarao K, Jiang S, Dimitrov DS (2007) Potent cross-reactive neutralization of SARS coronavirus isolates by human monoclonal antibodies. Proc Natl Acad Sci USA 104:12123–12128

Chapter 18
SARS Coronavirus Spike Protein Expression in HL-CZ Human Promonocytic Cells: Monoclonal Antibody and Cellular Transcriptomic Analyses

T. Narasaraju, P.L. Soong, J. ter Meulen, J. Goudsmit, and Vincent T.K. Chow

Abstract The SARS coronavirus (CoV) spike protein is a target of intensive research, as it is a major virulence factor. Transfection of SARS-CoV spike into Vero E6, HEK293T and HL-CZ cells leads to strong expression of the glycosylated spike protein, as shown by Western blot analyses and immunofluorescent imaging using spike-specific human monoclonal antibodies, indicating the potential utility of these antigens and antibodies as diagnostic reagents. Furthermore, we employed cDNA microarray analysis to probe the changes in host gene transcription attributed to transfection of a codon-optimized spike construct into the HL-CZ cell line of monocyte lineage that is linked to immunological responses. A diverse representation of 100 genes displayed altered transcriptional patterns in response to SARS-CoV spike expression, with 61 upregulated and 39 downregulated genes. Genes involved in intracellular trafficking, signaling, modulation or transcription were generally upregulated. In contrast, genes involved in cell metabolism and cytoskeleton formation were notably downregulated. The transcripts of other functional categories exhibited varied responses to SARS-CoV spike transfection. Collectively, our analyses elucidate numerous and complex transcriptomic events that occur in response to spike protein expression and that contribute towards SARS-CoV pathogenesis.

V.T.K. Chow (✉)
Infectious Diseases Program, Department of Microbiology, Yong Loo Lin School of Medicine, National University Health System, National University of Singapore, Kent Ridge, Singapore 117597
e-mail: micctk@nus.edu.sg

18.1 SARS Coronavirus: A Lethal Zoonotic Virus with Future Potential for Reemergence

In November 2002, the outbreak of the newly emerging severe acute respiratory syndrome (SARS) in many parts of the world, especially in South-East Asia and China, resulted in more than 8,000 cases, with a high fatality rate of about 10%. Infection is characterized by acute flu-like symptoms, causing severe atypical pneumonia with diffuse alveolar damage, and death within two weeks in severely infected patients. This global outbreak of SARS wreaked havoc on public health and socioeconomic stability, but was finally brought under control in July 2003.

Intensive collaboration identified the etiological agent as a new member of the genus Coronavirus. Analyses of the complete nucleotide sequence of the novel SARS-associated coronavirus (SARS-CoV) reveal significant similarities to the other coronaviruses, with the genes encoding the spike (S), envelope (E), membrane (M) and nucleocapsid (N) proteins possessing similar sequence frames. In addition, the novel SARS-CoV contains several unique genes (Poon et al. 2004).

Similar viruses with considerable homology to human SARS-CoV have been isolated from animals such as palm civets and horseshoe bats, suggesting zoonotic reservoirs and the possible reemergence of this deadly virus (Shi and Hu 2008). Hence, the mechanisms of SARS-CoV pathogenicity are the subject of extensive scientific research (Satija and Lal 2007). Furthermore, it is highly imperative that effective and safe vaccines are developed to counter a potential SARS epidemic in the future (He et al. 2004a, b). Candidate SARS-CoV vaccines have been successfully tested in animal models and in clinical trials. The lack of specific therapies against SARS necessitates molecular approaches for selecting attractive targets to facilitate the design of novel vaccines and antiviral agents against SARS-CoV infections (Groneberg et al. 2005; Han et al. 2006; van den Brink et al. 2005; ter Meulen et al. 2004, 2006; Poh et al. 2009).

18.2 SARS-CoV Spike Glycoprotein Contributes to Virulence and Pathogenesis

Comprising 1,255 amino acid residues, the SARS-CoV spike is the first protein to interact with the host membrane, culminating in a cascade of infection events. Given that this glycoprotein plays a crucial role in viral virulence and pathogenesis, its functions have been extensively characterized, and it is probably the most important viral target for vaccine design (Du et al. 2009). Containing intracellular, extracellular and transmembrane domains, the spike protein is classified as a type 1 transmembrane glycoprotein that is responsible for virus binding, fusion and entry (Simmons et al. 2004; Zhang et al. 2004; Zhou et al. 2004). Of particular interest and importance is the S1 unit which constitutes one of the two subunits of the extracellular domain. This domain is responsible for binding of the virus to its main receptor, angiotensin-converting enzyme 2 (ACE2), present on susceptible cells (Li et al. 2003; Babcock et al. 2004). Moreover, CD209L (L-SIGN) and DC-SIGN also facilitate cell entry of SARS-CoV (Jeffers et al. 2004; Shih et al. 2006). SARS-CoV spike is selectively recognized by pulmonary surfactant D and activates macrophages (Leth-Larsen et al. 2007).

Vero E6 cells are derived from the African green monkey kidney clone E6, and are the most permissive for SARS-CoV infection and replication. Human cell lines that are susceptible to SARS-CoV infection include LoVo, HepG2 and Huh-7. The lytic infection produced in Vero E6 cells mimics the apoptosis observed in the respiratory epithelial cells of severely ill SARS patients (Leong et al. 2005; Chow and Leong 2008). In contrast, persistent infection of LoVo cells does not exhibit visible cytopathic effect (CPE).

18.3 Generation of Human Monoclonal Antibodies by Selection from Recombinant Single-Chain Antibody-Phage Libraries Constructed from the Lymphocytes of a Convalescent SARS Patient

Antibody phage display is an established technology for the isolation of antibodies from large libraries of these molecules (Pini and Bracci 2000). Antibody-binding regions (Fab or scFv) are cloned as fusions of the phage M13 coat proteins. The procedure results in the isolation of genes encoding antibody-binding regions specific for the target antigen. These genes can be used for the construction of fully human immunoglobulin molecules, i.e., phage libraries can be either naïve or immune. The latter library is generated using antibody genes extracted from the B cells of an immunized donor or from a convalescent patient. Both methods may yield antibodies that can be employed for therapeutic applications (Hoogenboom 2002).

An immune scFv phage library was constructed from the lymphocytes of a convalescent Singapore SARS patient with high neutralizing antibody titers. Briefly, several hundred scFv clones were checked for specific binding to inactivated SARS-CoV. After further selection, binding affinities of monovalent scFv were estimated, and scFv-binding epitopes were mapped. Finally, V regions were cloned into IgG expression vectors for the expression of human IgG_1 molecules in the human PER.C6 cell line.

Thus, human monoclonal antibodies against the SARS-CoV spike glycoprotein have been successfully generated, and shown to completely prevent lung pathology and abolish pharyngeal shedding of SARS-CoV in infected ferrets (ter Meulen et al. 2004, 2006).

18.4 Expression of SARS-CoV Spike Protein in Mammalian Cell Lines and Specific Detection by Human Monoclonal Antibodies: Potential Application as Diagnostic Reagents

The human monoclonal antibodies (16 designated pgG103-014C03P01, pgG103-019C03P01, pgG103-020C03P01, pgG103-022C03P01, pgG103-023C03P01, pgG103-023aC03P01, pgG103-024C03P01, pgG103-031C03P01, pgG103-046C03P01, pgG103-050C03P01, pgG103-053C03P01, pgG103-054C03P01, pgG103-055C03P01, pgG103-056C03P01, pgG103-057C03P01 and pgG103-059C03P01) were analyzed by Western blotting and immunocytochemistry on human cell lines (HL-CZ and HEK293T) transfected with spike construct or with the pcDNA3.1 vector alone as control.

18.4.1 Western Blot Characterization of HL-CZ Cells Transfected with Codon-Optimized SARS-CoV Spike Construct

For expression of the spike protein, the pCon-S-Harv-Myc-His plasmid (incorporating the extracellular domain of the spike protein) was constructed on a pcDNA3.1 vector backbone with a double tag of Myc and His (kindly provided by Dr Michael Farzan, Harvard University). The construction of this codon-optimized SARS-CoV spike ectodomain is more effective for spike protein expression during transfection in mammalian cells by replacing the natural codons with optimum codons (Babcock et al. 2004).

HL-CZ is a suspension cell line derived from human promonocytic leukemia that possesses CD15, CD34 and CD11b markers, and is permissive for the propagation of dengue virus, influenza virus type A and HIV-1 (Lai et al. 2007). Twenty-four hours after cell seeding, one batch of HL-CZ cells was transiently transfected with pCon-S-Harv-Myc-His using TransFectin (Bio-Rad), while another batch was mock-transfected with pcDNA3.1 only as control. Protein samples of both batches were then analyzed by SDS-PAGE and immunoblotting using the human monoclonal antibodies to SARS-CoV spike. Among the 16 antibodies, the antibody designated pgG103-057C03P01 detected a single distinct band of ~180 kDa compatible with the glycosylated spike ectodomain in the total cell lysate and pellet of HL-CZ cells transfected with SARS-CoV spike, but not in the corresponding samples of cells transfected with pcDNA3.1 control. With the pgG103-057C03P01 antibody, Western blotting also revealed intense SARS-CoV spike bands in both the lysate and supernatant of spike-transfected human embryonic kidney HEK293T cells. These immunoblot results were further confirmed by using spike polyclonal antibody IMG-541 (Imgenex) which detected a weaker spike protein band (Fig. 18.1).

Fig. 18.1 Western blot analyses of SARS-CoV spike protein in transfected HL-CZ and HEK293T cells. Lanes 1–7 were probed with anti-spike human monoclonal antibody pgG103-057C03P01 followed by secondary antihuman antibody. SARS-CoV spike bands of ~180 kDa were detected in cells transfected with spike construct, i.e., lanes 2 (HL-CZ lysate), 4 (HL-CZ pellet), 6 (HEK293T lysate) and 7 (HEK293T supernatant). A similar but weak band was observed in lane 8 (pellet of spike-transfected HL-CZ cells but probed with commercial antibody). This protein was absent in cells transfected with pcDNA3.1 vector alone, i.e., lanes 1 (HL-CZ lysate), 3 (HL-CZ pellet) and 5 (HEK293T lysate). The blots were developed using an ECL kit (Amersham Pharmacia). Lane M depicts the prestained Kaleidoscope Protein ladder markers (Bio-Rad).

18.4.2 Microscopic Analysis of SARS-CoV Spike Expression in HL-CZ Cells by Indirect Immunofluorescence Labeling

Prior to transfection, HL-CZ cells were cultured on glass coverslips coated with Cell-Tak (BD Biosciences). To visualize the intracellular distribution of the spike protein, immunofluorescence assay (IFA) was performed on cells transfected for 24 h with either pCon-S-Harv-Myc-His or control cells transfected with the pcDNA3.1 vector alone. Monoclonal antibody pgG103-057C03P01 exhibited the strongest staining of spike protein with predominant cytoplasmic localization (Fig. 18.2a–c). Antibodies pgG103-023C03P01 and pgG103-054C03P01 also displayed positive but relatively less staining, mainly in the cytoplasm. However, antibodies such as pgG103-055C03P01 did not demonstrate any significant fluorescence (data not shown).

Transfected cells were also incubated with primary rabbit polyclonal antibody against SARS-CoV spike and secondary antirabbit antibody labeled with rhodamine. DAPI counterstaining was performed to highlight cell nuclei. Confocal microscopy revealed the wide distribution of spike in the cytoplasm of Vero E6 and HL-CZ cells transfected with pCon-S-Harv-Myc-His. The intense immunostaining in the cytoplasm of HL-CZ cells may be attributed to spike localization in smooth membrane vesicles or smooth endoplasmic reticulum. Spike staining is prominent in Vero E6 cells, which may partly be explained by their adherent property (Fig. 18.2e–f).

While spike expression has also been localized to the cell membrane, confocal microscopy studies show that SARS-CoV spike protein subunit C localizes in the cytoplasm of Sf-9 cells. The detection sensitivity of the spike protein-based IFA is reflected by its ability to detect SARS-CoV specific antibody at a lower dilution compared to commercial IFAs (Manopo et al. 2005).

Our results indicate that the pgG103-057C03P01 human monoclonal antibody is highly specific for detecting SARS-CoV spike antigen. Furthermore, transfected

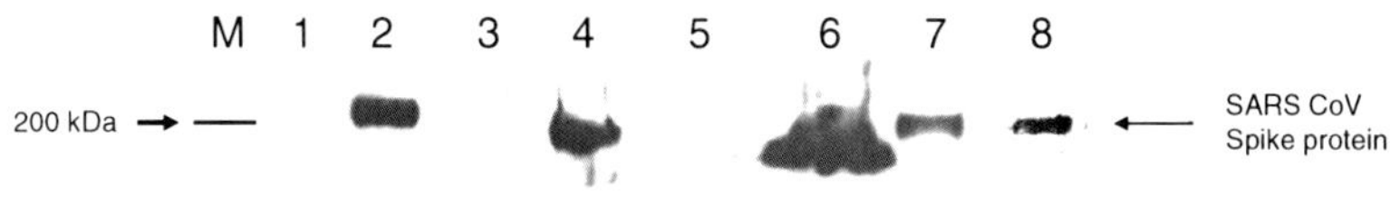

Fig. 18.2 Intracellular distribution of spike protein in HL-CZ cells transiently transfected for 24 h with (**a–c**) pCon-Spike-Harv-Myc-His construct or (**d**) pcDNA3.1 vector only, and incubated with anti-spike human monoclonal antibody pgG103-057C03P01 followed by secondary FITC-labeled antihuman antibody. (**e**) Vero and (**f**) HL-CZ cells were stained with DAPI to display nuclei, and with anti-spike polyclonal antibody followed by secondary antibody labeled with rhodamine. The superimposed images localize the spike protein to the cytoplasm giving rise to characteristic "ring" structures (*arrows*)

cell lines expressing the SARS-CoV spike extracellular domain can also serve as highly specific reagents for serodiagnosis of clinical specimens by Western blotting or immunocytochemistry.

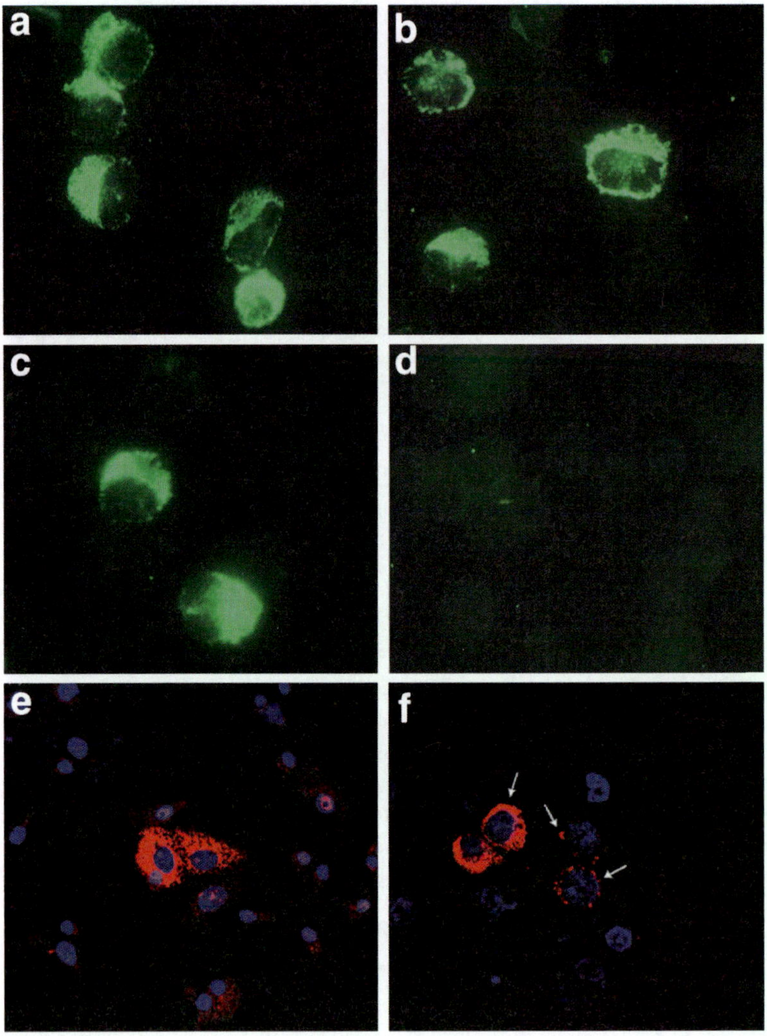

18.5 Transcriptomic Analysis Reveals Differentially Regulated Genes in HL-CZ Cells Following SARS-CoV Spike Transfection

The interactions between host and virus involve an elaborate and intricate interplay between molecular pathways initiated by the host in response to the infection and the mechanisms triggered by the virus to successfully infect the cell. Unraveling the transcriptomic events that ensue from the host–virus interactions can enhance our understanding of viral pathogenesis and disease manifestations, and can provide clues for designing preventive and therapeutic strategies to counter the infection (Leong and Chow 2006; Liew and Chow 2006). For example, multifarious and intriguing changes in genetic expression occur during infection of Vero E6 cells with whole live SARS-CoV (Leong et al. 2005). Compared with HL-CZ cells, Vero E6 cells are highly permissive to SARS-CoV infection and exhibit extensive CPE.

In order to gain insights into the role of the SARS-CoV spike protein in the human genetic responses and pathophysiology of SARS, we employed cDNA microarray to investigate gene expression changes in HL-CZ promonocytic cells following spike transfection. The various downstream transcriptomic changes that occur following this interaction will offer a clearer understanding of the role of spike during SARS-CoV pathogenesis.

Total cellular RNAs were extracted from HL-CZ cells transfected with pCon-S-Harv-Myc-His construct or with pcDNA3.1 alone using the SV Total RNA isolation kit (Promega). Labeling of first-strand cDNA with amino allyl-dUTP and purification of the CyDye-labeled cDNA prior to hybridization were performed with the CyScribe Post-Labeling kit and CyScribe GFX Purification kit (Amersham Biosciences). In both slides of set A of the gene array, Cy3 dye was incorporated into the spike-transfected cellular RNA, and Cy5 dye into vector-transfected cellular RNA. For set B, Cy5 and Cy3 dyes were incorporated into total RNA from cells transfected with spike and vector alone, respectively. This allows comparison of Cy-dye uptake efficiency as well as authentication of the microarray data.

Gene expression studies were performed using the Atlas Human 3.8I and 3.8II cDNA microarray panel (BD Biosciences) which comprises 7,600 human genes spotted onto two slides each consisting of 3,800 genes. Following microarray hybridization, the slides were scanned for Cy3 and Cy5 with a Genepix 4000B dual color image scanner (Axon Instruments).

Using the Acuity analysis software, low quality spots (e.g., extreme unevenness in intensity) were flagged off, but spots that exhibited sufficient signals were retained. Log values obtained from the scanning of labeled spike-transfected cellular cDNA were then normalized against values from labeled vector-transfected cDNA using locally weighted linear regression (LOWESS) analysis. Significant scan log ratios (spike versus vector transfection) of transcripts were then translated to actual fold changes in gene expression. By defining the threshold to be equivalent to greater than 1.96 standard deviations from the local mean based on the duplicated set, a total of 100 genes were identified at the 95% confidence level to be differentially expressed greater than two-fold. Of these, 61 transcripts were significantly upregulated, while 39 genes displayed reduced expression. Table 18.1 classifies the transcripts that were significantly altered at 24 h after spike transfection according to their known functional roles, including mRNAs of unknown function.

The most distinguishable genes to be downregulated by spike transfection were those involved in cell metabolism and cytoskeletal framework, while genes that mediate protein trafficking were generally upregulated. This was accompanied by a mixed response of transcripts that encode cell cycle and other proteins, suggesting an interesting interplay of host transcriptional alterations in response to spike protein expression.

18.5.1 Immune and Inflammatory Genes

Patients infected by SARS-CoV exhibit high plasma levels of IL1, IL6, IL8, IL12, monocyte chemoattractant protein-1 and IFN-γ inducible protein. High levels of IL8 and IL6 are associated with lung lesions in SARS patients at the acute stage. Elevations of IL8 and Th1-related cytokines induce hyperinflammatory innate responses following SARS-CoV invasion of the respiratory tract (Wang et al. 2007; Thiel and Weber 2008). Recombinant baculovirus expressing amino acids 17–688 of the spike protein induces AP-1 activation through MAP kinases and AP-1, with the domains spanning residues 324–488 and 609–688 being crucial for the induction of IL8 release in lung cells (Chang et al. 2004). IL8 induction can be specifically inhibited by antibody against ACE2. The IL-8 receptor binds to IL8 with high affinity, also binds to chemokine (C–X–C motif) ligand 1 (CXCL1/MGSA), and mediates neutrophil migration to sites of inflammation.

From our microarray data, the sharp upregulation of IL8 receptor transcript β (~12-fold increase) implies that spike actively induces expression of this receptor which has a stronger affinity for IL8. This finding concurs with Versteeg et al. (2007) who documented IL8 receptor induction by SARS-CoV spike. Interestingly, our previous study showed that the expression of IL8 receptor A or IL8RA (which is a low affinity receptor for IL8) is also upregulated in Vero E6 cells infected with SARS-CoV (Leong et al. 2005). Overproduction of specific inflammatory cytokines is a typical hallmark of SARS-CoV infections, and overactivity of the IL-8 receptor may be partly mediated by spike protein as an inflammatory response.

Another upregulated gene is the major histocompatibility complex (MHC) class I A. MHC class I molecules are expressed in nearly all cells, and play a central role in the immune system by presenting peptides derived from the endoplasmic reticulum lumen. Given that HL-CZ cells are derived from adult T cells on which MHC class I is typically expressed, upregulation of this gene is not surprising. In addition, the B-cell-specific transcription factor mRNA is increased ~9-fold upon spike transfection, indicating B-lymphocyte activation in response to SARS-CoV infection.

Table 18.1 Categories of human genes showing altered transcription following SARS-CoV spike transfection in HL-CZ cells

GenBank accession no.	Genes and their classification	Fold change in transcription	Up or down regulation
Cell cycle and development			
NM_006527	Stem-loop (histone) binding protein	2.18	↓
NM_003495	Histone 1, H4i	2.14	↑
NM_005474	Histone deacetylase 5	2.18	↑
NM_002895	Retinoblastoma-like 1 (p107)	2.74	↑
AF059617	Serum-inducible kinase	2.78	↓
U66838	Cyclin A1	3.95	↑
M26708	Prothymosin alpha	6.37	↑
Intracellular modulators			
NM_006762	Lysosomal associated multispanning membrane protein 5	3.25	↓
NM_016201	Angiomotin-like 2	2.15	↓
NM_004822	Netrin 1	2.10	↑
NM_004078	Cysteine and glycine-rich protein 1	2.19	↑
NM_016084	RAS, dexamethasone-induced 1	4.13	↑
U43148	Patched homolog	2.53	↓
X65293	Protein kinase C-epsilon	2.12	↓
U20537	Cysteine protease Mch2 isoform beta	2.11	↓
X74210	Adenylyl cyclase 2	2.11	↑
L13698	Gas1	2.30	↑
M68520	Cdc2-related protein kinase	2.51	↑
M55983	DNase I	2.69	↑
U02082	Guanine nucleotide regulatory protein	2.80	↑
D45887	Calmodulin	3.45	↑
M27544	Insulin-like growth factor	4.78	↑
X00351	Beta-actin	7.69	↑
Cell metabolism			
NM_016327	Ureidopropionase, beta	3.46	↓
NM_016203	Protein kinase, AMP-activated, gamma 2 non-catalytic subunit	2.40	↓
M61856	Cytochrome P450, subfamily IIC (mephenytoin 4-hydroxylase), polypeptide 18	2.12	↓
Cytoskeleton			
NM_003461	Zyxin	44.72	↑
NM_007127	Villin 1	4.06	↓
X03212	Mesothelial type II keratin K7	2.18	↓
X07695	Cytokeratin 4 C-terminal region	2.09	↓
M99063	Cytokeratin 2	2.05	↓
Signal transduction			
NM_004734	Doublecortin and CaM kinase-like 1	4.59	↓
NM_002730	Protein kinase, cAMP-dependent, catalytic, alpha	16.53	↑
L29511	GRB2 isoform	3.59	↓
L26584	Ras protein-specific guanine nucleotide-releasing factor 1	4.68	↑
U13667	G protein-coupled receptor	10.39	↑
X14034	Phospholipase C	10.64	↑
Protein trafficking			
NM_015930		2.07	↓

(*continued*)

Table 18.1 (continued)

GenBank accession no.	Genes and their classification	Fold change in transcription	Up or down regulation
	Transient receptor potential cation channel, subfamily V, member 2		
NM_007033	RER1 retention in endoplasmic reticulum 1 homolog	2.01	↓
NM_005085	Nucleoporin 214 kDa	2.07	↑
NM_016601	Potassium channel, subfamily K, member 9	2.15	↑
NM_007263	Coatomer protein complex, subunit epsilon	2.27	↑
NM_015994	ATPase, H + transporting, lysosomal 34 kDa, V1 subunit D	2.38	↑
NM_013245	Vacuolar protein sorting 4A	3.32	↑
M77235	Cardiac tetrodotoxin-insensitive voltage-dependent sodium channel alpha subunit	2.39	↑
X91788	Icln protein	5.07	↑
S70609	Glycine transporter type 1b	2.70	↑
Protein translation			
NM_016199	LSM7 homolog, U6 small nuclear RNA associated	2.55	↓
NM_005520	Heterogeneous nuclear ribonucleoprotein H1	2.07	↑
NM_016024	RNA binding motif protein, X-linked 2	3.05	↑
Transcriptional regulators			
NM_003428	Zinc finger protein 84	2.58	↑
NM_003425	Zinc finger protein 45	3.03	↑
NM_003426	Zinc finger protein 74	5.29	↑
NM_003446	Zinc finger protein 157	10.41	↑
NM_003430	Zinc finger protein 91	46.05	↑
NM_004865	TBP-like 1	6.05	↓
Cell receptors			
NM_016240	Scavenger receptor class A, member 3	2.14	↑
NM_002116	Major histocompatibility complex, class I, A	2.40	↑
NM_001557	Interleukin 8 receptor, beta	12.44	↑
S85655	Prohibitin	3.68	↓
L04947	Receptor tyrosine kinase	2.06	↓
Miscellaneous proteins			
NM_005625	Syndecan binding protein (syntenin)	221.40	↑
NM_003462	Dynein, axonemal, light intermediate polypeptide 1	2.45	↑
NM_015878	Antizyme inhibitor 1	4.51	↓
NM_002450	Metallothionein 1 L	3.48	↓
NM_004882	CBF1 interacting corepressor	3.22	↓
NM_016316	REV1-like	2.43	↓
NM_005822	Down syndrome critical region gene 1-like 1	2.07	↓
NM_002780	Pregnancy specific beta-1-glycoprotein 4	2.01	↓
NM_016229	Cytochrome b5 reductase 2	2.06	↑
NM_006303	JTV1	2.36	↑
NM_016243	Cytochrome b5 reductase 1	2.53	↑
NM_016023	CGI-77 protein	3.22	↑
NM_016528	Hydroxyacid oxidase 3 (medium-chain)	4.02	↑
NM_016227	Chromosome 1 open reading frame 9	7.58	↑
NM_006982	Cartilage paired-class homeoprotein 1	13.85	↑
U02081	Guanine nucleotide regulatory protein	2.67	↓
Z29083	5 T4 oncofetal antigen	2.52	↓
M38690	CD9 antigen	2.43	↓

(*continued*)

Table 18.1 (continued)

GenBank accession no.	Genes and their classification	Fold change in transcription	Up or down regulation
M81882	Glutamate decarboxylase	2.33	↓
M37033	CD53 glycoprotein	2.20	↓
U91618	Proneurotensin/proneuromedin N	2.09	↓
Y00978	Dihydrolipoamide acetyltransferase	2.07	↓
AB011539	MEGF6	2.24	↑
Z24459	MTCP1	2.31	↑
U49089	Neuroendocrine-dlg	3.13	↑
NM_003460	Zona pellucida glycoprotein 2 (sperm receptor)	2.33	↑
Hypothetical proteins			
NM_015919	Zinc finger protein 226	2.18	↓
NM_004793	Protease, serine, 15	2.02	↓
NM_012118	CCR4 carbon catabolite repression 4-like	2.33	↑
U08853	Transcription factor LCR-F1	2.07	↑
AF060515	Cyclin K	2.34	↑
U08191	R kappa B	2.77	↑
M83221	I-Rel	3.12	↑
AF015950	Telomerase reverse transcriptase	4.08	↑
D28118	DB1	8.47	↑
M96944	B-cell specific transcription factor	9.19	↑
NM_016385	Cylindromatosis (turban tumor syndrome)	3.27	↓
NM_016627	Archaemetzincins-2	2.78	↓
NM_015913	Thioredoxin domain containing 12 (endoplasmic reticulum)	2.46	↑

18.5.2 Cell Cycle Genes

Infectious bronchitis coronavirus infection induces cell cycle perturbations with downstream effects on viral replication (Dove et al. 2006). SARS-CoV spike expression in HL-CZ cells is also accompanied by differential responses of transcripts that encode cell cycle-related proteins. Following spike transfection, the cyclin A1 transcript is upregulated ~4-fold. Cyclin A1 binds to and regulates cdk1 and cdk2 which are critical factors during the S and G_2 phases of the cell cycle. Consequently, modification of cyclin A1 transcription has an indirect impact on virus maturation and life cycle. In HIV-1 progeny formation, cyclin A1 binds Rb family members, and the p21/waf1 family of endogenous cdk inhibitors, as well as the E2F-1 transcriptional factor, all of which are important in regulating cell cycle progression (Liang et al. 2005).

Also upregulated is the retinoblastoma-like 1 (p107) (RBL1) transcript variant 1 whose protein is similar in sequence and possibly function to the retinoblastoma 1 (RB1) gene product. RB1 is a tumor suppressor protein involved in cell cycle regulation, being phosphorylated in the S to M phase transition, but is dephosphorylated in the G_1 phase.

Prothymosin α is an abundant acidic nuclear protein with a role in supporting cell proliferation. It is present in all mammalian tissues, and is usually proportional to the proliferative activity of the tissue. Although this protein does not directly regulate cell division, it is required for entry from the G_2 to M phase, and its mRNA is induced in normal human lymphocytes when stimulated by mitogens (Gómez-Márquez and Rodríguez 1998). Spike transfection upregulates prothymosin α (6.4-fold), compatible with extensive studies establishing that viral infections transmit proliferation signals to otherwise terminally differentiated cells (Vareli et al. 1995). Cells that overexpress this transcript also exhibit higher levels of histone-H1 depleted chromatin, by interacting with histone H1, leading to the decondensation and remodeling of chromatin fibers. Such changes in chromatin organization are important in essential cellular processes such as recombination, replication, transcription and chromosome packaging. Microarray analysis demonstrated upregulation of the histone H1 transcript concurrent with that of prothymosin α.

Another notable upregulated transcript (~2.2-fold) is histone deacetylase 5 (HDAC5) which also plays an important role in chromatin remodeling. Whether a cell is permissive for viral infection may be dependent on the state of cellular differentiation. For example, permissiveness for human cytomegalovirus (HCMV) infection is linked to repression of the viral major immediate early promoter (MIEP). In HCMV, monocytes represent a site of viral latency in HCMV carriers, and reactivated virus is only observed upon differentiation into macrophages. Histone deacetylases (HDACs) are involved in MIEP repression in nonpermissive cells, since inhibition of HDACs induces viral permissiveness and increases MIEP activity (Murphy et al. 2002). HDACs modulate MIEP transcriptional activity by affecting the acetylation of transcriptional factors as well as histones. A related protein, HDAC6, regulates formation of the HIV-mediated fusion pore. Binding of the HIV gp120 protein to CD4-positive permissive cells increases the level of acetylated α-tubulin. Overexpression of active HDAC6 inhibits tubulin acetylation, thus preventing HIV-1 envelope-dependent cell fusion as well as infection without affecting the expression of other HIV-1 receptors (Valenzuela-Fernández et al. 2005).

18.5.3 Cytoskeletal Genes

SARS-CoV-infected cells exhibit diminished expression of genes associated with the maintenance of cytoskeletal structure. Many cellular processes depend on cytoskeletal rearrangements. Villin 1 is a structural protein regulated by increased intracellular calcium that in turn binds to actin, and contributes to the formation of microvilli in the small intestine. Cytokeratins are part of the family of keratins consisting of about 20 different proteins that comprise intermediate filaments in almost all epithelial cells. These filaments form the internal infrastructure of cells and are vital for maintaining cellular integrity. Disruption of such features affects cellular flexibility, tensile strength and interaction with other cellular components (Glass et al. 2006). Infection of animal cells by a number of viruses generally results in CPE. In lytic viral infections, there is a profound impairment of cell structural integrity, and inhibition of host protein synthesis related to widespread destruction of the intermediate filament network, e.g. proteolysis of cytokeratins K7 and K18 by the adenovirus late-acting L3 proteinase (Zhang and Schneider 1994). The collapse of this network (including vimentin and lamin), coupled with suppression of new keratin synthesis to restore the damaged filament network, contributes to CPE. In SARS-CoV infection, these microtubule networks may become disrupted, making it conducive for subsequent viral release, partially attributed to the spike protein.

Following SARS-CoV spike transfection, the mRNA encoding zyxin is greatly upregulated ~45-fold. Zyxin enhances cell motility through rearrangement of the actin cytoskeleton, and it facilitates the formation of the molecular complex that promotes the assembly of actin-rich structures targeted particularly to the inner face of the plasma membrane (Beckerle 1997). Correct positioning of zyxin within the cell is also critical for its physiological function. Mislocation of zyxin affects cell migration and spreading, influences behavior of the cell edge, and causes irregular distribution of important proteins that promote actin assembly (Drees et al. 1999). Furthermore, the β-actin transcript is upregulated ~7.7-fold, suggesting the complex interplay between the various elements that contribute to structural integrity within the cell.

18.5.4 *Trafficking and Transport Genes*

The vacuolar protein sorting 4A gene transcript is upregulated upon spike transfection. In many viral infections, viral proteins must recruit a variety of cellular factors, including members of the vacuolar protein sorting family, for subsequent viral budding and egress, failing which the viral particles would be nearly assembled but remain tethered to the cell surface (Sherer et al. 2003).

Another noteworthy upregulated transcript (~5-fold) is ICln mRNA which encodes a 42-kDa chloride channel regulatory protein essential in regulating cell volume, and which interacts with the specific platelet integrin $\alpha_{IIb}\beta_3$ through the KVGFFKR motif on platelet membranes (Larkin et al. 2004). Treatment with the specific channel protein inhibitory agent, acyclovir, culminates in inhibition of platelet aggregation and integrin activation. It is also noteworthy that the SARS-CoV protein 3a upregulates fibrinogen expression in lung epithelial cells, congruent with observations of SARS patients with thrombocytopenia, and suggesting dysfunctional coagulation and activation of the fibrinogen pathway (Tan et al. 2005).

18.6 Conclusions and Future Prospects

This is the first report documenting SARS-CoV expression in the human HL-CZ cell line that triggers a cascade of host transcriptional responses involving a wide repertoire of genes belonging to key functional classes including inflammation, immunity, cell signaling and trafficking. Intriguingly, most of the transcriptional factors upregulated in response to spike transfection are zinc finger proteins. In contrast to live SARS-CoV infection, this study provides comparative insights into the host transcriptomic effects of the spike protein when expressed alone in relevant human cells, thus shedding light on its specific roles during viral infection. These data indicate that SARS-CoV spike expression significantly modifies the regulation of numerous genes in HL-CZ cells, reiterate its importance in SARS pathogenesis, and further justify it as a key target for vaccine development. Such studies can enhance our understanding of the "infectomics" of viral infections in different human and mammalian cell types. More in-depth analyses of the individual genes may be conducted by quantitative real-time RT-PCR to verify the microarray data, and by using gene-knockout animal models to further explore their specific roles in SARS-CoV infection in vivo.

Acknowledgments The authors thank J. de Kruif, H.N. Leong, B. Liu and W.M. Yeo for their valuable assistance. This study was funded by the Biomedical Research Council, Singapore and the Microbiology Vaccine Initiative, National University of Singapore.

References

Babcock GJ, Esshaki DJ, Thomas WD, Ambrosino DM (2004) Amino acids 270 to 510 of the severe acute respiratory syndrome coronavirus spike protein are required for interaction with receptor. J Virol 78:4552–4560
Beckerle MC (1997) Zyxin: zinc fingers at sites of cell adhesion. Bioessays 19:949–957
Chang YJ, Liu CY, Chiang BL, Chao YC, Chen CC (2004) Induction of IL-8 release in lung cells via activator protein-1 by recombinant baculovirus displaying severe acute respiratory syndrome-coronavirus spike proteins: identification of two functional regions. J Immunol 173:7602–7614
Chow VT, Leong WF (2008) Severe acute respiratory syndrome coronavirus induces differential host gene expression responses associated with pathogenesis. In: Yang D (ed) RNA viruses: host gene response to infection. World Scientific, New Jersey, pp 295–320
Dove B, Brooks G, Bicknell K, Wurm T, Hiscox JA (2006) Cell cycle perturbations induced by infection with the coronavirus infectious bronchitis virus and their effect on virus replication. J Virol 80:4147–4156
Drees EB, Andrews KM, Beckerle MC (1999) Molecular dissection of zyxin function reveals its involvement in cell motility. J Cell Biol 147:1549–1559
Du L, He Y, Zhou Y, Liu S, Zheng BJ, Jiang S (2009) The spike protein of SARS-CoV – a target for vaccine and therapeutic development. Nat Rev Microbiol 7:226–236
Glass C, Kim KH, Fuchs E (2006) Sequence and expression of a human type II mesothelial keratin. J Cell Biol 101:2366–2373
Gómez-Márquez J, Rodríguez P (1998) Prothymosin α is a chromatin-remodelling protein in mammalian cells. Biochem J 333:1–3
Groneberg DA, Poutanen SM, Low DE, Lode H, Welte T, Zabel P (2005) Treatment and vaccines for severe acute respiratory syndrome. Lancet Infect Dis 5:147–155
Han DP, Penn-Nicholson A, Cho MW (2006) Identification of critical determinants on ACE2 for SARS-CoV entry and development of a potent entry inhibitor. Virology 350:15–25
He Y, Zhou Y, Liu S, Kou Z, Li W, Farzan M, Jiang S (2004a) Receptor-binding domain of SARS-CoV spike protein induces highly potent neutralizing antibodies: implication for developing subunit vaccine. Biochem Biophys Res Commun 324:773–781

He Y, Zhou Y, Siddiqui P, Jiang S (2004b) Inactivated SARS-CoV vaccine elicits high titers of spike protein-specific antibodies that block receptor binding and virus entry. Biochem Biophys Res Commun 325:445–452

Hoogenboom HR (2002) Overview of antibody phage-display technology and its applications. Methods Mol Biol 178:1–37

Jeffers SA, Tusell SM, Gillim-Ross L, Hemmila EM, Achenbach JE, Babcock GJ, Thomas WD Jr, Thackray LB, Young MD, Mason RJ, Ambrosino DM, Wentworth DE, Demartini JC, Holmes KV (2004) CD209L (L-SIGN) is a receptor for severe acute respiratory syndrome coronavirus. Proc Natl Acad Sci USA 101:15748–15753

Lai CC, Jou MJ, Huang SY, Li SW, Wan L, Tsai FJ, Lin CW (2007) Proteomic analysis of up-regulated proteins in human promonocyte cells expressing severe acute respiratory syndrome coronavirus 3C-like protease. Proteomics 7:1446–1460

Larkin D, Murphy D, Reilly DF, Cahill M, Sattler E, Harriott P, Cahill DJ, Moran N (2004) ICln, a novel integrin $\alpha_{IIb}\beta_3$-associated protein, functionally regulates platelet activation. J Biol Chem 279:27286–27293

Leong WF, Chow VT (2006) Transcriptomic and proteomic analyses of rhabdomyosarcoma cells reveal differential cellular gene expression in response to enterovirus 71 infection. Cell Microbiol 8:565–580

Leong WF, Tan HC, Ooi EE, Koh DR, Chow VT (2005) Microarray and real time RT-PCR analyses of differential human gene expression patterns induced by severe acute respiratory syndrome (SARS) coronavirus infection of Vero E6 cells. Microbes Infect 7:248–259

Leth-Larsen R, Zhong F, Chow VT, Holmskov U, Lu J (2007) The SARS coronavirus spike glycoprotein is selectively recognized by lung surfactant protein D and activates macrophages. Immunobiology 212:201–211

Li W, Moore MJ, Vasilieva N, Sui J, Wong SK, Berne MA, Somasundaran M, Sullivan JL, Luzuriaga K, Greenough TC, Choe H, Farzan M (2003) Angiotensin-converting enzyme 2 is a functional receptor for the SARS coronavirus. Nature 426:450–454

Liang WS, Maddukuri A, Teslovich TM, de la Fuente C, Agbottah E, Dadgar S, Kehn K, Hautaniemi S, Purnfery A, Stephan DA, Kashanchi F (2005) Therapeutic targets for HIV-1 infection in the host proteome. Retrovirology 2:20

Liew KJ, Chow VT (2006) Microarray and real-time RT-PCR analyses of a novel set of differentially expressed human genes in ECV304 endothelial-like cells infected with dengue virus type 2. J Virol Methods 131:47–57

Manopo I, Lu L, He Q, Chee LL, Chan SW, Kwang J (2005) Evaluation of a safe and sensitive Spike protein-based immunofluorescence assay for the detection of antibody responses to SARS-CoV. J Immunol Methods 296:37–44

Murphy JC, Fischle W, Verdin E, Sinclair JH (2002) Control of cytomegalovirus lytic gene expression by histone acetylation. EMBO J 21:1112–1120

Pini A, Bracci L (2000) Phage display of antibody fragments. Curr Protein Pept Sci 1:155–169

Poh WP, Narasaraju T, Pereira NA, Zhong F, Phoon MC, Macary PA, Wong SH, Lu J, Koh DR, Chow VT (2009) Characterization of cytotoxic T-lymphocyte epitopes and immune responses to SARS coronavirus spike DNA vaccine expressing the RGD-integrin-binding motif. J Med Virol 81:1131–1139

Poon LL, Guan Y, Nicholls JM, Yuen KY, Peiris JS (2004) The aetiology, origins, and diagnosis of severe acute respiratory syndrome. Lancet Infect Dis Rev 4:663–671

Satija N, Lal SK (2007) The molecular biology of SARS coronavirus. Ann N Y Acad Sci 1102:26–38

Sherer NM, Lehmann MJ, Jemenez-Soto LF, Ingmundson A, Horner SM, Cicchetti G, Allen PG, Pypaert M, Cunningham JM, Mothes W (2003) Visualization of retroviral replication in living cells reveals budding into multivesicular bodies. Traffic 4:785–801

Shi Z, Hu Z (2008) A review of studies on animal reservoirs of the SARS coronavirus. Virus Res 133:74–87

Shih YP, Chen CY, Liu SJ, Chen KH, Lee YM, Chao YC, Chen YM (2006) Identifying epitopes responsible for neutralizing antibody and DC-SIGN binding on the spike glycoprotein of the severe acute respiratory syndrome coronavirus. J Virol 80:10315–10324

Simmons G, Reeves JD, Rennekamp AJ, Amberg SM, Piefer AJ, Bates P (2004) Characterization of severe acute respiratory syndrome-associated coronavirus (SARS-CoV) spike glycoprotein-mediated viral entry. Proc Natl Acad Sci 101:4240–4245

Tan YJ, Tham PY, Chan DZ, Chou CF, Shen S, Fielding BC, Tan TH, Lim SG, Hong W (2005) The severe acute respiratory syndrome coronavirus 3a protein up-regulates expression of fibrinogen in lung epithelial cells. J Virol 79:10083–10087

ter Meulen J, Bakker AB, van den Brink EN, Weverling GJ, Martina BE, Haagmans BL, Kuiken T, de Kruif J, Preiser W, Spaan W, Gelderblom HR, Goudsmit J, Osterhaus AD (2004) Human monoclonal antibody as prophylaxis for SARS coronavirus infection in ferrets. Lancet 363:2139–2141

ter Meulen M, van den Brink EN, Poon LL, Marissen WE, Leung CS, Cox F, Cheung CY, Bakker AQ, Bogaards JA, van Deventer E, Preiser W, Doerr HW, Chow VT, de Kruif J, Peiris JS, Goudsmit J (2006) Human monoclonal antibody combination against SARS coronavirus: synergy and coverage of escape mutants. PLoS Med 3:e237

Thiel V, Weber F (2008) Interferon and cytokine responses to SARS-coronavirus infection. Cytokine Growth Factor Rev 19:121–132

Valenzuela-Fernández A, Álvarez S, Gordon-Alonso M, Barrero M, Ursa Á, Cabrero JR, Fernández G, Naranjo-Suárez S, Yáñez-Mo M, Serrador JM, Muñoz-Fernández MÁ, Sánchez Madrid F (2005) Histone deacetylase 6 regulates human immunodeficiency virus type 1 infection. Mol Biol Cell 16:5445–5454

van den Brink EN, Ter Meulen J, Cox F, Jongeneelen MA, Thijsse A, Throsby M, Marissen WE, Rood PM, Bakker AB, Gelderblom HR, Martina BE, Osterhaus AD, Preiser W, Doerr HW, de Kruif J, Goudsmit J (2005) Molecular and biological characterization of human monoclonal antibodies binding to the spike and nucleocapsid proteins of severe acute respiratory syndrome coronavirus. J Virol 79:1635–1644

Vareli K, Lazaridis MF, Tsolas O (1995) Prothymosin α mRNA levels vary with c-myc expression during tissue proliferation, viral infection and heat shock. FEBS Lett 371:337–340

Versteeg GA, van de Nes PS, Bredenbeek PJ, Spaan WJ (2007) The coronavirus spike protein induces endoplasmic reticulum stress and upregulation of intracellular chemokine mRNA concentrations. J Virol 81:10981–10990

Wang W, Ye L, Ye L, Li B, Gao B, Zeng Y, Kong L, Fang X, Zheng H, Wu Z, She Y (2007) Up-regulation of IL-6 and TNF-alpha induced by SARS-coronavirus spike protein in murine macrophages via NF-kappaB pathway. Virus Res 128:1–8

Zhang Y, Schneider RJ (1994) Adenovirus inhibition of cell translation facilitates release of virus particles and enhances degradation of the cytokeratin network. J Virol 68:2544–2555

Zhang H, Wang G, Li J, Nie Y, Shi X, Lian G, Wang W, Yin X, Zhao Y, Qu X, Ding M, Deng H (2004) Identification of an antigenic determinant on the S2 domain of the severe acute respiratory syndrome coronavirus spike glycoprotein capable of inducing neutralizing antibodies. J Virol 78:6938–6945

Zhou T, Wang H, Luo D, Rowe T, Wang Z, Hogan RJ, Qiu S, Bunzel RJ, Huang G, Mishra V, Voss TG, Kimberly R, Luo M (2004) An exposed domain in the severe acute respiratory syndrome coronavirus spike protein induces neutralizing antibodies. J Virol 78:7217–7226

Chapter 19
Signaling Pathways of SARS-CoV In Vitro and In Vivo

Tetsuya Mizutani

Abstract Severe acute respiratory syndrome (SARS) is a respiratory illness with variable symptoms that was recognized as the first near-pandemic infectious disease of the twenty-first century. A novel human coronavirus, named SARS coronavirus (SARS-CoV), derived from SARS patients was reported as the etiologic agent of SARS. Studying the signaling pathways of SARS-infected cells is key to understanding the molecular mechanism of SARS viral infection. Cell death is observed in cultured Vero E6 cells after SARS-CoV infection. From SARS-CoV infection to cell death, p38 mitogen-activated protein kinase (MAPK) is a key participant in the determination of cell death and survival. Two signaling pathways comprising signal transducer and activator of transcription 3 (STAT3) and p90 ribosomal S6 kinase (p90RSK) are downstream of p38 MAPK. AKT and JNK (Jun NH$_2$-terminal kinase) signaling pathways are important to establish persistent infection of SARS-CoV in Vero E6 cells. Expression studies of SARS-CoV proteins indicate that the viral proteins are able to activate signaling pathways of host cells. The study of signaling pathways in SARS-CoV patients is difficult to perform compared with in vitro studies due to the effects of the human immune system. This review highlights recent progress in characterizing signal transduction pathways in SARS-CoV-infected cells in vitro and in vivo.

19.1 Introduction

Severe acute respiratory syndrome (SARS) is a respiratory illness with variable flu-like symptoms and pneumonia, which is caused by the SARS coronavirus (SARS-CoV) (Drosten et al. 2003; Ksiazek et al. 2003; Peiris et al. 2003a, 2003b; Poutanen

T. Mizutani
Department of Virology I, National Institute of Infectious Diseases, Gakuen 4-7-1, Musashimurayama, Tokyo208-0011, Japan
e-mail: tmizutan@nih.go.jp

et al. 2003; Tsang et al. 2003). SARS was first recognized in China in November 2002 and subsequently spread to 29 other countries, thus emerging as the first near-pandemic infectious disease of the twenty-first century. A worldwide total of 8,096 cases of SARS, of which 774 (9.6%) resulted in death, was reported by the World Health Organization (WHO) (http://www.who.int/csr/sars/country/table2004_04_21/en/index.html).

SARS-CoV belongs to the *Coronaviridae* family (order *Nidovirales*) of enveloped single-stranded positive RNA viruses (Marra et al. 2003; Rota et al. 2003; Thiel et al. 2003). The SARS-CoV genome is approximately 30 kb in length and is the longest known amongst the RNA virus genomes. The SARS-CoV genomic RNA has a cap structure and a poly-A tail at the 5' and 3' ends, respectively. SARS-CoV genome replication occurs in the cytoplasm. During viral replication, a full-length genomic negative-stranded RNA is transcribed from genomic positive-stranded RNA by the viral RNA polymerase that is initially translated from genomic RNA. Approximately 60% of SARS-CoV genomic RNA encodes viral polymerase and its related proteins. The mRNA transcription of coronavirus is unique, because all mRNAs have a nested set structure. The mRNAs have a 5' leader sequence of approximately 70 nucleotides and poly-A tails at the 3' end. Mouse hepatitis virus (MHV), which is a prototype of coronavirus, has seven mRNAs, whereas SARS-CoV has at least nine mRNAs. The leader RNA is transcribed from the 3' end of full-length genomic negative-stranded RNA. There is strong evidence that the leader RNA is transcribed as small sized-RNA, which is approximately 70 bases in length. The leader RNA binds to intragenic initiation sites on negative-stranded RNA, and then viral RNA polymerase starts to transcribe mRNA at the site. The SARS-CoV viral genomic RNA comprises 14 open reading frames (ORFs), and eight of the encoded proteins are unique compared with other coronaviruses. These unique proteins are thought to be involved in the pathogenetic mechanism of SARS-CoV.

Large overlapping polyproteins (1a and 1b) encoded by approximately 60% of the SARS-CoV viral genome are processed into 16 nonstructural proteins including polymerase and proteases (chymotrypsin-like cysteine protease and papain-like protease). These proteins are thought to be essential in viral replication and transcription. The viral particle of SARS-CoV mainly consists of four structural proteins, spike (S), membrane (M), envelope (E), and nucleocapsid (N) (Fig. 19.1). The viral particle may also comprise viral accessory proteins that bind to the structural proteins. The S protein binds to the viral receptor of host cells and enables the virus to enter the cytoplasm by endocytosis.

SARS-CoV has the potential to cause respiratory illness in human patients. Cytokine storm occurs in SARS-CoV-infected patients and is one of the observed pathologic mechanisms of SARS-CoV infection. On the other hand, apoptotic cell death is observed in vitro when SARS-CoV-sensitive cultured cells such as Vero E6 cells are used (Mizutani et al. 2004c). Various signaling pathways are activated during the entire process of viral infection, from S protein–ACE2 (Angiotensin-converting enzyme-2) binding for internalization into the host cells to apoptotic cell death (Mizutani 2007). The most common signaling pathways are mitogen-activated

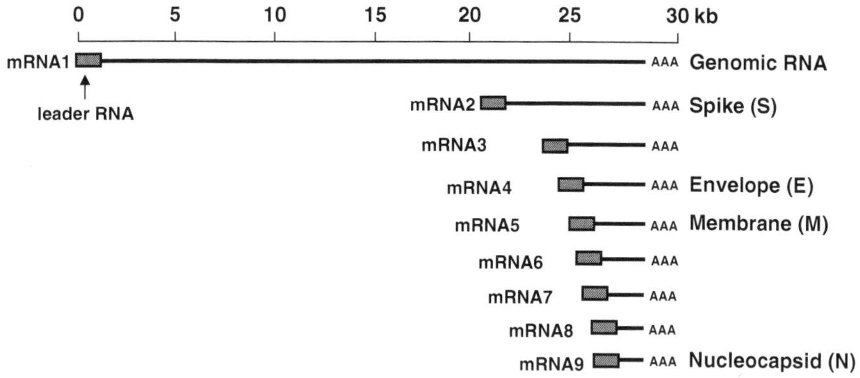

Fig. 19.1 SARS-CoV mRNAs in infected cells. Structural proteins are indicated

protein kinase (MAPK) pathways, which include Jun NH_2-terminal kinase (JNK), extracellular signal-regulated kinase (ERK), and p38 MAPK. These three major MAPKs are highly conserved in a wide range of species from yeast to mammals and are regulatory proteins of cell death and cell survival in living cells. Thus, MAPKs are key to the process of apoptosis (Garrington and Johnson 1999). MAPKK kinase (MAPKKK) activates MAPK kinase (MAPKK), and then MAPKK activates MAPK. Generally, the ERK signaling pathway promotes cell survival and proliferation, and JNK and p38 MAPK induce apoptosis. However, the role of each MAPK varies depending on cell type and stimulation. Many signaling pathways are activated in virus-infected cells, and cross-talk activation between signaling pathways occurs. Thus, signaling pathways regulating cell death and survival in virus-infected cells is highly complex.

Analysis of activated signaling pathways in SARS-CoV-infected cells and patients is required for understanding the pathogenesis of SARS. This review highlights recent progress in characterizing signal transduction pathways induced by SARS-CoV infection in vivo and in vitro.

19.2 p38 MAPK Signaling Pathway in Viral Infection

The p38 MAP kinase is expressed in response to stressors, and viral infection generally induces activation of p38 MAPK. The roles of p38 MAPK in viral infection/replication have been researched recently as described below.

Environmental stresses, such as UV irradiation, oxidative stimuli and pro-inflammatory cytokines, are able to induce activation of p38 MAPK. There are at least four isoforms of p38 MAPK: p38α, p38β, p38γ, and p38δ. 1999; Platanias 2003; Lee et al. 2004), but these isoforms are generally not distinguished in the field of virology. However, these four isoforms exhibit different properties and have different cellular functions. The p38α and p38β MAPKs have more than 70% similarity at the amino acid sequence level, and their functions are inhibited by the pyridinyl imidazole inhibitor SB203580 [4-(4-fluorophenyl)-2-(4-methylsulfinylphenyl)-5-(4-pyridyl) imidazole]. Conversely, p38γ and p38δ MAPKs, which have 60% similarity to p38α, are not inhibited by SB203580. Furthermore, p38α and p38β MAPKs are widely expressed in tissues, whereas the expression of p38γ and p38δ MAPKs is tissue-specific. In the field of virology, because SB203580 is generally used as an inhibitor of the p38 MAPK signaling pathway, it can be used in studying the role of p38α and/or p38β MAPKs in SARS-CoV infection. The kinases upstream of p38 MAPK are MKK3 (MAPK kinase 3) and MKK6 (MAPK kinase 6), which are known to phosphorylate and activate p38 MAPK. Interestingly, MKK6 has affinity to PKR in the presence of double-stranded RNA, poly(rI;rC) and PKR is able to activate MKK6, but not MKK3. This result indicated that interaction of MKK6 and PKR provides a mechanism to regulate activation of p38 MAPK (Silva et al. 2004). In hepatitis C virus-core-expressing cells, PKR has an important role in cell cycle arrest and was shown to interact strongly with p38 MAPK (Spaziani et al. 2006). As upstream of p38MAPK, transforming growth factor (TGF)-β-activating kinase (TAK1), apoptosis signal-regulatory kinase (ASK1), and MAPKKK4 are known as MAPKKKs.

Downstream targets of p38 MAPK are well-known as mitogen and stress-activated protein kinase 1 (MSK1), MAP kinase-interacting kinase 1 (MNK1), and MAPK-activated protein kinase 2 and 3 (MAPKAPK 2 and 3) (Freshney et al. 1994). These signaling pathway proteins have further downstream targets. For example, MNK1 activates the eukaryotic translation initiation factor 4E (eIF4E) (Gingras et al. 1999). MAPKAPK2 activates heat shock protein 27 (HSP27), cAMP response element-binding protein (CREB), and transcription factor-1 (ATF-1) (Garrido et al. 2003; Tan et al. 1996). Mouse hepatitis virus (MHV) A59 strain induces interleukin-6 (IL-6) production via eIF4E phosphorylation under p38 MAPK activation (Banerjee et al. 2002). Inhibitors of p38 MAPK inhibit transcription of viral mRNA and production of viral protein, indicating that p38 MAPK is utilized to promote viral protein synthesis. Conversely, p38 MAPK enhances transcription of CHOP (C/EBP homologous protein) encoded by the growth arrest- and DNA damage-inducible gene 153 (GADD153) (Wang et al. 1996). The p38 MAPK-induced apoptosis via activation of the CHOP pathway occurs in cells infected with Japanese encephalitis virus (JEV) (Su et al. 2002). The H5N1 subtype of influenza virus induces tumor necrosis factor alpha (TNF-α) expression via activation of p38 MAPK (Lee et al. 2005). The p38 MAPK signaling pathway is thought to primarily induce apoptosis in virus-infected cells. As described in HCV infection, p38 MAPK is able to promote cell survival. Although p38 MAPK is activated in many cases of viral infection, viral proteins sometimes negatively regulate p38 MAPK. The ORF 61 protein of varicella-zoster virus (VZV) is known to repress phosphorylation of p38 MAPK for negative regulation of cellular proinflammatory responses (Rahaus et al. 2005). Thus, activation or inactivation of p38 MAPK occurs in the pathogenesis of disease caused by viral infection.

19.3 p38 MAPK Signaling Pathway in SARS-CoV-Infected Cells

The p38 MAPK signaling pathway takes part in cell death, as previously described. Apoptosis is an active and physiologic type of cell death and is a host cell's protective mechanism for preventing the spread of viral particles before production of viral particles. Vero and Vero E6 cells, which are monkey kidney cells, are widely used in SARS-CoV research because of their high susceptibilities to infection due to lack of interferon genes. Apoptosis has been shown to be inducible by infection with SARS-CoV (Mizutani et al. 2004c; Yan et al. 2004). Cytopathic effects (CPEs), defined as focal cell rounding and DNA fragmentation typical of apoptosis, are observed in SARS-CoV-infected Vero E6 cells at 24 h post-infection (h.p.i.) (Mizutani et al. 2004c). Activated caspase 3, which has an essential role in apoptosis, was detected at peak levels at 24 h.p.i. On the other hand, the phosphorylation level of p38 MAPK reached a maximum at 18 h.p.i. in virus-infected cells. The phosphorylated p38 MAPK was active, as shown by using an in vitro kinase assay. The CPE observed in SARS-CoV-infected cells is slightly inhibited by SB203580, and therefore p38 MAPK activation is thought to induce CPE of virus-infected cells. However, DNA fragmentation is not inhibited by the inhibitor. Apoptosis and CPE are thought to be linked, and activation of p38 MAPK is a promoter of cell death in Vero E6 cells infected by SARS-CoV. However, SB203580 treatment of Vero E6 cells indicates that there is no requirement for p38 MAPK activation in SARS-CoV replication. The p38 MAPK signaling pathway perhaps has other roles in SARS-CoV-infected cells.

The downstream targets of p38 MAPK are phosphorylated in SARS-CoV-infected cells. The level of phosphorylated eIF4E is increased in SARS-CoV-infected cells (Mizutani et al. 2004c). However, the activated eIF4E does not regulate viral protein synthesis, as demonstrated by the similar kinetics of viral protein accumulation in infected Vero E6 cells in the presence and absence of SB203580. Both MAPKAPK-2 and its substrate HSP-27 are phosphorylated in virus-infected cells. HSP-27 is known as an anti-apoptotic protein as it inhibits apoptosome formation (Garrido et al. 2003). CREB is also known to mediate a survival signal under various conditions (Tan et al. 1996; Ginty et al 1994; von Knethen et al. 1998), and CREB is also phosphorylated in SARS-CoV-infected cells. The expression of SARS-N protein in transfected COS-1 cells induces phosphorylation of p38 MAPK, HSP-27, and CREB (Surjit et al. 2004), whereas the viral-N protein expression system of vaccinia virus (DIs-N) does not induce phosphorylation of p38 MAPK (Mizutani et al. 2006d). Activation of the p38 MAPK pathway induces actin reorganization in COS-1 cells devoid of growth factors (Surjit et al. 2004). Furthermore, the 7a protein of SARS-CoV induces apoptotic cell death and phosphorylation of p38 MAPK in 293 T cells (Kopecky-Bromberg et al. 2006). However, SB203580 does not prevent cell rounding, apoptosis, and chromatin condensation induced by the 7a protein. The differences in the results are most likely due to the use of different cell cultures and expression systems. Overall, phosphorylated proteins downstream of p38 MAPK have the potential to induce an anti-apoptotic environment in SARS-CoV-infected cells. However, activated p38 MAPK in SARS-CoV-infected cells is thought to be able to promote both cell death and survival. Perhaps there are other substrates of p38 MAPK that are inducible on cell death of Vero E6 cells caused by SARS-CoV infection, or perhaps there is cross-talk between the p38 MAPK signaling pathway and other signaling pathways inducing cell death.

19.4 Downstream Signaling Pathways of p38 MAPK

In Vero E6 cells, signal transducer and activator of transcription (STAT) 3 protein is constitutively phosphorylated at Tyr-705 and is slightly phosphorylated at Ser-727 (Mizutani et al. 2004a). SARS-CoV infection is able to induce dephosphorylation of STAT3 Tyr-705 after 18 h.p.i. On the other hand, Ser-727-phosphorylated STAT3 is slightly increased at the same point in time. The activity of STAT transcription factors is induced by phosphorylation of a single tyrosine residue, leading to dimerization via an intermolecular SH2 phosphotyrosine interaction (Shuai et al. 1993, 1994; Schindler et al. 1992a, 1992b). STAT3 is known to be activated in response to interleukin-6 (IL-6) and IL-10, and is thought to act as an anti-apoptotic transcription factor (Rajan and Mckay 1998; Grandis et al. 2000; Mora et al. 2002). Tyr-705 phosphorylation of STAT is necessary for its activation (Shuai et al. 1993, 1994; Schindler et al. 1992b), suggesting that SARS-CoV infection leads to a decrease in STAT3 activation. Furthermore, STAT3 does not act as a transcriptional enhancer in SARS-CoV-infected Vero E6 cells, as shown by the disappearance of Tyr-705-phosphorylated STAT3 from the nuclear fraction post SARS-CoV infection. The proteins upstream of STAT3 in the signaling pathway are Janus kinases (JAK1 and 2) and Tyk2, which are phosphorylated at low levels in mock-infected Vero E6 cells, even after virus infection. The signal transducing adaptor molecule 1 (STAM1), which is known to be associated with Jak2 and Jak3 via the immunoreceptor tyrosine-based activation motif, is upregulated in SARS-CoV-infected Vero E6 cells (Leong et al. 2005). Therefore, Tyr-705 dephosphorylation of STAT3 in virus-infected cells is independent of its upstream kinases, and there may be other signaling pathways regulating STAT3 phosphorylation and dephosphorylation. Two inhibitors of p38 MAPK (SB203580 and SB202190) partially inhibit dephosphorylation of STAT3 at Tyr-705, indicating that the p38 MAPK signaling pathway is upstream of Tyr-705 dephosphorylation of STAT3 in SARS-CoV-infected Vero E6 cells. Inactivation of STAT3 via p38 MAPK activation may induce cell death in SARS-CoV-infected cells. However, the kinetics of STAT3 after SARS-CoV infection varies for different cell types. The suppressors of cytokine signaling-3 (SOCS3) mRNA are suppressed in SARS-CoV-infected Caco-2 cells (Okabayashi et al. 2006), leading to continuous activation of STAT3.

A serine/threonine kinase, p90 ribosomal S6 kinase (RSK), belongs to another signaling pathway, which is regulated by p38 MAPK. Generally, p90RSK is phosphorylated at Thr-573 by ERK (Gavin and Nebreda 1999; Smith et al. 1999), and this phosphorylation induces autophosphorylation at Ser-380, and then PDK1 (phosphoinositide-dependent kinase 1) phosphorylates at Ser-221 (Frödin et al. 2000; Jensen et al. 1999; Richards et al. 1999). No significant differences are observed in phosphorylation levels of p90RSK at Ser-221 and Thr-573 in SARS-CoV-infected Vero E6 cells (Mizutani et al. 2006a). However, Ser-380 of p90RSK is phosphorylated in virus-infected confluent cells. Thus, phosphorylation of p90RSK Ser-380 is upregulated without upregulation of Thr-573 in SARS-CoV-infected cells. The phosphorylation of Ser-380 is decreased in SB203580-treated virus-infected cells, indicating that p38 MAPK can induce phosphorylation of Ser-380. Furthermore, p90RSK phosphorylates CREB (Frodin and Gammeltoft 1999). In SARS-CoV-infected Vero E6 cells, p38 MAPK activation induces phosphorylation of p90RSK Ser-380, and then CREB is thought to be phosphorylated by activated p90RSK. Thus, p90RSK may have anti-apoptotic activity in SARS-CoV-infected cells.

19.5 ERK1/2 Activation by SARS-CoV Infection

The SARS-CoV S protein is able to induce phosphorylation of ERK1/2 in HEK293T cells (Liu et al. 2007). The S-induced protein kinase C (PKC)/ERK signaling pathway promotes nuclear factor-kappa B (NF-κB) binding to the cyclo-oxygenase-2 (COX-2) promoter. Similar results have been reported using the N protein of SARS-CoV (Yan et al. 2006). SARS-CoV S protein expression induces release of interleukin-8 (IL-8) via ERK and p38 MAPK signaling pathways including activator protein 1 (AP-1) in A549 cells (Chang et al. 2004). On the contrary, phosphorylation of ERK1/2 is downregulated in N protein-expressing COS-1 cells in the absence of serum (Surjit et al. 2004). Thus, viral proteins can potentially up- or downregulate phosphorylation of ERK1/2. ERK1/2 is observed to be phosphorylated in SARS-CoV-infected Vero E6 cells (Mizutani et al. 2004a). After treatment with MAPK/ERK kinase 1 and 2 (MEK1/2)-specific inhibitor (PD98059), SARS-CoV-infected Vero E6 cells exhibit no significant changes in activated caspase-3 or caspase-7. Thus, activation of ERK1/2 is not sufficient to prevent cell death by SARS-CoV infection. Furthermore, activation of ERK1/2 is not necessary to establish persistent infection of SARS-CoV in Vero E6 cells (Mizutani et al. 2005).

19.6 JNK Activation by SARS-CoV

The SARS-CoV S protein induces CREB binding to COX-2 promoter mediated via the phosphatidylinositol 3-kinase (PI3K)/PKC/JNK pathway in HEK293T cells (Liu et al. 2007). Expression of the SARS-CoV N protein induces phosphorylation of JNK in Vero E6 cells (Mizutani et al. 2006d) and in COS-1 cells in the absence of serum (Surjit et al. 2004). The phosphorylation level of Jun, which is dependent upon activation of JNK, also increases in the absence of serum. The SARS-CoV N protein can activate AP-1, which is composed of homodimers and heterodimers of Fos, Jun, CREB, and activating transcription factor (ATF) subunits, in Vero and Huh7 cells (He et al. 2003). The viral accessory proteins, 3a and 7a, phosphorylate JNK1 and JNK3 in HEK293T cells (Kanzawa et al. 2006). Overall, viral proteins are able to induce phosphorylation of JNK in several cell lines. SARS-CoV infection induces phosphorylation of JNK in Vero E6 cells after at least 12 h.p.i. (Mizutani et al. 2004a). The Vero E6 cells begin to show rounding at 24 h.p.i and persistently infected cells are observed after 48 h.p.i (Mizutani et al. 2005). At this time, JNK, Akt, and p38 MAPK are phosphorylated in virus-infected cells. Treatment with an inhibitor of JNK (SP600125), and PI3K (LY294002), inhibits the establishment of persistence, whereas treatment with an inhibitor of MEK1/2 (PD98059) and p38 MAPK (SB203580) does not inhibit persistence of infection (Mizutani et al. 2005). Thus, two different signaling pathways of JNK and PI3K/Akt are important for the establishment of persistently infected Vero E6 cells (Mizutani et al. 2006d, 2007).

19.7 PI3K/Akt Activation by SARS-CoV

Akt, which is also known as protein kinase B (PKB), is phosphorylated at both Ser-473 and Thr-308 residues via the PI3K signaling pathway upon stimulation by growth factors, insulin, and hormones (Toker 2000; Brazil and Hemmings 2001; Scheid and Woodgett 2003; Welch et al. 1998). The main role of Akt is inhibition of apoptosis via phosphorylation of the forkhead transcription factor (FKHR) family, glycogen synthase kinase-3β (GSK-3β), caspase-9, and Bcl-associated death protein (Bad) (Cardone et al. 1998; Cross et al. 1995; Datta et al. 1997). Interestingly, GSK-3 regulates phosphorylation of N protein (Wu et al. 2009). The M protein of SARS-CoV induces apoptosis in both HEK293T cells and transgenic *Drosophila* (Chan et al. 2007). The M protein-induced apoptosis involves mitochondrial release of cytochrome *c* protein. In SARS-CoV-infected Vero E6 cells, Ser-473 of Akt is phosphorylated at 8 h.p.i. and maximal phosphorylation is observed at 18 h.p.i. (Mizutani et al. 2004b), after which Akt is dephosphorylated. Thr-308 phosphorylation has not been detected in Vero E6 cells. The phosphorylation of Ser-473 of Akt by viral infection is inhibited by LY294002, which is an inhibitor of the PI3K signaling pathway. An in vitro kinase activity assay of Akt in SARS-CoV-infected cells indicated that Akt is highly phosphorylated only at serine residues, but Akt activity is low. Therefore, weak activation of Akt cannot prevent apoptosis induced by SARS-CoV infection in Vero E6 cells. The phosphorylation of Akt in virus-infected cells is necessary to establish persistence, but Akt is not phosphorylated after establishing persistent cell lines (Mizutani et al. 2005, 2006d), suggesting that activation of PI3K/Akt is essential for the establishment of persistent infection with

SARS-CoV at points in time before cell death. The above characterizations of Akt in SARS-CoV-infected Vero E6 cells are mainly derived from experiments using confluent cells. When subconfluent Vero E6 cells are infected by SARS-CoV, cell proliferation is inhibited (Mizutani et al. 2006c). SARS-CoV infection induces dephosphorylation of a serine residue of Akt without phosphorylation in subconfluent cultures. Thus, downregulation of Akt activity in SARS-CoV-infected cells prevents cell proliferation.

19.8 NF-κB Activation and Inhibition by SARS-CoV Proteins

The SARS-CoV N protein is able to activate NF-κB in Vero E6 cells (Liao et al. 2005). As described above, the S- and N-induced PKC/ERK signaling pathway promotes NF-κB binding to the COX-2 promoter (Liu et al. 2007; Yan et al. 2006). SARS-CoV S and N proteins may cause inflammation of the lungs by activating *COX-2* gene expression. The 3a and 7a viral accessory proteins enhance NF-κB mediated transcription in HEK293T cells (Kanzawa et al. 2006). In contrast, the N protein inhibits interferon production in 293 T cells via inhibition of NF-κB (Kopecky-Bromberg et al. 2007). The M protein also suppresses NF-κB activity (Fang et al. 2007). Growth arrest and apoptosis via caspase-3 and caspase-9 activities are induced in SARS-CoV 3C-like protease ($3CL^{pro}$)-expressing human promonocyte HL-CZ cell line (Lin et al. 2006). The SARS-CoV $3CL^{pro}$ may increase activation of NF-κB and upregulate cytochrome *c* oxidase and downregulate Hsp-70, inducing mitochondrial-mediated apoptosis (Lai et al. 2007). Viral papain-like protease (PLP) regulates antagonism of IRF3 and NF-κB signaling pathways (Frieman et al. 2009).

19.9 Inhibitory Effects of Viral Proteins on the Cell Cycle

The 3a protein of SARS-CoV has the potential to inhibit cell cycle progression at the G_1 phase in HEK293, COS-7, and Vero cells (Yuan et al. 2005, 2007). The C-terminal region of the 3a protein, which includes a potential ATPase motif, is essential to inhibit the cell cycle. The 3a protein expression reduces cyclin D3 level and inhibits retinoblastoma (Rb) phosphorylation. The p53 phosphorylation is increased by 3a expression. The 7a protein expression also blocks cell cycle progression at the G_0/G_1 phase in HEK293, COS-7, and Vero cells by mechanisms similar to those of the 3a protein (Yuan et al. 2006). The N protein is a substrate of cyclin-dependent kinase (CDK) as well as GSK, MAPK, and casein kinase II (Surjit et al. 2005). The N protein directly binds to cyclin D and inhibits activity of the cyclin D–CDK4 complex. The N protein also inhibits CDK2 activity by direct binding to the CDK2–cyclin complex, resulting in blocking the S phase progression in COS-7 and Huh7 cells (Surjit et al. 2006). Therefore, proteins of SARS-CoV may have the ability to inhibit the progression of the host cell cycle, but further detailed analysis is required in SARS-CoV-infected cells.

19.10 Apoptotic Signaling Pathway

SARS-CoV infection induces apoptotic cell death in Vero E6 cells, via dephosphorylation of STAT3 by p38 MAPK activation, and inactivation of Akt, as previously described. Recent study suggest that SARS-CoV triggers apoptosis via protein kinase R (PKR) (Krähling et al. 2009). Overexpression of SARS-CoV proteins can induce apoptosis in variable cell lines. Induction of apoptosis by various viral proteins may occur at different stages of the infection cycle. SARS-CoV $3CL^{pro}$ expression in HL-CZ cells induces apoptosis via caspase-3 and caspase-9 (Lin et al. 2006). Furthermore, $3CL^{pro}$ expression in HL-CZ cells upregulates proteins located in the mitochondria, but downregulates Hsp-70, which antagonizes apoptosis-inducing factor (Lai et al. 2007). The SARS-CoV 8a protein, localized in the mitochondria of infected cells, increases mitochondrial transmembrane potential, reactive oxygen species production, and caspase-3 activation, resulting in inducing apoptosis in Vero, HEK293, and Huh7 cells (Chen et al. 2007). ORF 6 induces apoptosis via caspase-3 mediated, ER stress and JNK-dependent pathways (Ye et al. 2008). SARS-CoV N protein modulates the TGF-β signaling pathway to block apoptosis of SARS-CoV-infected host cells (Zhao et al. 2008). In the absence of serum, the SARS-CoV N protein can induce apoptosis by activating the mitochondrial pathway (Zhang et al. 2007), and/or by downregulating ERK and Akt signaling pathways (Surjit et al. 2004) in COS-1 cells, but not in Hep-G2 and Huh-7 cells (Zhang et al. 2007). The SARS-CoV S protein and its C-terminal domain (S2) induce apoptosis in Vero E6 cells, but the S1, E, M, and N proteins are not able to induce apoptosis in Vero E6 cells (Chow et al. 2005). In contrast, the SARS-CoV M and N proteins can induce apoptosis in human pulmonary fibroblast (HPF) cells (Zhao et al. 2006). The M protein induces apoptosis through modulation of the Akt pathway and mitochondrial cytochrome *c* release in HEK293T cells and transgenic *Drosophila* [85]. Overexpression of SARS-CoV 3a protein in Vero E6 cells induces apoptosis, mediated through a caspase-8-dependent pathway or p38 MAPK (Law et al. 2005; Waye et al. 2005; Padhan et al. 2008). The 3a protein expression in *Drosophila* induces apoptosis, which could be modulated by cellular cytochrome *c* levels and caspase activity (Wong et al. 2005). The SARS-CoV 3b protein induces both necrosis and apoptosis in Vero E6 cells (Khan et al. 2006). The SARS-CoV 7a protein interacts with pro-survival proteins, basal cell lymphoma-extra large (Bcl-xL), B cell lymphoma 2 (Bcl-2), Bcl-w, A1, and myeloid cell leukaemia sequence 1 (Mcl-1), at the endoplasmic reticulum and the mitochondria, resulting in triggering apoptosis in HEK293T and Vero E6 cells (Tan et al. 2007). Interestingly, the 7a protein does not interact with the pro-apoptotic members, Bcl-2 associated X protein (Bax), Bcl-2 homologous killer (Bak), Bad, and Bcl-2 interacting domain (Bid). However, a mutant virus without the 7a/7b gene is able to induce extensive CPEs in the Vero cell line (Yount et al. 2005), suggesting that the 7a protein is not a dominant contributor to virus-induced cell death in this cell culture system. The SARS-CoV N protein downregulates the level of Bcl-2 in COS-1 cells (Surjit et al. 2004). The SARS-CoV E protein induces apoptosis in Jurkat T cells in the absence of growth factors, but apoptosis is inhibited by overexpression of Bcl-xL via interaction with the E protein (Yang et al. 2005). Apoptosis is also inhibited by overexpression of Bcl-2 in SARS-CoV-infected Vero cells (Bordi et al. 2006). In the virus-infected Vero cells, downregulation of Bcl-2 and upregulation of Bax are observed (Ren et al. 2005). Bcl-xL activation plays important roles in establishing persistent infection of SARS-CoV (Mizutani et al. 2006b). The N protein upregulates the Bcl-xL protein level (Mizutani et al. 2006d). These reports indicate that Bcl-xL activation is the key to preventing apoptosis due to SARS-CoV infection. The other viral proteins localized in the mitochondria of infected cells may also interact with Bcl-xL and other pro-survival proteins.

19.11 Signaling Pathways in SARS Patients

Western blots are used to analyze signaling pathway proteins of cultured cells infected with SARS-CoV or transfected with plasmids encoding viral proteins. Thus, the kinetics of phosphor-proteins regulating signaling pathways is important for understanding which signaling pathways are activated in virus-infected cells. However, in vivo analysis and amounts of mRNA from whole blood or tissues of SARS patients are primarily analyzed using DNA microarrays. Unfortunately, when the level of mRNA related to a signaling pathway increases in SARS patients, as measured by DNA microarray analysis, the results do not suggest activation of particular signaling pathways, due to the analysis being performed on a mixed population of cells. The roles of signaling pathways may be different amongst different cell types. Analyses of signaling pathways in virus-infected patients are still difficult to perform for these reasons. However, flow cytometric analysis of cell samples from virus-infected patients provides an improved method for the investigation of signaling pathways in vivo. Flow cytometric analysis of phospho-p38 indicated that augmented p38 MAPK phosphorylation of CD14 monocytes was associated with suppressed p38 MAPK phosphorylation of CD8 lymphocytes, suggesting that altered leukocyte p38 activation contributes to abnormal blood cytokine profiles in SARS patients (Lee et al. 2004).

Analysis of cell apoptosis in SARS patients is key to understanding the signaling pathways that regulate apoptosis. In SARS patients, lymphopenia caused by depletion of T lymphocytes by apoptosis is a common abnormality (Chen et al. 2006). Compared to healthy controls, SARS patients have significantly lower lymphocyte and platelet counts and have significantly higher vascular cell adhesion molecule-1 (sVCAM-1) levels and soluble Fas ligand (sFasL) levels, as determined using

ELISA (enzyme-linked immunosorbent assay). SARS patients also have intracellular activated caspase-3 fragment levels, as measured using flow cytometry (Peiris et al. 2003b). Liver impairment commonly occurs amongst patients with SARS, indicating that SARS-CoV may be localized in the liver (Chau et al. 2004). The pathologic features, perhaps due to apoptosis, are the presence of acidophilic bodies, ballooning of hepatocytes, and mild to moderate lobular activities. The thyroid glands of SARS-infected patients show extensive injury due to apoptosis of the follicular epithelial cells and the parafollicular cells, as measured using terminal deoxynucleotidyl transferase-mediated dUTP nick end-labeling assay (Wei et al. 2007). Necrosis is also observed in splenic lymphoid tissue and lymph nodes of SARS patients (Ding et al. 2003). MyD88-mediated innate immune signaling and inflammatory cell recruitment to the lung in BALB/c mice may be required for protection from lethal recombinant mouse-adapted SARS-CoV infection (Sheahan et al. 2008). Further detailed analysis of apoptosis in cells of SARS patients is required, but the initial reports indicate the activation of apoptotic signaling pathways in SARS patients.

19.12 Conclusion

Both pro-apoptotic and pro-survival signaling pathways are activated in SARS-CoV-infected cells (Fig. 19.2). The balance of activities of signaling pathways is important for determination of cell death or cell survival. In SARS patients, analysis of signaling pathways is further complicated because many cell types respond to viral infection. For example, immune cells infected by SARS-CoV produce and release cytokines, and the cytokines activate other cells. Thus, in SARS patients, many types of cells are infected by SARS-CoV, compared with one type of cell used for in vitro experiments. In addition, the viral proteins that interact with cellular proteins in signaling pathways must be further clarified to understand the

Fig. 19.2 Signaling pathways in cells infected with SARS-CoV. Because each report in this chapter used different cultured cells, this figure is shown based on our experiments using Vero E6 cells

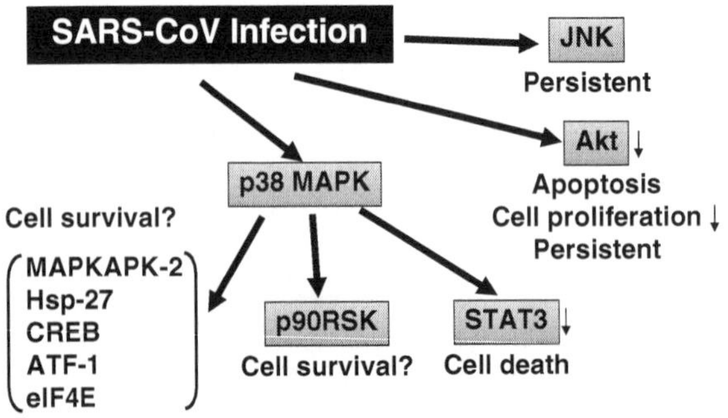

molecular mechanisms of SARS-CoV infection. It is particularly important to determine the viral proteins that are necessary and sufficient to fully activate signaling pathways leading to apoptotic cell death. Determining the SARS-CoV-induced signaling pathways in SARS patients will enable the development of therapeutic reagents that can inhibit the pathways of apoptotic cell death and production of cytokines.

Acknowledgments I am grateful to Drs. Shuetsu Fukushi, Masayuki Saijo, Momoko Ogata, Kouji Sakai, Ichiro Kurane, and Shigeru Morikawa (National Institute of Infectious Diseases) for their comments on this manuscript. This work was supported, in part, by a grant-in-aid from the Ministry of Health, Labor, and Welfare of Japan and Japan Society for the Promotion of Science.

References

Banerjee S, Narayanan K, Mizutani T, Makino S (2002) Murine coronavirus replication-induced p38 mitogen-activated protein kinase activation promotes interleukin-6 production and virus replication in cultured cells. J Virol 76:5937–5948

Bonni A, Brunet A, West AE, Datta SR, Takasu MA, Greenberg ME (1999) Cell survival promoted by the Ras-MAPK signaling pathway by transcription-dependent and -independent mechanisms. Science 286:1358–1362

Bordi L, Castilletti C, Falasca L, Ciccosanti F, Calcaterra S, Rozera G, Di Caro A, Zaniratti S, Rinaldi A, Ippolito G, Piacentini M, Capobianchi MR (2006) Bcl-2 inhibits the caspase-dependent apoptosis induced by SARS-CoV without affecting virus replication kinetics. Arch Virol 151:369–377

Brazil DP, Hemmings BA (2001) Ten years of protein kinase B signalling: a hard Akt to follow. Trends Biochem Sci 26:657–664

Cardone MH, Roy N, Stennicke HR, Salvesen GS, Franke TF, Stanbridge E, Frisch S, Reed JC (1998) Regulation of cell death protease caspase-9 by phosphorylation. Science 282:1318–1321

Chan CM, Ma CW, Chan WY, Chan HY (2007) The SARS-Coronavirus Membrane protein induces apoptosis through modulating the Akt survival pathway. Arch Biochem Biophys 459 (2):197–207

Chang YJ, Liu CY, Chiang BL, Chao YC, Chen CC (2004) Induction of IL-8 release in lung cells via activator protein-1 by recombinant baculovirus displaying severe acute respiratory syndrome-coronavirus spike proteins: identification of two functional regions. J Immunol 173:7602–7614

Chau TN, Lee KC, Yao H, Tsang TY, Chow TC, Yeung YC, Choi KW, Tso YK, Lau T, Lai ST, Lai CL (2004) SARS-associated viral hepatitis caused by a novel coronavirus: report of three cases. Hepatology 39:302–310

Chen RF, Chang JC, Yeh WT, Lee CH, Liu JW, Eng HL, Yang KD (2006) Role of vascular cell adhesion molecules and leukocyte apoptosis in the lymphopenia and thrombocytopenia of patients with severe acute respiratory syndrome (SARS). Microbes Infect 8:122–127

Chen CY, Ping YH, Lee HC, Chen KH, Lee YM, Chan YJ, Lien TC, Jap TS, Lin CH, Kao LS, Chen YM (2007) Open reading frame 8a of the human severe acute respiratory syndrome coronavirus not only promotes viral replication but also induces apoptosis. J Infect Dis 196:405–415

Chow KY, Yeung YS, Hon CC, Zeng F, Law KM, Leung FC (2005) Adenovirus-mediated expression of the C-terminal domain of SARS-CoV spike protein is sufficient to induce apoptosis in Vero E6 cells. FEBS Lett 579:6699–6704

Cross DA, Alessi DR, Cohen P, Andjelkovich M, Hemmings BA (1995) Inhibition of glycogen synthase kinase-3 by insulin mediated by protein kinase B. Nature 378:785–789

Datta SR, Dudek H, Tao X, Masters S, Fu H, Gotoh Y, Greenberg ME (1997) Akt phosphorylation of BAD couples survival signals to the cell-intrinsic death machinery. Cell 91:231–241

Ding Y, Wang H, Shen H, Li Z, Geng J, Han H, Cai J, Li X, Kang W, Weng D, Lu Y, Wu D, He L, Yao K (2003) The clinical pathology of severe acute respiratory syndrome (SARS): a report from China. J Pathol 200:282–289

Drosten C, Gunther S, Preiser W, van der Werf S, Brodt HR, Becker S, Rabenau H, Panning M, Kolesnikova L, Fouchier RA et al (2003) Identification of a novel coronavirus in patients with severe acute respiratory syndrome. N Engl J Med 348:1967–1976

Fang X, Gao J, Zheng H, Li B, Kong L, Zhang Y, Wang W, Zeng Y, Ye L (2007) The membrane protein of SARS-CoV suppresses NF-kappaB activation. J Med Virol 79:1431–1439

Freshney NW, Rawlinson L, Guesdon F, Jones E, Cowley S, Hsuan J, Saklatvala J (1994) Interleukin-1 activates a novel protein kinase cascade that results in the phosphorylation of Hsp27. Cell 78:1039–1049

Frieman M, Ratia K, Johnston RE, Mescar AD, Baric RS (2009) SARS Coronavirus papain-like protease ubiquitin-like domain and catalytic domain regulate antagonism of IRF3 and NFkB signaling. J Virol 83:6689–6705

Frodin M, Gammeltoft S (1999) Role and regulation of 90 kDa ribosomal S6 kinase (RSK) in signal transduction. Mol Cell Endocrinol 151:65–77

Frödin M, Jensen CJ, Merienne K, Gammeltoft S (2000) A phosphoserine-regulated docking site in the protein kinase RSK2 that recruits and activates PDK1. EMBO J 19:2924–2934

Garrido C, Schmitt E, Candé C, Vahsen N, Parcellier A, Kroemer G (2003) HSP27 and HSP70: potentially oncogenic apoptosis inhibitors. Cell Cycle 2:579–584

Garrington TP, Johnson GL (1999) Organization and regulation of mitogen-activated protein kinase signaling pathways. Curr Opin Cell Biol 11:211–218

Gavin AC, Nebreda AR (1999) A MAP kinase docking site is required for phosphorylation and activation of p90rsk/MAPKAP kinase-1. Curr Biol 9:281–284

Gingras AC, Raught B, Sonenberg N (1999) eIF4 initiation factors: effectors of mRNA recruitment to ribosomes and regulators of translation. Annu Rev Biochem 68:913–963

Ginty DD, Bonni A, Greenberg ME (1994) Nerve growth factor activates a Ras-dependent protein kinase that stimulates c-fos transcription via phosphorylation of CREB. Cell 77:713–725

Grandis JR, Drenning SD, Zeng Q, Watkins SC, Melhem MF, Endo S, Johnson DE, Huang L, He Y, Kim JD (2000) Constitutive activation of Stat3 signaling abrogates apoptosis in squamous cell carcinogenesis in vivo. Proc Natl Acad Sci U S A 97:4227–4232

He R, Leeson A, Andonov A, Li Y, Bastien N, Cao J, Osiowy C, Dobie F, Cutts T, Ballantine M, Li X (2003) Activation of AP-1 signal transduction pathway by SARS coronavirus nucleocapsid protein. Biochem Biophys Res Commun 311:870–876

Jensen CJ, Buch MB, Krag TO, Hemmings BA, Gammeltoft S, Frodin M (1999) 90-kDa ribosomal S6 kinase is phosphorylated and activated by 3-phosphoinositide-dependent protein kinase-1. J Biol Chem 274:27168–27176

Kanzawa N, Nishigaki K, Hayashi T, Ishii Y, Furukawa S, Niiro A, Yasui F, Kohara M, Morita K, Matsushima K, Le MQ, Masuda T, Kannagi M (2006) Augmentation of chemokine production by severe acute respiratory syndrome coronavirus 3a/X1 and 7a/X4 proteins through NF-kappaB activation. FEBS Lett 580:6807–6812

Khan S, Fielding BC, Tan TH, Chou CF, Shen S, Lim SG, Hong W, Tan YJ (2006) Overexpression of severe acute respiratory syndrome coronavirus 3b protein induces both apoptosis and necrosis in Vero E6 cells. Virus Res 122:20–27

Kopecky-Bromberg SA, Martinez-Sobrido L, Palese P (2006) 7a Protein of Severe Acute Respiratory Syndrome Coronavirus Inhibits Cellular Protein Synthesis and Activates p38 Mitogen-Activated Protein Kinase. J Virol 80:785–793

Kopecky-Bromberg SA, Martínez-Sobrido L, Frieman M, Baric RA, Palese P (2007) Severe acute respiratory syndrome coronavirus open reading frame (ORF) 3b, ORF 6, and nucleocapsid proteins function as interferon antagonists. J Virol 81:548–557

Krähling V, Stein DA, Spiegel M, Weber F, Mühlberger E (2009) Severe acute respiratory syndrome coronavirus triggers apoptosis via protein kinase R but is resistant to its antiviral activity. J Virol 83:2298–2309

Ksiazek TG, Erdman D, Goldsmith CS, Zaki SR, Peret T, Emery S, Tong S, Urbani C, Comer JA, Lim W et al (2003) A novel coronavirus associated with severe acute respiratory syndrome. N Engl J Med 348:1953–1966

Lai CC, Jou MJ, Huang SY, Li SW, Wan L, Tsai FJ, Lin CW (2007) Proteomic analysis of up-regulated proteins in human promonocyte cells expressing severe acute respiratory syndrome coronavirus 3C-like protease. Proteomics 7:1446–1460

Law PT, Wong CH, Au TC, Chuck CP, Kong SK, Chan PK, To KF, Lo AW, Chan JY, Suen YK, Chan HY, Fung KP, Waye MM, Sung JJ, Lo YM, Tsui SK (2005) The 3a protein of severe acute respiratory syndrome-associated coronavirus induces apoptosis in Vero E6 cells. J Gen Virol 86:1921–1930

Lee CH, Chen RF, Liu JW, Yeh WT, Chang JC, Liu PM, Eng HL, Lin MC, Yang KD (2004) Altered p38 mitogen-activated protein kinase expression in different leukocytes with increment of immunosuppressive mediators in patients with severe acute respiratory syndrome. J Immunol 172:7841–7847

Lee DC, Cheung CY, Law AH, Mok CK, Peiris M, Lau AS (2005) p38 mitogen-activated protein kinase-dependent hyperinduction of tumor necrosis factor alpha expression in response to Avian Influenza Virus H5N1. J Virol 79:10147–10154

Leong WF, Tan HC, Ooi EE, Koh DR, Chow VT (2005) Microarray and real-time RT-PCR analyses of differential human gene expression patterns induced by severe acute respiratory syndrome (SARS) coronavirus infection of Vero cells. Microbes Infect 7:248–259

Liao QJ, Ye LB, Timani KA, Zeng YC, She YL, Ye L, Wu ZH (2005) Activation of NF-kappaB by the full-length nucleocapsid protein of the SARS coronavirus. Acta Biochim Biophys Sin 37:607–612

Lin CW, Lin KH, Hsieh TH, Shiu SY, Li JY (2006) Severe acute respiratory syndrome coronavirus 3C-like protease-induced apoptosis. FEMS Immunol Med Microbiol 46:375–380

Liu M, Yang Y, Gu C, Yue Y, Wu KK, Wu J, Zhu Y (2007) Spike protein of SARS-CoV stimulates cyclooxygenase-2 expression via both calcium-dependent and calcium-independent protein kinase C pathways. FASEB J 21:1586–1596

Marra MA, Jones SJ, Astell CR, Holt RA, Brooks-Wilson A, Butterfield YS, Khattra J, Asano JK, Barber SA, Chan SY et al (2003) The genome sequence of the SARS-Associated Coronavirus. Science 300:1399–1404

Mizutani T (2007) Emerging infectious diseases. Lai SK (eds) Signal transduction in SARS-CoV-infected cells. Ann N Y Acad Sci 1102:86–95

Mizutani T, Fukushi S, Murakami M, Hirano T, Saijo M, Kurane I, Morikawa S (2004a) Tyrosine dephosphorylation of STAT3 in SARS coronavirus-infected Vero E6 cells. FEBS Lett 577:187–192

Mizutani T, Fukushi S, Saijo M, Kurane I, Morikawa S (2004b) Importance of Akt signaling pathway for apoptosis in SARS-CoV-infected Vero E6 cells. Virology 327:169–174

Mizutani T, Fukushi S, Saijo M, Kurane I, Morikawa S (2004c) Phosphorylation of p38 MAPK and its downstream targets in SARS coronavirus-infected cells. Biochem Biophys Res Commun 319:1228–1234

Mizutani T, Fukushi S, Saijo M, Kurane I, Morikawa S (2005) JNK and PI3K/Akt signaling pathways are required for establishing persistent SARS-CoV infection in Vero E6 cells. Biochim Biophys Acta 1741:4–10

Mizutani T, Fukushi S, Saijo M, Kurane I, Morikawa S (2006a) Regulation of p90RSK phosphorylation by SARS-CoV infection in Vero E6 cells. FEBS Lett 580:1417–1424

Mizutani T, Fukushi S, Saijo M, Kurane I, Morikawa S (2006b) Characterization of persistent SARS-CoV infection in Vero E6 cells. In Perlman S, Holmes K (eds) The Nidoviruses. The control of SARS and other Nidovirus diseases, vol 581. Wiley, New York, pp 323–326

Mizutani T, Fukushi S, Iizuka D, Inanami O, Kuwabara M, Takashima H, Yanagawa H, Saijo M, Kurane I, Morikawa S (2006c) Inhibition of Cell Proliferation by SARS-CoV infection in Vero E6 cells. FEMS Immunol Med Microbiol 46:236–243

Mizutani T, Fukushi S, Ishii K, Sasaki Y, Kenri T, Saijo M, Kanaji Y, Shirota K, Kurane I, Morikawa S (2006d) Mechanisms of establishment of persistent SARS-CoV-infected cells. Biochem Biophys Res Commun 347:261–265

Mizutani T, Fukushi S, Kenri T, Sasaki Y, Ishii K, Endoh D, Zamoto A, Saijo M, Kurane I, Morikawa S (2007) Enhancement of cytotoxicity against Vero E6 cells persistently infected with SARS-CoV by *Mycoplasma fermentans*. Arch Virol 152:1019–1025

Mora LB, Buettner R, Seigne J, Diaz J, Ahmad N, Garcia R, Bowman T, Falcone R, Fairclough R, Cantor A, Muro-Cacho C, Livingston S, Karras J, Pow-Sang J, Jove R (2002) Constitutive activation of Stat3 in human prostate tumors and cell lines: direct inhibition of Stat3 signaling induces apoptosis of prostate cancer cells. Cancer Res 62:6659–6666

Okabayashi T, Kariwa H, Yokota S, Iki S, Indoh T, Yokosawa N, Takashima I, Tsutsumi H, Fujii N (2006) Cytokine regulation in SARS coronavirus infection compared to other respiratory virus infections. J Med Virol 78:417–424

Padhan K, Minakshi R, Towheed MA, Jameel S (2008) Severe acute respiratory syndrome coronavirus 3a protein activates the mitochondrial death pathway through p38 MAP kinase activation. J Gen Virol 89:1960–1969

Peiris JS, Lai ST, Poon LL, Guan Y, Yam LY, Lim W, Nicholls J, Yee WK, Yan WW, Cheung MT, Cheng VC, Chan KH, Tsang DN, Yung RW, Ng TK, Yuen KY, SARS study group (2003a) Coronavirus as a possible cause of severe acute respiratory syndrome. Lancet 361:1319–1325

Peiris JS, Yuen KY, Osterhaus AD, Stohr K (2003b) The severe acute respiratory syndrome. N Engl J Med 349:2431–2441

Platanias LC (2003) p38 mitogen-activated protein kinase pathway and its role in interferon signaling. Pharmacol Ther 98:129–142

Poutanen SM, Low DE, Henry B, Finkelstein S, Rose D, Green K, Tellier R, Draker R, Adachi D, Ayers M et al (2003) Identification of severe acute respiratory syndrome in Canada. N Engl J Med 348:1995–2005

Rahaus M, Desloges N, Wolff MH (2005) ORF61 Protein of Varicella-Zoster Virus Influences JNK/SAPK and p38/MAPK Phosphorylation. J Med Virol 76:424–433

Rajan P, McKay RD (1998) Multiple routes to astrocytic differentiation in the CNS. J Neurosci 18:3620–3629

Ren L, Yang R, Guo L, Qu J, Wang J, Hung T (2005) Apoptosis induced by the SARS-associated coronavirus in Vero cells is replication-dependent and involves caspase. DNA Cell Biol 24:496–502

Richards SA, Fu J, Romanelli A, Shimamura A, Blenis J (1999) Ribosomal S6 kinase 1 (RSK1) activation requires signals dependent on and independent of the MAP kinase ERK. Curr Biol 9:810–820

Rota PA, Oberste MS, Monroe SS, Nix WA, Campagnoli R, Icenogle JP, Peñaranda S, Bankamp B, Maher K, Chen MH et al (2003) Characterization of a novel coronavirus associated with severe acute respiratory syndrome. Science 300:1394–1399

Scheid MP, Woodgett JR (2003) Unravelling the activation mechanisms of protein kinase B/Akt. FEBS Lett 546:108–112

Schindler C, Shuai K, Prezioso VR, Darnell JE Jr (1992a) Activation of transcription by IFN-gamma: tyrosine phosphorylation of a 91-kD DNA binding protein. Science 258:1808–1812

Schindler C, Shuai K, Prezioso VR, Darnell JE Jr (1992b) Interferon-dependent tyrosine phosphorylation of a latent cytoplasmic transcription factor. Science 257:809–813

Sheahan T, Morrison TE, Funkhouser W, Uematsu S, Akira S, Baric RS, Heise MT (2008) MyD88 is required for protection from lethal infection with a mouse-adapted SARS-CoV. PLoS Pathog E1000240

Shuai K, Stark GR, Kerr IM, Darnell JE Jr (1993) A single phosphotyrosine residue of Stat91 required for gene activation by interferon-gamma. Science 261:1744–1746

Shuai K, Horvath CM, Huang LH, Qureshi SA, Cowburn D, Darnell JE Jr (1994) Interferon activation of the transcription factor Stat91 involves dimerization through SH2-phosphotyrosyl peptide interactions. Cell 76:821–828

Silva AM, Whitmore M, Xu Z, Jiang Z, Li X, Williams BR (2004) Protein kinase R (PKR) interacts with and activates mitogen-activated protein kinase kinase 6 (MKK6) in response to double-stranded RNA stimulation. J Biol Chem 279:37670–37676

Smith JA, Poteet-Smith CE, Malarkey K, Sturgill TW (1999) Identification of an extracellular signal-regulated kinase (ERK) docking site in ribosomal S6 kinase, a sequence critical for activation by ERK in vivo. J Biol Chem 274:2893–2898

Spaziani A, Alisi A, Sanna D, Balsano C (2006) Role of p38 MAPK and RNA-dependent protein kinase (PKR) in Hepatitis C virus core-dependent nuclear delocalization of Cyclin B1. J Biol Chem 281:10983–10989

Su HL, Liao CL, Lin YL (2002) Japanese encephalitis virus infection initiates endoplasmic reticulum stress and an unfolded protein response. J Virol 76:4162–4171

Surjit M, Liu B, Jameel S, Chow VTK, Lal SK (2004) The SARS coronavirus nucleocapsid (N) protein induces actin reorganization and apoptosis in COS-1 cells. Biochem J 383:13–18

Surjit M, Kumar R, Mishra RN, Reddy MK, Chow VT, Lal SK (2005) The severe acute respiratory syndrome coronavirus nucleocapsid protein is phosphorylated and localizes in the cytoplasm by 14-3-3-mediated translocation. J Virol 79:11476–11486

Surjit M, Liu B, Chow VT, Lal SK (2006) The nucleocapsid protein of severe acute respiratory syndrome-coronavirus inhibits the activity of cyclin-cyclin-dependent kinase complex and blocks S phase progression in mammalian cells. J Biol Chem 281:10669–10681

Tan Y, Rouse J, Zhang A, Cariati S, Cohen P, Comb MJ (1996) FGF and stress regulate CREB and ATF-1 via a pathway involving p38 MAP kinase and MAPKAP kinase-2. EMBO J 15:4629–4642

Tan YX, Tan TH, Lee MJ, Tham PY, Gunalan V, Druce J, Birch C, Catton M, Fu NY, Yu VC, Tan YJ (2007) Induction of apoptosis by the severe acute respiratory syndrome coronavirus 7a protein is dependent on its interaction with the Bcl-XL protein. J Virol 81:6346–6355

Thiel V, Ivanov KA, Putics A, Hertzig T, Schelle B, Bayer S, Weissbrich B, Snijder EJ, Rabenau H, Doerr HW et al (2003) Mechanisms and enzymes involved in SARS coronavirus genome expression. J Gen Virol 84:2305–2315

Toker A (2000) Protein kinases as mediators of phosphoinositide 3-kinase signaling. Mol Pharmacol 57:652–658

Tsang KW, Ho PL, Ooi GC, Yee WK, Wang T, Chan-Yeung M, Lam WK, Seto WH, Yam LY, Cheung TM et al (2003) A cluster of cases of severe acute respiratory syndrome in Hong Kong. N Engl J Med 348:1977–1985

von Knethen A, Lotero A, Brune B (1998) Etoposide and cisplatin induced apoptosis in activated RAW 264.7 macrophages is attenuated by cAMP-induced gene expression. Oncogene 17:387–394

Wang XZ, Lawson B, Brewer JW, Zinszner H, Sanjay A, Mi LJ, Boorstein R, Kreibich G, Hendershot LM, Ron D (1996) Signals from the stressed endoplasmic reticulum induce C/EBP-homologous protein (CHOP/GADD153). Mol Cell Biol 16:4273–4280

Waye YM, Law WP, Wong CH, C Au T, Chuck CP, Kong SK, S Chan P, To KF, I Lo A, W Chan J, Suen YK, Edwin Chan HY, Fung KP, Y Sung J, Dennis Lo YM, W Tsui S. (2005) The 3a protein of SARS-coronavirus induces apoptosis in Vero E6 cells. Conf Proc IEEE Eng Med Biol Soc 7:7482–7485

Wei L, Sun S, Xu CH, Zhang J, Xu Y, Zhu H, Peh SC, Korteweg C, McNutt MA, Gu J (2007) Pathology of the thyroid in severe acute respiratory syndrome. Hum Pathol 38:95–102

Welch H, Eguinoa A, Stephens LR, Hawkins PT PT (1998) Protein kinase B and rac are activated in parallel within a phosphatidylinositide 3OH-kinase-controlled signaling pathway. J Biol Chem 273:11248–11256

Wong SL, Chen Y, Chan CM, Chan CS, Chan PK, Chui YL, Fung KP, Waye MM, Tsui SK, Chan HY (2005) In vivo functional characterization of the SARS-Coronavirus 3a protein in Drosophila. Biochem Biophys Res Commun 337:720–729

Wu CH, Yeh SH, Tsay YG, Shieh YH, Kao CL, Chen YS, Wang SH, Kuo TJ, Chen DS, Chen PJ (2009) Glycogen synthase kinase-3 regulates the phosphorylation of severe acute respiratory syndrome coronavirus nucleocapsid protein and viral replication. J Biol Chem 284:5229–5239

Yan H, Xiao G, Zhang J, Hu Y, Yuan F, Cole DK, Zheng C, Gao GF (2004) SARS coronavirus induces apoptosis in Vero E6 cells. J Med Virol 73:323–331

Yan X, Hao Q, Mu Y, Timani KA, Ye L, Zhu Y, Wu J (2006) Nucleocapsid protein of SARS-CoV activates the expression of cyclooxygenase-2 by binding directly to regulatory elements for nuclear factor-kappa B and CCAAT/enhancer binding protein. Int J Biochem Cell Biol 38:1417–1428

Yang Y, Xiong Z, Zhang S, Yan Y, Nguyen J, Ng B, Lu H, Brendese J, Yang F, Wang H, Yang XF (2005) Bcl-xL inhibits T cell apoptosis induced by expression of SARS coronavirus E protein in the absence of growth factors. Biochem J 392:135–143

Ye Z, Wong CK, Li P, Xie Y (2008) A SARS-CoV protein, ORF-6, induces caspase-3 mediated, ER stress and JNK-dependent apoptosis. Biochim Biophys Acta 1780(12):1383–1387

Yount B, Roberts RS, Sims AC, Deming D, Frieman MB, Sparks J, Denison MR, Davis N, Baric RS (2005) Severe acute respiratory syndrome coronavirus group-specific open reading frames encode nonessential functions for replication in cell cultures and mice. J Virol 79:14909–14922

Yuan X, Shan Y, Zhao Z, Chen J, Cong Y (2005) G0/G1 arrest and apoptosis induced by SARS-CoV 3b protein in transfected cells. Virol J 2:66

Yuan X, Wu J, Shan Y, Yao Z, Dong B, Chen B, Zhao Z, Wang S, Chen J, Cong Y (2006) SARS coronavirus 7a protein blocks cell cycle progression at G0/G1 phase via the cyclin D3/pRb pathway. Virology 346:74–85

Yuan X, Yao Z, Wu J, Zhou Y, Shan Y, Dong B, Zhao Z, Hua P, Chen J, Cong Y (2007) G1 phase cell cycle arrest induced by SARS-CoV 3a protein via the cyclin D3/pRb pathway. Am J Respir Cell Mol Biol 37:9–19

Zhang L, Wei L, Jiang D, Wang J, Cong X, Fei R (2007) SARS-CoV nucleocapsid protein induced apoptosis of COS-1 mediated by the mitochondrial pathway. Artif Cells Blood Substit Immobil Biotechnol 35:237–253

Zhao G, Shi SQ, Yang Y, Peng JP (2006) M and N proteins of SARS coronavirus induce apoptosis in HPF cells. Cell Biol Toxicol 22:313–322

Zhao X, Nicholls JM, Chen YG (2008) Severe acute respiratory syndrome-associated coronavirus nucleocapsid protein interacts with Smad3 and modulates transforming growth factor-beta signaling. J Biol Chem 283:3272–3280

Index

Necrosis, 157, 186, 231, 232, 239–241
NendoU, 88
Neutralization assay, 282
Neutralizing antibodies, 8
Neutralizing antibody titers, 282
NFkB, 137
NF-kB, 313
N-glycosylation, 178
NK cells, 264
NL63, 283
NMR, 52, 53
Non-structural proteins, 78, 117
N protein, 56, 251
nsp1, 56, 79
nsp2, 80
nsp3, 80
nsp4, 84
nsp5, 84
nsp6, 84
nsp7, 85
nsp8, 55, 85
nsp9, 55, 85
nsp10, 85
nsp11, 86
nsp12, 86
nsp13, 87
nsp14, 87
nsp15, 88
nsp16, 89
Nuclear, 157
Nucleocapsid, 32, 36
Nucleocapsid protein, 129–131
 acetylation, 131
 apoptosis, 139
 B23 interaction, 140
 cellular kinases, 132
 cellular localization, 134
 cyclin-CDK complex, 136
 multimer, 135
 phosphorylation, 132
 RNA binding domains, 134
 sumoylation, 132

O

Octanucleotide, 54
O-glycosylation, 4
Open reading frame, 32
ORF6, 155, 157–158
ORF8, 160, 162
ORF3a, 155–157, 162
 in other human coronavirus, 170–172
 of SARS-CoV, 167–169
ORF7a, 155, 158–160
ORF8a, 155, 160–162, 177–187
ORF8ab, 160, 161, 177–187
ORF8ab, 155
ORF3b, 155–157
ORF7b, 155, 158–160, 162
ORF8b, 155, 160–161, 177–187
ORF9b, 155, 162
ORF9b of SARS-CoV, 172–173

P

Packaging signals, 48
Papain-like protease (PLpro), 124
 catalytic triad, 124
 tertiary structure, 124
Peptide, 33
Phosphatidylinositol 3-kinase (PI3K), 311, 312
Phylogeny, 180
Plaque reduction assay, 285
Plasma membrane, 156
Plasminogen activator inhibitor 1 (PAI-1), 249
p38 MAPK, 307–311, 315
Poly(ADP-ribose), 81
Poly(A) binding protein (PABP), 56
Polymerase, 78, 85
Polymerase co-factors, 102, 103, 106, 108
Polyprotein, 78, 116
Polypyrimidine tract-binding (PTB) protein, 56, 57
Poly(A) tail, 54
pp1a/pp1ab, 78
p90 ribosomal S6 kinase (RSK), 311
Primase, 78, 85
Proof-reading, 87
Protease, 78, 81, 84, 117–124
 main, 117
 main protease, 78, 84
 papain-like, 124
 papain-like protease, 78, 81
 specificity, 81
 structure, 80, 84

Proteases, 70
Proteasome, 183, 185
14-3-3 Protein, 133
Protein kinase R (PKR), 314
Proteolysis, 33
Proteolytic cleavage, 12–13
Proteolytic processing, 78
Pseudoknot, 48, 54, 55, 68

R

RDRP, 67
Receptor binding domain, 4–5
Receptor binding motif (RBM), 8
Receptors, 31, 39, 41
 binding, 31, 35, 39–41
Refolding, 39
Regulated intramembrane proteolysis, 27
Renilla luciferase, 284
Rening-angiotensin system (RAS), 7, 11, 24, 207–209
 ACE2, 205, 207–210, 213
Replicase gene
 expression, 79
 organization, 78
Replicase-transcriptase complex, 75, 91
Reporter genes, 280
Retroviral gag–pol construct, 281
Retroviral pseudotypes, 280
RFP, 284
Ribosome leaky scanning, 160, 162
RNA-binding domain, 78, 83, 85
RNA 5'-cap formation, 87, 89, 90
5' RNA cap structure, 89, 90
RNA-protein interaction, 134
RNA 5'-triphosphatase, 87

S

SARS coronavirus (SARS-CoV), 47, 49–52, 54, 57, 247, 307–317
SARS-CoV accessory proteins, 173–174
SARS-CoV accessory proteins ORF3a and 9b, 167–174
SARS-CoV Evolution, 196, 207
 Chinese horseshoe bat, 197
 Himalayan palm civet, 197
SARS-CoV vaccines, 279
SARS pseudotypes, 282

SARS spike (S) envelope gene construct, 280
S1 domain, 33, 34, 36
Secondary structural model, 51
S

Transcriptomic analysis, 295–301
Transforming growth factor-b1 (TGF-β1), 249
Transmembrane domain, 78, 83
Transmembrane protein, 24
Transmissible gastroenteritis virus (TGEV), 49, 50, 53, 55
Transmission, 9, 180
Type II pneumocytes, 7–8
Type I transmembrane protein, 4

U

Ubiquitin, 183
3′ untranslated regions (UTRs), 48, 52, 54–56
5′ untranslated regions (UTRs), 48, 51, 52, 56
Untranslated regions (UTRs), 47
U-turn motif, 52

V

Vaccine development, 211, 213–215
 immunopotentiation of disease, 217–219
 immunosenescence, 213, 216
Vaccines, 270

Viral evasive strategies, 265–266
Viral inactivation, 33
Viral replication, 177, 185–188
Virions, 31, 33, 156, 160
Virus, 158–160
Virus-like particles, 156, 158, 159
Virus neutralization, 279

W

Wet markets, 285
WHO, 283

X

X domain, 81
X-ray crystallography, 37

Y

Yeast-two-hybrid, 184
YFP, 284
YNMG-like tetraloop, 53

Z

Zinc-binding domain, 81, 83, 87